Industrial and Applied Mathematics

Industrial and Applied Mathematics, a Scopus-indexed book series, publishes high-quality research-level monographs, lecture notes and contributed volumes focusing on areas where mathematics is used in a fundamental way, such as industrial mathematics, bio-mathematics, financial mathematics, applied statistics, operations research and computer science.

Avishek Adhikari • Erik Trell • Ashis SenGupta
Editors

Quantitative Frontiers

Exploring Applications of Mathematics and Statistics

Springer

Editors
Avishek Adhikari
Department of Mathematics
Presidency University
Kolkata, West Bengal, India

Erik Trell
Department of Health, Medicine and Caring Sciences
Linköping University
Linköping, Sweden

Ashis SenGupta
Department of Population Health Sciences
Augusta University
Augusta, GA, USA

ISSN 2364-6837 ISSN 2364-6845 (electronic)
Industrial and Applied Mathematics
ISBN 978-981-96-9434-1 ISBN 978-981-96-9435-8 (eBook)
https://doi.org/10.1007/978-981-96-9435-8

Mathematics Subject Classification: 70G65, 46A13, 60Axx, 32A55, 83C35, 47Nxx, 26A42, 62Axx, 11B39, 94A60

This Springer imprint is published by the registered company Springer Nature Singapore Pte Ltd.
The registered company address is: 152 Beach Road, #21-01/04 Gateway East, Singapore 189721, Singapore

Dedicated to the Enduring Memory of Professor Dr. M. R. Adhikari (January 2, 1944 to June 29, 2021)

This book is respectfully and lovingly dedicated to the late ***Professor Dr. Mahima Ranjan Adhikari****,* ***Founder President*** *of the Institute for Mathematics, Bioinformatics, Information Technology and Computer Science* ***(IMBIC)****,* ***which has branches in Japan and Sweden****. He was the former Secretary of the* ***Calcutta Mathematical Society****. Professor Adhikari was not merely an academic; he was a* ***foundational pillar*** *of mathematical research and education in India, whose legacy is defined by brilliance, dedication, and extraordinary kindness. A true scholar, Professor Adhikari achieved*

First Class First *in the M.Sc. Examination in Pure Mathematics at Calcutta University in 1966, an accomplishment for which he was awarded a* **Gold Medal**. *Throughout his long and distinguished career, which included roles as a Professor at both* **University of Calcutta** *and* **Burdwan University** *and Assistant Professor at* **Presidency College, Kolkata**, *he shaped the intellectual landscape of his field. His influence was global, marked by invitations to institutions across the USA, Russia, UK, Greece, Switzerland, Italy, Japan, Sweden, France, China, Turkey, Bangladesh, Malaysia,* **and Nepal**. *Most profoundly, he was a* **dedicated mentor**, *successfully guiding* **twelve Ph.D. students** *across nine diverse areas of mathematics. Professor Adhikari's commitment to scholarship continued until his final days, producing the three-volume* **Basic Topology** *series, which was published by Springer. He authored and co-authored* **nine textbooks** *and edited* **two Research Monographs**, *with his published works being cited internationally. Professor Adhikari's greatest contribution, however, was his spirit. He will forever remain in the heart of all the people he touched owing to his* **ever-supportive, encouraging, and loving nature**. *His life was an equation of excellence, mentorship, and profound human warmth.*

Preface

This volume is *commemoration volume in memory of Professor M R Adhikari*, the founder President of the Institute for Mathematics, Bioinformatics, Information Technology and Computer Science (IMBIC). IMBIC is an International Research Institute in India having branches in Japan and Sweden. It was established in 2006 with the avowed object for promotion of advanced studies and research in Mathematics, Statistics, Bioinformatics, Computer Science, Information Technology, Nanotechnology, and their closely allied areas. This Institute is an autonomous, voluntary, non-government organization registered under the West Bengal Societies Registration Act, 1961.

This volume presents a collection of research papers from esteemed scholars across the globe, including the USA, Russia, Norway, Sweden, Nigeria, and India. The book delves into a wide range of subjects, from the theoretical foundations of mathematical physics to the practical applications of statistics, engineering, and computer science.

The 15 chapters offer a fascinating blend of theoretical and applied research. From the exploration of gravitational waves and the intricacies of color theory to the mathematical foundations of atomic structure and the statistical analysis of complex systems, this book caters to a broad audience of researchers and students. These chapters are being recommended by internationally reputed reviewers, not only restricted within the Program Committee of the 17th and 18th International Conference 2023 MSAST 2024, but also by subject experts from all over the globe.

A notable aspect of this book is the interdisciplinary nature of its content. For instance, the chapter on the "Linguistics Semiology and Geometric Lie Algebra Foundation of Atomic Structure and Periodic System" bridges the gap between linguistics and physics, offering a novel perspective on the fundamental building blocks of matter. Similarly, the chapter on "Fibonacci generation of exact distribution of Fibonacci summands" explores the mathematical properties of Fibonacci numbers, a topic with applications in various fields, including computer science and cryptography.

Mathematical applications span the first 12 chapters. Chapter 1 deals with a novel mathematical approach to generate gravitational waves in laboratory conditions by leveraging the amplification effect of gravitational waves at the difference frequency of electromagnetic waves. Chapter 2 explores the concept of harmonic vibrations, comparing traditional methods (vibrating strings, Chladni plates) with a nonlinear, golden ratio-based distribution known as Jacob's Ladder. Chapter 3 proposes a semiotic mathematical approach to the periodic table, linking it to linguistics, geometry, and physics, and suggests this framework as a potential path to understanding the mechanism of superconductivity. Chapter 4 proposes a new conceptualization of time as "clock time" and "temporal frequency" to address Einstein's twin paradox and time dilation near black holes, considering them opposite extremes of a unified time framework. Chapter 5 describes a comprehensive study using Lie algebra to investigate the Standard Model, extending its application to the periodic table, atomic structure, and eventually modeling real-world structures, chemical compounds, and compositions at various scales. Chapter 6 investigates the resilience of secret sharing schemes against leakage attacks, where an adversary can obtain partial information about individual shares. It re-examines static and adaptive leakage models for general access structures, building upon previous work by Srinivasan-Vasudevan and Kumar-Meka-Sahai. Chapter 7 investigates the approximation behavior of the Picard singular integral for functions in generalized Hölder spaces, quantifying the rate of approximation using a generalized norm and modulus of continuity. A numerical investigation using Matlab's bvp4c solver has been conducted in Chapter 8 to analyze the influence of heat and mass transfer on hydromagnetic nanofluid flow over an exponentially stretching surface, considering effects like chemical reaction, viscous dissipation, radiation, and slip conditions. Chapter 9 explores the Fibonacci-based structure of the distribution of Fibonacci summands, revealing a connection to Pascal's triangle and suggesting a reinterpretation of the latter as a Fibonacci-generated epi-structure. Chapter 10 introduces set-valued Riemann integrals on time scales, providing definitions, integrability criteria, and properties. It explores the connection between set-valued and ordinary Riemann integrals, convexity, and support functions. Chapter 11 provides a comprehensive analysis of real projective space $\mathbb{RP}^m$, particularly focusing on dimensions 1, 2, and 3. It explores multiple formal representations of $\mathbb{RP}^m$ and utilizes topological surgery theory to offer clear visualizations, highlighting its significance in both physics and mathematics. Chapter 12 analyzes the compatibility of decay laws obtained from similarity methods and fixed point methods for high Reynolds number homogeneous isotropic turbulence, focusing on the Von Karman-Howarth equation and the energy spectrum equation.

Statistical applications span in next three chapters. Chapter 13 discusses the adaptation of circular and axial statistical models to handle asymmetric data, along with the development of bivariate extensions for analyzing dependent variables, with a focus on mixture models with independent coordinate variables. Chapter 14 investigates classification and regression error bounds for non-identically distributed data, providing theoretical guarantees for universal learning rules and k-nearest neighbor regression under non-stationary noise conditions. The findings are applied to wireless network scenarios. The final Chap. 15 proposes a probabilistic approach to efficient data storage in the presence of feature selection and under-sampling, providing theoretical bounds and practical applications in data science.

In summary, this edited volume offers a valuable resource for researchers, students, and practitioners interested in the latest advancements in various scientific disciplines. By bringing together experts from around the world, the book fosters interdisciplinary collaboration and promotes the exchange of ideas. This volume is dedicated to the memory of Professor M R Adhikari, the founder President of the institute IMBIC, whose continuous encouragement and support will always be remembered by the mathematical community.

We extend our sincere gratitude to all contributors to this volume, particularly the corresponding authors, for their cooperation and timely submission of articles and revisions. We are also deeply indebted to the following reviewers for their invaluable assistance in reviewing these articles:

Sergey Chervon (Russia), Alexandr Balakin (Russia), Godwin Joseph Ibeh (Nigeria), Emmanuel E. Etim (Nigeria), Stein Johansen (Norway), Partha Sarathi Roy (Australia), Sandip Jana (India), Hiranmoy Mondal (India), Satyajit Roy (India), Md. Nasim Akhtar (India), Kuntal Banerjee (India), Indranil Ghosh (USA), Nitin Gupta (India), and Dietrich von Rosen (Sweden). We are thankful to Dr. Chandan Goswami, former Research Scholar at the Department of Mathematics, Presidency University, Kolkata for all kinds of editorial support without which this book would not have been possible.

We are grateful to Springer for publishing this volume and to all individuals who have extended their support and cooperation in bringing the project of IMBIC for publishing this commemorative volume to fruition.

Kolkata, India — Avishek Adhikari
Augusta, GA, USA — Ashis SenGupta
Malmo, Sweden — Erik Trell

About the Book

This edited volume presents a diverse collection of research papers from esteemed scholars across the globe, including the USA, Russia, Norway, Sweden, Nigeria, and India. The book delves into a wide range of subjects, from the theoretical foundations of mathematical physics to the practical applications of statistics, engineering, and computer science.

The 15 chapters offer a fascinating blend of theoretical and applied research. From the exploration of gravitational waves and the intricacies of color theory to the mathematical foundations of atomic structure and the statistical analysis of complex systems, this book caters to a broad audience of researchers and students.

A notable aspect of this book is the interdisciplinary nature of its content. For instance, the chapter on the "Linguistics Semiology and Geometric Lie Algebra Foundation of Atomic Structure and Periodic System" bridges the gap between linguistics and physics, offering a novel perspective on the fundamental building blocks of matter. Similarly, the chapter on "Fibonacci generation of exact distribution of Fibonacci summands" explores the mathematical properties of Fibonacci numbers, a topic with applications in various fields, including computer science and cryptography.

The book also includes several chapters focused on applied research, such as the numerical analysis of fluid flow and the development of machine learning algorithms for data analysis. These chapters highlight the practical implications of theoretical research and demonstrate how mathematical and statistical techniques can be used to solve real-world problems.

Overall, this edited volume provides a valuable resource for researchers and students interested in a wide range of scientific disciplines. It offers a glimpse into the cutting-edge research being conducted around the world and showcases the power of interdisciplinary collaboration.

In summary, this edited volume offers a valuable resource for researchers, students, and practitioners interested in the latest advancements in various scientific disciplines. By bringing together experts from around the world, the book fosters interdisciplinary collaboration and promotes the exchange of ideas. This volume is dedicated to the memory of Professor M R Adhikari, the founder President of the institute IMBIC, whose continuous encouragement and support will always be remembered by the mathematical community.

Salient features of the book

- **Interdisciplinary Focus:** As applications of mathematics and statistics, the book covers a wide range of topics, from physics to computer science and engineering, highlighting the interconnectedness of these fields.
- **Global Perspective:** Contributions from renowned scholars from diverse countries like the USA, Russia, Norway, Sweden, Nigeria, and India offer a global perspective on cutting-edge research.
- **Theoretical Depth and Practical Relevance:** The book delves into both fundamental theoretical concepts and practical applications, making it suitable for researchers and practitioners alike.
- **Innovative Research:** The chapters present original research, exploring novel ideas and techniques in various fields.
- **Potential Impact:** The findings and insights presented in the book have the potential to influence future research and technological advancements.
- **High-Quality Scholarship:** The contributions from esteemed institutions and experts ensure the quality and rigor of the research.

Contents

Editors and Contributors

About the Editors

Avishek Adhikari, Professor and HOD, Department of Mathematics, Presidency University, Kolkata, India. M.Sc. (Gold Medalist), a recipient of the President of India Medal, ISCA Young Scientist Award, and NANUM fund by International Mathematical Union (IMU), received his Ph.D. Degree from the Indian Statistical Institute. Professor Adhikari was a faculty member at the Department of Pure Mathematics, University of Calcutta (2006–2019). He is the founder secretary of IMBIC, India (having Branches in Sweden and Japan), Treasurer of Cryptology Research Society of India, and was the Former Eastern Zonal Coordinator for NBHM MSc and Ph.D. Scholarship Examination. He was a Post-Doctoral fellow at INRIA-Rocquencourt, France, a visiting scientist at Linkoping University, Sweden, and a visiting scientist at ISI, Kolkata. His academic visit includes Japan (7 times, funded by JSPS, Japan and JST, Japan), Sweden, France, England, Switzerland, South Korea, Russia, Malaysia, Bangladesh, and Singapore. He has published six basic textbooks on mathematics including *Basic Modern Algebra with Applications* (Springer, 2014), *Basic Topology, Volume 1* (Springer, 2022), *Basic Topology, Volume 2* (Springer, 2022), and edited two research monographs including Springer Monograph (2017) on *Mathematical and Statistical Applications in Life Sciences and Engineering*. He was one of the Program Co-Chairs of Indocrypt 2021. He has published more than 75 research articles in reputed international journals, conference proceedings and contributed volumes. He is/was one of the investigators of 12 Sponsored Research Projects funded by the agencies like DRDO, WESEE (Ministry of Defense), DST, and NBHM including two International Collaborative Projects supported by DST-JSPS (Indo-Japan Project) and DST-JST (Indo-Japan). Ten of his Ph.D. students have already obtained Ph.D. Degree. Currently, four of the Ph.D. scholars are doing their research works under his supervision. His research interest includes Algebra, Graph Theory, and Cryptology.

Erik Trell, Professor Emeritus, Faculty of Health Sciences, University of Linköping, Sweden. Associate professor in Internal Medicine, Malmö General Hospital, University of Lund, 1968–1982. Ph.D. thesis 1972 on Pulmonary Hypertension. From 1973 to 1982 I planned, designed, and run the new Department of Preventive Medicine at the Medical Clinic, reaching international recognition in its innovative span both construction; adding of alcohol disorders, prediabetes, and tobacco-related cancer; inclusion of the female birth year cohorts and the mammographic unit; and development of the world's first online interactive computer system for the monitoring.

All this reported in many theses and "firsts" in high-impact Journals and contributed to appointment in 1986 as professor and chief of the new Department of Family Medicine, Faculty of Health Sciences (Health University), of Linköping (retiring in 2006 but staying on as professor emeritus and continuing to serve as "district doctor" in Karlstad).

The Health University carried me on over a range of forefront projects like planning and organization of the new Medical Faculty of Heraklion University, the Mediterranean Health Research Program and the PHC services on Crete; advancing online decentralized maternal and child health care computer system regionally and in the Bhorugram desert community in India; transferring the Malmö Preventive Program and related data banks to user-operated "Healthometer" personal "HPS" (Health Positioning System) instrument in optional, e.g., smartphone media; and applying the Health for All-inspired Östergötland county Health Policy Program.

Now stopping to exemplify, much similar could be specified on education, research, advisory and consultancy functions, conferences, etc. domestically and over the world but I may conclude that it took me to a, rarer and rarer, kind of guru of the game. Summing up, I have some 450 medical and 40 mathematical/physical publications, 156 of the medical ones searchable in PubMed and 219, including some of the mathematical, in ResearchGate.

Ashis SenGupta, Emeritus Scientist, Council of Scientific and Industrial Research, Ministry of Science and Technology, Govt. of India; Advisor/Consultant at the Indian Statistical Institute, India; Adjunct Professor, Augusta University, Georgia, USA; and Distinguished Professor, Middle East Technical University, Turkey. Advisor, Analythium, Canada; Chairman/Expert of various Committees, SERB-DEST, Govt. of India.

Visiting Professor worldwide including at Stanford University, University of Wisconsin-Madison, Concordia University, Canada; Belgium, Sweden, Germany, Israel, etc.

Ph.D. from Ohio State Univ, USA. Seventeen Ph.D. scholars including ongoing ones in India, Turkey and the USA. More than 120 publications including 13 books and volumes in Multivariate Analysis, Directional Statistics, Statistical Machine Learning, Financial Statistics, and Reliability Inference.

Editor, Gujarat Journal of Statistics and Data Science. Ex Editor-in-Chief of Environmental and Ecological Statistics (Springer, USA), and of Jr. of ISPS (India); Assoc. Editor, Book series FIM, Springer.

Principal Investigator of three Mega projects from Govt. of India. Recipient of several International and National recognitions, including two Lifetime Achievements and Distinguished Statistician Awards.

Ex Vice-President of International Forum for Interdisciplinary Mathematics and Ex President of International Indian Statistical Association (India chap) successively for several terms. Elected President of Mathematical Sciences section of Indian Science Congress in 2011–2012. Elected Member of International Statistical Institute; Fellow of the National Academy of Sciences, India, Fellow of ISPS, Fellow of AIIA, and Fellow of the American Statistical Association—currently the only working statistician in India with this recognition.

Contributors

Godfrey E. Akpojotor Delta State University, Abraka, Nigeria
International Centre for Basic Research, Abuja, Nigeria

Alexander O. Animalu International Centre for Basic Research, Abuja, Nigeria
University of Nigeria, Nsukka, Enugu State, Nigeria

Barry C. Arnold University of California, Riverside, CA, USA

Hemen Bharali Assam Don Bosco University Kamarkuchi, Sonapur, Assam, India

Sabyasachi Dutta SRM University - AP, Amaravati, Andhra Pradesh, India

Samuel Edeagu International Centre for Basic Research, Abuja, Nigeria
University of Nigeria, Nsukka, Enugu State, Nigeria
University of Kent, Canterbury, UK

Igor Fomin Bauman Moscow State Technical University, Moscow, Russia

Ghurumuruhan Ganesan University of Bristol, Bristol, UK

Vladimir Gladyshev Bauman Moscow State Technical University, Moscow, Russia

O'Paul Isikaku-Ironkwe Bernd T. Matthias Superconductivity Lab., RTSD Technologies, San Diego, CA, USA

Stein E. Johansen Norwegian University of Science and Technology, Trondheim, Norway

Vladimir Kauts Bauman Moscow State Technical University, Moscow, Russia

Amit Kumar Laha Amity University Kolkata, Newtown, West Bengal, India

Tanmay Saha Central University of Kerala, Periye, Kerala, India

Vikuozonuo Sekhose Assam Don Bosco University Kamarkuchi, Sonapur, Assam, India

Ashis SenGupta Augusta University, Augusta, GA, USA
Middle East Technical University, Ankara, Turkey

Abhay Pratap Singh IIT Roorkee, Roorkee, Uttarakhand, India

Uaday Singh IIT Roorkee, Roorkee, Uttarakhand, India

Carl Swopes Swopes Inc., Color Research, Las Cruces, NM, USA

Erik Trell University of Linköping, Linköping, Sweden

Krishnandan Verma Maryam Ajmal Women's College of Science and Technology, Hojai, Assam, India

Chapter 1
Laboratory Gravitational Waves: Sources and Detection

Vladimir Gladyshev, Vladimir Kauts, and Igor Fomin

Abstract Laboratory sources of gravitational waves are considered based on the interaction of intense laser radiation and matter and by means of electromagnetic radiation itself. It is shown that the perturbations of matter induced by powerful electromagnetic radiation with optical frequencies correspond to the high-frequency gravitational waves with the amplitude of the order $h_0 \sim 10^{-40}$. Also, a system of standing electromagnetic waves in electromagnetic resonators is considered as a source of gravitational waves. It is shown that in this case the amplitude of gravitational waves depends on the difference in the frequency of standing electromagnetic waves as $h_0 \sim \tilde{\omega}^{-2}$. This effect implies a significant increase in the amplitude of low-frequency gravitational waves induced by electromagnetic waves. Also, the obtained characteristics of the gravitational waves are used to assess the possibility of their registration by existing and prospective detectors.

Keywords Gravitational waves · Electromagnetic waves · Electromagnetic resonators · Detectors of gravitational waves

1.1 Introduction

Currently, various astrophysical processes [1] and cosmological perturbations in the early Universe [1, 2] are considered to be the sources of gravitational wave radiation. The LIGO and Virgo experiments [3, 4] used the interference method proposed in [5] to detect gravitational waves resulting from the mergers of black holes and neutron stars. Primordial gravitational waves of cosmological genesis have not been detected as of now.

V. Gladyshev (✉) · V. Kauts · I. Fomin
Bauman Moscow State Technical University, Moscow, Russia
e-mail: gladyshev@bmstu.ru; kauts@bmstu.ru; fomin_iv@bmstu.ru

A. Adhikari et al. (eds.), *Quantitative Frontiers*, Industrial and Applied Mathematics, https://doi.org/10.1007/978-981-96-9435-8_1

Various laboratory sources of gravitational waves are also considered, both of the type based on interaction of electromagnetic radiation and matter [6, 7] and the type treating an electromagnetic field as a source of gravitational waves [8–14].

In other words, gravitational wave fluctuations of the spacetime metric may be induced by both medium perturbations and an alternating electromagnetic field $\delta T_{\mu\nu} = \delta T_{\mu\nu}^{(\text{medium})} + \delta T_{\mu\nu}^{(\text{em})}$, while it is the specific character of gravitational wave generation that determines whether the first or the second factor dominates.

According to the estimates given in [6, 7], a short pulse of powerful laser radiation interacting with matter $\delta T_{\mu\nu}^{(\text{medium})} \gg \delta T_{\mu\nu}^{(\text{em})}$ makes it possible to generate high-frequency gravitational waves with an amplitude of the order $h_0 \sim 10^{-40}$ in terms of realistic parameters for this experiment, which is much lower than the sensitivity limit of modern detectors $h_0 \sim 10^{-21} - 10^{-22}$ [3, 4]. Nevertheless, works [15–19] show that taking into account the processes underlying conversion of intense laser radiation into a gravitational wave in nonlinear dielectric media posits the relevance of these methods of generating and detecting high-frequency gravitational waves. Moreover, papers [20–23], which analyze the processes of generating gravitational wave radiation by means of specially configured electromagnetic fields in vacuum $\delta T_{\mu\nu}^{(\text{medium})} = 0$, dielectric media $\delta T_{\mu\nu}^{(\text{medium})} \neq 0$, and an external magnetic field, imply that it should be possible in laboratory conditions to obtain gravitational waves with controlled characteristics, that is, potentially amplifiable.

Another source of gravitational waves to consider would be two standing plane electromagnetic waves with frequencies ω_1 and ω_2 in a Fabry-Pérot resonator [21]. The approach proposed considers an amplification factor governing the gravitational waves to be their amplitude as a function of the difference frequency of standing electromagnetic waves $\tilde{\omega} = \omega_1 - \omega_2$ of the form $h_0 \sim \tilde{\omega}^{-2}$, which is a consequence of the gravitational field equations within the scope of Einstein's linearized theory of gravity.

1.2 Gravitational Field Equations in the Linearized Theory of Gravity

Gravitational field equations as per Einstein's theory $G_{\mu\nu} = \kappa T_{\mu\nu}$ may also be written as follows [1]:

$$R_{\mu\nu} = \kappa \left(T_{\mu\nu} - \frac{1}{2} g_{\mu\nu} T \right), \tag{1.1}$$

where $\kappa = \frac{8\pi G}{c^4}$ is Einstein's constant, $G_{\mu\nu} = R_{\mu\nu} - \frac{1}{2} g_{\mu\nu} R$ is the Einstein tensor, $R_{\mu\nu}$ is the Ricci tensor, $R = R_\mu{}^\mu = g^{\mu\nu} R_{\nu\mu}$ is the scalar curvature, G is the gravitational constant, c is the speed of light in vacuum, $T_{\mu\nu}$ is the stress-energy tensor of the gravitational field source $T = T_\rho{}^\rho = g^{\rho\sigma} T_{\rho\sigma}$, satisfying the conservation laws [1]:

$$\frac{\partial T^{\mu\nu}}{\partial x^\nu} = 0, \tag{1.2}$$

where $g_{\mu\nu}$ is the metric tensor defining the geometry of spacetime, the Greek indices range over 0, 1, 2, 3, while the Latin indices range over 1, 2, 3.

Within the scope of linearized general relativity theory, weak gravitational fields are considered to be small perturbations of Minkowski space

$$g_{\mu\nu} = \eta_{\mu\nu} + h_{\mu\nu}, \mid h_{\mu\nu} \mid \ll 1, \tag{1.3}$$

and for harmonic calibration

$$\partial^\mu \left(h_{\mu\nu} - \frac{1}{2}\eta_{\mu\nu} h_\rho{}^\rho \right) = 0, \tag{1.4}$$

where $\eta_{\mu\nu}$ is the metric tensor of Minkowski space, $\partial^\mu \equiv \frac{\partial}{\partial x^\mu}$ and $h_\rho{}^\rho = h_{\rho\sigma}\eta^{\rho\sigma}$ are the gravitational field equations to be reduced to the following form:

$$\Delta h_{\mu\nu} - \frac{1}{c^2}\frac{\partial^2 h_{\mu\nu}}{\partial t^2} = 2\kappa \left(T_{\mu\nu} - \frac{1}{2}\eta_{\mu\nu} T \right). \tag{1.5}$$

Wave solutions of Eq. (1.5) are gravitational waves, which, in the case of propagation along a designated direction $x^1 = x$, can be divided into three types according to the non-zero components of the tensor $h_{\mu\nu}$, specifically [1]:

1. h_{22}, h_{23}, h_{33}—transverse-transverse (TT)
2. $h_{12}, h_{13}, h_{20}, h_{30}$—longitudinal-transverse (LT)
3. h_{11}, h_{10}, h_{00}—longitudinal-longitudinal (LL)

For the case of a plane gravitational wave propagating in the direction $x^1 = x$, its polarization is described by only two non-zero components of the tensor $h_{\mu\nu}$: $h_{23} = h_{32}$ or $h_{22} = -h_{33}$, while the remaining non-zero components are gauge artifacts which depend on the choice of the coordinate system [1].

Hence, a tensor $h_{\mu\nu}$ is considered to function as vacuum gravitational wave solution to the gravitational field Eq. (1.5), the components of said tensor satisfying the conditions of a transverse-traceless (TT) calibration [1]:

$$h_{0\mu} = 0, h_{ii} = 0, \partial^i h_{ij} = 0. \tag{1.6}$$

In a transverse-traceless calibration, solutions to the gravitational field equations may be written as follows:

$$h_{ij}^{TT}(t, \boldsymbol{x}) = \frac{4G}{rc^4}\Lambda_{ij,kl}(\boldsymbol{n}) \int d^4x' T_{kl}\left(t - \frac{r}{c} + \frac{\boldsymbol{x}' \cdot \boldsymbol{n}}{c}, \boldsymbol{x}' \right). \tag{1.7}$$

where $r = |\boldsymbol{x} - \boldsymbol{x}'|$, coordinates $x^\mu = (ct, \boldsymbol{x})$, tensor

$$\begin{aligned}\Lambda_{ij,kl}(\boldsymbol{n}) = P_{ik}P_{jl} - \frac{1}{2}P_{ij}P_{kl} = \delta_{ik}\delta_{jl} - \frac{1}{2}\delta_{ij}\delta_{kl} - n_j n_l \delta_{ik} - n_i n_k \delta_{jl} \\ + \frac{1}{2}n_k n_l \delta_{ij} + \frac{1}{2}n_i n_j \delta_{kl} + \frac{1}{2}n_i n_j n_k n_l,\end{aligned} \tag{1.8}$$

and n_k is the unit vector in the direction of the gravitational wave propagation.

Within the scope of the quadrupole approximation and in the transverse-traceless calibration, the gravitational field is determined as follows:

$$h_{ij}^{TT}(t, \boldsymbol{x}) = \frac{1}{r}\frac{2G}{c^4}\Lambda_{ij,kl}(\boldsymbol{n})\ddot{Q}^{kl}\left(t - \frac{r}{c}\right), \tag{1.9}$$

where

$$Q^{ij} = M^{ij} - \frac{1}{3}\delta^{ij}M_{kk} \equiv \int d^3x\rho\left(x^i x^j - \frac{1}{3}r^2\delta^{ij}\right), \tag{1.10}$$

$$M^{ij} = \int d^3x\rho(t, \boldsymbol{x})x^i x^j, \tag{1.11}$$

where $\rho(t, \boldsymbol{x})$ is the density of matter distribution.

Thus, as a rule, evaluation of characteristics of transverse gravitational waves in linearized gravity is based upon the expressions (1.9), (1.10), and (1.11) [1].

1.3 Computing Characteristics of Laboratory Gravitational Waves According to the Quadrupole Formula

Irradiating matter by short pulses of powerful laser radiation is investigated as a potential mechanism for generating high-frequency gravitational waves, suggesting that the characteristics of gravitational wave radiation should increase with the power of modern lasers [6].

Let us consider the shock wave model proposed in the paper [6] to describe the waves which arise when a powerful laser pulse acts on a thin plate of the target material. Shock waves generate respective gravitational waves, which means that this model may be used to determine their characteristics. Next, we will consider a model of an ionic crystal interacting with an electromagnetic wave. We will also find the characteristics of gravitational waves associated with this process, compare them to the results of the previous model, and consider the prospects of detecting gravitational waves obtained by these methods.

1.4 Shock Wave Model

The paper [6] examined shock waves arising from irradiating matter with short pulses of powerful laser radiation as sources of high-frequency gravitational waves. The shock wave front coordinate may be written as follows $z_S = v_S t$, where the shock wave propagation velocity may be defined as

$$v_S \cong \sqrt{\frac{p_S}{\rho_0}}. \tag{1.12}$$

In this expression, p_S is the pressure of the shock wave, ρ_0 is the density of the medium.

The density of the shock wave in the direction of propagation z is defined as

$$\rho_S = \begin{cases} 4\rho_0, \frac{3}{4}z_S < z < z_S, \\ \rho_0, z_S < z < L, \\ 0, \text{in the rest of space,} \end{cases} \tag{1.13}$$

where L is the dimension characterizing the material in the direction of the shock wave propagation.

The pressure of the shock wave in the medium p_S and the intensity of the laser radiation I_L are linked as follows [6]:

$$p_S = \left(\rho_c I_L^2\right)^{\frac{1}{3}}. \tag{1.14}$$

The critical density ρ_c may be defined as:

$$\rho_c = \frac{\varepsilon_0 m_e m_i}{Ze^2}\frac{(2\pi c)^2}{\lambda_L^2}, \tag{1.15}$$

where ε_0 is the relative permittivity of the vacuum, m_e is the electron rest mass, m_i is the mass of an ion, e is the electron charge, λ_L is the laser wavelength, Z is the atomic number.

The gravitational wave amplitude h_{zz} and the power of the gravitational radiation P_g induced by the shock waves are determined as follows [6]:

$$h_{zz} = \frac{7}{2}\frac{GP_L\tau}{rc^4}\sqrt{\frac{\rho_c}{\rho_0}} = 2.89 \times 10^{-44} \times \sqrt{\frac{\rho_c}{\rho_0}}\left(\frac{P_L\tau}{r}\right), \tag{1.16}$$

$$P_g = \frac{147}{160}\frac{GS^2\rho_0^2 v_S^6}{c^5} = \frac{147}{160}\frac{GP_L^2\rho_c}{c^5\rho_0} = 2.53 \times 10^{-53} \times \left(\frac{\rho_c}{\rho_0}\right)P_L^2, \tag{1.17}$$

where P_L is the laser power, τ is the duration of the laser pulse.

For the following parameters: $Z = 6$ (carbon), $\lambda_L = 351$ nm, $P_L = 5 \times 10^{14}$ W, $\tau = 1$ ns, $r = 10$ m, the following values for the amplitude and power of gravitational waves are obtained from the expressions (1.16), (1.17) [6]

$$h_{zz} \cong 5.1 \times 10^{-40}, \tag{1.18}$$

$$P_g \cong 7.86 \times 10^{-25}\text{W}. \tag{1.19}$$

The amplitude obtained is almost 20 orders of magnitude lower than the amplitudes of gravitational waves from astrophysical sources detected in the LIGO and VIRGO experiments [3, 4].

According to the estimates presented, it is necessary to develop fundamentally new methods for generating gravitational waves in laboratory conditions. An alternative direction may follow from potentially increasing the amplitude of gravitational waves via the material of the medium $\sqrt{\frac{\rho_c}{\rho_0}}$ (taking into account the wavelength λ_L) and the energy of laser radiation $W_L\tau$.

1.5 Dielectrics in the Electromagnetic Wave Field

Now we shall consider another approach to analyzing gravitational wave generation via interaction of matter and electromagnetic waves. To this end, consider the mechanism of gravitational wave excitation in an ionic crystal as a result of an intense electromagnetic wave passing through it [7].

When a plane electromagnetic wave passes through the crystal, the ions in each cell will perform forced oscillations around the equilibrium position, leading to the appearance of an additional dipole moment in this cell, said moment periodically varying in time. The total contribution of all such periodically variable dipole moments from all crystal cells is manifested on the macroscale as the effect of medium polarization in the field of an external electromagnetic wave.

The displacement vector for crystal ions in the electromagnetic wave field may be determined as follows [7]:

$$\vec{\xi} = \left(\vec{r}_+ - \vec{r}_-\right)\sqrt{\rho} = \frac{\omega_T\sqrt{\varepsilon_\infty - \varepsilon_0}}{2\sqrt{\pi}\left(\omega_T^2 - \omega^2\right)}\vec{E}, \tag{1.20}$$

where ω is the frequency of the electromagnetic wave, ω_T is the natural frequency of the elastic transverse oscillations in the crystal, ε_0, ε_∞ is the relative permittivity of the crystal at extremely low and extremely high electromagnetic wave frequencies, ρ is the density of the reduced mass $\mu = \frac{m_+ m_-}{(m_+ + m_-)}$.

Further, we calculate the amplitude of gravitational waves emitted by a system of two ions during the passage of an electromagnetic wave. To this end, consider the oscillations of two crystal cell ions in a designated direction x near the center of mass of this system, assuming that the wave propagates along the axis z.

In this case, we state the displacements of the crystal lattice ions with respect to their center of mass as

$$r_+ = x_1 = \frac{\mu l_0}{(m_+ + m_-)} + \frac{m_+}{\sqrt{\rho}\,(m_+ + m_-)}\left[\frac{\omega_T\sqrt{\varepsilon_\infty - \varepsilon_0}}{2\sqrt{\pi}\left(\omega_T^2 - \omega^2\right)}E_z(t)\right], \tag{1.21}$$

$$r_- = x_2 = \frac{\mu l_0}{(m_+ + m_-)} + \frac{m_-}{\sqrt{\rho}\,(m_+ + m_-)}\left[\frac{\omega_T\sqrt{\varepsilon_\infty - \varepsilon_0}}{2\sqrt{\pi}\left(\omega_T^2 - \omega^2\right)}E_z(t)\right], \tag{1.22}$$

is the average position of the ions with respect to the center of mass $\pm\frac{l_0}{2}$, where l_0 is the initial distance between the crystal lattice ions, and the intensity of the plane electromagnetic wave is defined as

$$E_z(t) = E_0 \exp\left[i\,(kz - \omega t)\right]. \tag{1.23}$$

The quadrupole moment is defined as follows [1]:

$$Q^{ij}(t) = \frac{1}{c^2}\int T^{00}\left(t, x^n\right) x^i x^j dx^3, \tag{1.24}$$

where $T^{00}(t, x^n)$ is the first component of the stress-energy tensor of the system, that is, the energy density of the system.

In this case, we write the non-zero component of the quadrupole moment as:

$$\begin{aligned} Q^{xx} &= \int_V m_+\delta\,(x - x_1)\,x^2\delta(y)dy\delta(z)dz + \int_V m_-\delta\,(x - x_2)\,x^2\delta(y)dy\delta(z)dz \\ &= m_+x_1^2 + m_-x_2^2, \end{aligned} \tag{1.25}$$

where $Q^{xx} = Q_{xx}$.

The corresponding non-zero components of the tensor P_{ij} characterizing the direction of propagation z or the vector $n_i = (0, 0, 1)$ of the gravitational wave are equal to $P_{11} = P_{22} = 1$.

The components of the gravitational wave tensor in transverse-traceless (TT) gauge are stated as follows:

$$\bar{h}_{i0}^{TT}\,(t, z) = 0, \tag{1.26}$$

$$\bar{h}_{ij}^{TT}\,(t, z) = \frac{2G}{zc^4}\left[\frac{d^2}{dt^2}Q_{ij}^{TT}\left(t - \frac{z}{c}\right)\right], \tag{1.27}$$

where:

$$Q_{ij}^{TT} = \left(P_{kl} P_{jm} - \frac{1}{2} P_{kn} P_{lm} \right) Q_{lm}. \tag{1.28}$$

The non-zero components of the gravitational wave tensor are hence defined as:

$$\bar{h}_{xx}^{TT} = -\bar{h}_{yy}^{TT} = \frac{G}{zc^4} \left[\frac{d^2}{dt^2} Q_{xx} \left(t - \frac{z}{c} \right) \right]. \tag{1.29}$$

Next, consider the frequency of the electromagnetic wave:

$$\omega = \omega_T + \Delta\omega, \, \Delta\omega < \omega_T, \tag{1.30}$$

close to the oscillation frequency of the crystal lattice, which leads us to deriving the expression

$$\frac{\omega_T}{\omega_T^2 - \omega^2} \approx -\frac{1}{2\Delta\omega}. \tag{1.31}$$

Assuming that the ion masses are approximately equal to $m_+ \approx m_- = m_i$, neglecting the small terms in Eq. (1.29), Eqs. (1.21), (1.22), and (1.25) yield the expressions for transverse gravitational waves:

$$\begin{aligned} \bar{h}_{xx}^{TT} = -\bar{h}_{yy}^{TT} = -\frac{Gm_i \left(\varepsilon_\infty - \varepsilon_0 \right)}{2\pi z \rho c^4} \left(\frac{\omega_T}{\Delta\omega} \right)^2 E_0^2 \\ \times \mathrm{Re} \left[\exp \left\{ \frac{2i}{c} \left[ckz - \omega_T \left(ct - z \right) \right] \right\} \right] \end{aligned} \tag{1.32}$$

For N-ions of the crystal, taking into account that $\sum m_i = Nm_i$, and assuming $\gamma_\omega \equiv \frac{\varepsilon_\infty - \varepsilon_0}{2\pi} = \mathrm{const}$ as the dispersion coefficient, we obtain:

$$\bar{h}_{xx}^{TT} = -\bar{h}_{yy}^{TT} = -\frac{G}{zc^4} \left(\frac{\omega_T}{\Delta\omega} \right)^2 \gamma_\omega E_0^2 V \times \mathrm{Re} \left[\exp \left\{ \frac{2i}{c} \left[ckz - \omega_T \left(ct - z \right) \right] \right\} \right], \tag{1.33}$$

where V is the volume of the dielectric affected by the electromagnetic radiation.

These gravitational wave fluctuations of the spacetime metric may be written as

$$\bar{h}_{\alpha\beta}^{TT} = \begin{pmatrix} -h_+ & 0 & 0 \\ 0 & h_+ & 0 \\ 0 & 0 & 0 \end{pmatrix}_{\alpha\beta} \times \mathrm{Re} \left[\exp \left\{ \frac{2i}{c} \left[ckz - \omega_T \left(ct - z \right) \right] \right\} \right], \tag{1.34}$$

where $\alpha, \beta = 1, 2, 3$, and

$$h_+ = \frac{G}{zc^4}\left(\frac{\omega_T}{\Delta\omega}\right)^2 \gamma_\omega E_0^2 V = \frac{Gm_i}{z\rho c^4}\left(\frac{\omega_T}{\Delta\omega}\right)^2 \gamma_\omega E_0^2 N, \tag{1.35}$$

where h_+ is the "plus" amplitude of the gravitational wave polarization.

The frequency of natural transverse oscillations of ionic crystals $\omega_T \sim 10^{13} s^{-1}$ [7] is close to the optical frequency range $\omega_{opt} \sim 10^{14} - 10^{15} s^{-1}$; therefore, the condition $\Delta\omega < \omega_T$ is satisfied for laser radiation, and Eq. (1.35) may be used in this case to find the gravitational wave amplitude.

Now we shall estimate the maximum gravitational wave amplitude h_+, taking into account the critical ion displacement $\xi_{\max}$ approaching the dissolution of the crystal lattice.

Equation (1.20), taking into account (1.31), yields:

$$(E_0)_{\max} = -2\xi_{\max}\Delta\omega = \mp 2\left(\frac{l_{\max}}{2}\right)\sqrt{\rho}\Delta\omega \tag{1.36}$$

Substituting (1.36) into (1.35), we obtain:

$$h_+ = \frac{G}{zc^4}\gamma_\omega l_{\max}^2 \omega_T^2 m_i N \tag{1.37}$$

For ordinary ionic crystals, the dispersion coefficient is $\gamma_\omega \sim 1$ [7], the model parameters are $\omega_T \sim 10^{13} s^{-1}$, $l_{\max} \sim 10^{-9}$ m, $z = 10$ m, $m_i \sim 10^{-26}$ kg.

For these values, Eq. (1.37) yields:

$$h_+ \cong 10^{-63} \times N \tag{1.38}$$

For $N \sim 10^{23} h_+ \cong 10^{-40}$, which corresponds to the result (1.18) that was derived from the shock wave model for the same distance of $z = 10$ m.

Next, we estimate the maximum value of the gravitational radiation power [1]:

$$P_g = -\frac{dE_g}{dt} = \frac{G}{5c^5}\left\langle \dddot{Q}_{ij}\dddot{Q}_{ij}\right\rangle = \frac{G}{40c^5}\frac{m_i^2\gamma_\omega^2\omega_T^6}{\rho^2(\Delta\omega)^4}E_{0\max}^4 N^2 = \frac{G}{40c^5}m_i^2\gamma_\omega^2\omega_T^6 l_{\max}^4 N^2 \tag{1.39}$$

For the same parameters of the ionic crystal, Eq. (1.39) yields:

$$P_g \cong 10^{-64} \times N^2 \simeq 10^{-18}\ \text{W}. \tag{1.40}$$

Therefore, for the model describing interaction of dielectrics with electromagnetic waves, the estimate of the gravitational wave radiation power is higher than in the shock wave model (1.19) at the same amplitude.

Note also that the estimates derived depend on the parameters of the crystals, and thus it is possible to improve the characteristics of gravitational waves by selecting specific dielectric media.

However, the amplitude of gravitational waves induced in these processes is too small to use existing methods of high-frequency gravitational wave detection [22].

1.6 Gravitational Waves Induced by Electromagnetic Waves

In the framework of linearized general relativity, weak gravitational fields are considered as small perturbations of Minkowski space (1.3), while for a certain selected direction $x^1 = x$ the gravitational field Eq. (1.1) are explicitly written as follows [18–21]:

$$-h_{11,00} + 2h_{01,01} - h_{00,11} + h_{22,00} + h_{33,00} = 2\kappa T_{00}, \tag{1.41}$$

$$h_{11,00} - 2h_{01,01} + h_{00,11} + h_{22,11} + h_{33,11} = 2\kappa T_{11}, \tag{1.42}$$

$$h_{22,01} + h_{33,01} = 2\kappa T_{01}, h_{22,11} - h_{22,00} = 2\kappa T_{22}, h_{33,11} - h_{33,00} = 2\kappa T_{33}, \tag{1.43}$$

where $h_{\mu\nu,\sigma\rho} = \frac{\partial^2 h_{\mu\nu}}{\partial x^\sigma \partial x^\rho}$, $\frac{\partial}{\partial x^0} = \frac{1}{c}\frac{\partial}{\partial t}$, $\frac{\partial}{\partial x^1} = \frac{\partial}{\partial x}$, and the stress-energy tensor $T_{\mu\nu}$ satisfies the conservation laws

$$\frac{\partial T^{\mu\nu}}{\partial x^\mu} = 0. \tag{1.44}$$

In general, the system of Eqs. (1.41)–(1.44) may possess solutions in the form of both transverse gravitational waves defined by the components h_{22} and h_{33}, and longitudinal gravitational waves h_{00}, h_{10}, h_{01}, h_{11} [18–21].

In vacuum ($T_{\mu\nu} = 0$), these equations have the following form:

$$-h_{11,00} + 2h_{01,01} - h_{00,11} + h_{22,00} + h_{33,00} = 0, \tag{1.45}$$

$$h_{11,00} - 2h_{01,01} + h_{00,11} + h_{22,11} + h_{33,11} = 0, \tag{1.46}$$

$$h_{22,01} + h_{33,01} = 0, h_{22,11} - h_{22,00} = 0, h_{33,11} - h_{33,00} = 0, \tag{1.47}$$

which implies the following condition:

$$h_{11,00} - 2h_{01,01} + h_{00,11} = 0. \tag{1.48}$$

The system of Eqs. (1.45)–(1.47) may hence be written as follows:

$$h_{22,00} + h_{33,00} = 0, h_{22,01} + h_{33,01} = 0, h_{22,01} + h_{33,01} = 0, \tag{1.49}$$

$$h_{22,11} - h_{22,00} = 0, h_{33,11} - h_{33,00} = 0. \tag{1.50}$$

and has solutions in the form of transverse plane gravitational waves [1]:

$$h_{22} = -h_{33} = h_0 \cos\left[\omega_g \left(t \mp \frac{x}{c}\right)\right], \tag{1.51}$$

where h_0 and ω_g are the amplitude and frequency of gravitational waves in empty space, the superscript corresponding to the case of wave propagation in the positive direction $x^1 = x$, the subscript indicating the opposite direction.

Therefore, the internal solutions of the system of Eqs. (1.41)–(1.43) for a given stress-energy tensor $T_{\mu\nu}$ correspond to the gravitational field associated with the source, and the solution (1.51) in the form of transverse plane gravitational waves is external. Consequently, comparing the characteristics of bound and free gravitational waves involves selecting a certain type of gravitational field sources to match gravitational wave solutions of Eqs. (1.41)–(1.43). Electromagnetic waves in various media and external fields may be considered as a possible source [18–21].

The relationship between internal and external solutions is determined from the condition of continuity for the energy flux density of gravitational waves [19–21].

$$ct^{01} = -\frac{c}{4\kappa}\left(\partial_0 h_{\alpha\beta} \partial_1 h^{\alpha\beta}\right), \alpha, \beta = 1, 2, 3, \tag{1.52}$$

at the interface separating the region filled with an electromagnetic field from empty space $(ct^{01})_{\text{in}} = (ct^{01})_{\text{out}}$ in the event of complete absorption of the electromagnetic wave by a surface.

Therefore, analysis of gravitational waves associated with an electromagnetic field is of interest for both surveying how they differ from free gravitational waves in empty space and assessing the possibility of developing sources of observable gravitational waves.

1.7 Generation of Gravitational Waves by a System of Standing Electromagnetic Waves

The paper [21] considered a source of gravitational waves in the form of standing electromagnetic waves in an electromagnetic resonator of length L with mirrors located at the points $x = -\frac{L}{2}$ and $x = \frac{L}{2}$ and the following non-zero components of the electric and magnetic fields:

$$E_y = E_1\left[\cos\left(\omega_1\left(t-\frac{x}{c}\right)\right)+\cos\left(\omega_1\left(t+\frac{x}{c}\right)\right)\right] \\ +E_2\left[\cos\left(\omega_2\left(t-\tfrac{x}{c}\right)\right)+\cos\left(\omega_2\left(t+\tfrac{x}{c}\right)\right)\right] \tag{1.53}$$

$$H_z = H_1\left[\cos\left(\omega_1\left(t-\frac{x}{c}\right)\right)-\cos\left(\omega_1\left(t+\frac{x}{c}\right)\right)\right] \\ +H_2\left[\cos\left(\omega_2\left(t-\frac{x}{c}\right)\right)-\cos\left(\omega_2\left(t+\frac{x}{c}\right)\right)\right] \tag{1.54}$$

where $E_1 = H_1$ and $E_2 = H_2$.

Non-zero components of the stress-energy tensor are:

$$T_{00} = T_{11} = \frac{E_1E_2}{4\pi}\left[\cos\left(\tilde{\omega}\left(t-\frac{x}{c}\right)\right)+\cos\left(\tilde{\omega}\left(t+\frac{x}{c}\right)\right)\right], \tag{1.55}$$

$$T_{01} = T_{10} = -\frac{E_1E_2}{4\pi}\left[\cos\left(\tilde{\omega}\left(t-\frac{x}{c}\right)\right)-\cos\left(\tilde{\omega}\left(t+\frac{x}{c}\right)\right)\right], \tag{1.56}$$

where $\tilde{\omega} = \omega_1 - \omega_2$ is the difference frequency, while this stress-energy tensor with its components satisfies the conservation laws (1.44).

Exact solutions of the gravitational field Eqs. (1.41)–(1.43) in this case are [21]:

$$h_{00} = h_{11} = -\frac{4GE_1E_2}{c^3\tilde{\omega}}\left(\frac{L}{2}+x\right)\sin\left(\tilde{\omega}\left(t-\frac{x}{c}\right)\right) \\ -\frac{4GE_1E_2}{c^3\tilde{\omega}}\left(\frac{L}{2}-x\right)\sin\left(\tilde{\omega}\left(t+\frac{x}{c}\right)\right) \tag{1.57}$$

$$h_{01} = h_{10} = \frac{4GE_1E_2}{c^3\tilde{\omega}}\left(\frac{L}{2}+x\right)\sin\left(\tilde{\omega}\left(t-\frac{x}{c}\right)\right) \\ -\frac{4GE_1E_2}{c^3\tilde{\omega}}\left(\frac{L}{2}-x\right)\sin\left(\tilde{\omega}\left(t+\frac{x}{c}\right)\right) \tag{1.58}$$

$$h_{22} + h_{33} = -\frac{4GE_1E_2}{c^2\tilde{\omega}^2}\cos\left(\tilde{\omega}\left(t-\frac{x}{c}\right)\right)-\frac{4GE_1E_2}{c^2\tilde{\omega}^2}\cos\left(\tilde{\omega}\left(t+\frac{x}{c}\right)\right), \tag{1.59}$$

The external gravitational wave solutions are written as follows [19]:

$$h_{22} = -h_{33} = \frac{2GE_1E_2}{c^2\tilde{\omega}^2}\cos\left(\tilde{\omega}\left(t\mp\frac{x}{c}\right)\right) = h_0\cos\left(\tilde{\omega}\left(t\mp\frac{x}{c}\right)\right). \tag{1.60}$$

Therefore, the spacetime metric in the vicinity of an electromagnetic resonator filled with two standing electromagnetic waves is defined as:

$$\begin{aligned} ds^2 = c^2dt^2 - dx^2 - \left[1 - h_0 \cos\left(\tilde{\omega}\left(t \mp \frac{x}{c}\right)\right)\right] dy^2 \\ - \left[1 + h_0 \cos\left(\tilde{\omega}\left(t \mp \frac{x}{c}\right)\right)\right] dz^2, \end{aligned} \quad (1.61)$$

that is, it contains only the components of one of the types of $h_+(x,t) = h_0 \cos\left(\tilde{\omega}\left(t \mp \frac{x}{c}\right)\right)$ gravitational wave polarization and it is diagonal.

In this case, the ratio of the amplitudes of the bound longitudinal and free transverse gravitational waves is:

$$r_{gw} \equiv \frac{h_0^{(c)}}{h_0} \simeq \frac{\tilde{\omega} L}{c} = 2\pi \left(\frac{L}{\lambda_{gw}}\right), \quad (1.62)$$

and for a small difference frequency, such as that $\left(\frac{\tilde{\omega} L}{c}\right) \ll 1$, we obtain $r_{gw} \ll 1$, meaning that, with further propagation of the gravitational wave after absorption of the electromagnetic one, most of the gravitational field is located outside the source in the form of low-frequency transverse gravitational waves with a frequency $\omega_{gw} = \tilde{\omega}$.

1.8 The Ratio of the Source Parameters to Those of the GW Detector

As part of the problem of generating and detecting gravitational waves, the controlled parameters of the source comprise the electric field strengths $E_1 = E_2 = E_0$ in the resonator and their difference frequency $\tilde{\omega}$, while the controlled parameter of the detector is the amplitude of the gravitational waves detected h_0.

Considering the generation of standing electromagnetic waves with components (1.53)–(1.54) by means of laser radiation in an electromagnetic resonator, the electric field strength inside the resonator is determined as follows:

$$E_0^2 = \frac{8\pi}{c} I_L \times Q = \frac{8\pi}{c} \left(\frac{P_L}{S}\right) \times Q, \quad (1.63)$$

where P_L is the pumping laser power in the resonator, S is the cross-sectional area of the beam, Q is the resonator Q-factor, and, taking into account $f_{gw} = \tilde{f} = \frac{\tilde{\omega}}{2\pi}$.

Equation (1.60) yields expressions for estimating the difference frequency:

$$f_{gw} = \tilde{f} = \frac{2}{\sqrt{\pi}} \left(\frac{G}{c^3} \frac{P_L}{S} Q \right)^{\frac{1}{2}} \times h_0^{-\frac{1}{2}} \text{ [Hz]}, \tag{1.64}$$

and for assessing the laser radiation power:

$$P_L = \left(\frac{\pi c^3}{4G} \right) \frac{S h_0}{Q} \tilde{f}^2 \text{ [W]}. \tag{1.65}$$

Assuming the cross-sectional area of the beam to be $S \simeq 1\ \text{cm}^2 = 10^{-4}\ \text{m}^2$, and the resonator Q-factor to be $Q = 10^6$, Eq. (1.65) yields:

$$P_L \simeq 3 \times 10^{25} \times h_0 \times \left(\frac{f}{\text{Hz}} \right)^2 \text{ [W]}. \tag{1.66}$$

Expression (1.66) defines the relationship between the parameters of the source and the detector in an implementation of the Hertz experiment, that is, an experiment in generating and detecting gravitational waves with controlled characteristics of the source and the detector.

1.9 Conclusion: Parameter Estimates for the Source and the GW Detectors

We shall now consider the estimates of laser radiation power necessary for performing a GW experiment using certain methods for gravitational wave detection.

First, consider the application of the interferometric method of gravitational wave detection, said method based on observing how gravitational radiation affects the interference pattern in Michelson interferometers [3, 4].

In the case of low-frequency gravitational waves, the frequency of laser radiation in the interferometer significantly exceeds the gravitational wave frequency $\Omega \gg f_{gw}$, which means that the length of the gravitational wave significantly exceeds the interferometer arm length $\lambda_{gw} \gg L$.

For the aLIGO and KAGRA interferometers, the maximum sensitivity $h_0 \simeq 10^{-23}$ is reached at the frequency $f_{gw} \simeq 100 - 300$ Hz [23]. If these values are treated as optimal, Eq. (1.66) yields the laser power $P_L \simeq 3 \times (10^6 - 10^7)$ W, necessary to implement the Hertz experiment using the gravitational wave generation method proposed.

It should also be noted that the new generation of ground gravity-wave antennas in development aims at sensitivity $h_0 \simeq 10^{-25}$ at frequencies $f_{gw} \sim 100$ Hz (Einstein Telescope) [24] and $h_0 \simeq 10^{-26}$ at frequencies $f_{gw} \sim 10^3$ Hz (NEMO) [25]. The required power of the resonator pumping laser in this case is hence estimated to be $P_L \simeq 3 \times (10^4 - 10^5)$ W.

The paper [26] proposed to detect gravitational waves by measuring electric field in superconductors, said field occurring in response to a gravitational wave. The sensitivity of these detectors at the resonant frequency (corresponding to the natural frequency of the superconductor ion oscillations) $f_{gw} \simeq 10^3$ Hz was estimated as $h_0 \simeq 10^{-24}$. In this case, the laser power required to implement the Hertz experiment is equal to $P_L \simeq 3 \times 10^7$ W.

The papers [27, 28] estimated the effect that elastic deformations of a crystal lattice in the field of a plane gravitational wave had on the optical properties of the crystal. In [27], measuring the phase difference $\delta\varphi$ of two beams propagating in the crystal orthogonally to the direction of the gravitational wave propagation was investigated as a detection method, while the limiting sensitivity of these detectors was also estimated to be $h_0 \simeq 10^{-24}$.

In this case, the power of the source laser radiation is:

$$P_L \simeq 3 \times 10^7 \times \left(\frac{f}{\text{Hz}}\right)^2 [\text{W}], \tag{1.67}$$

that is, for the frequency $\tilde{f} = f_{gw} \simeq 10^3$ Hz we have a similar estimate of the power of laser radiation $P_L \simeq 3 \times 10^7$ W. However, unlike in the case of the previous method, it is possible to detect gravitational waves in a wider frequency range.

Therefore, a laser radiation source with a power $P_L \geq 3 \times 10^6$ W is required to implement the Hertz experiment based on the gravitational wave generation method proposed. Although modern lasers are capable of much higher power $P_L \simeq 10^{15}$ W, this radiation is generated in the form of short pulses $\tau \leq 1$ ns [29, 30], the analytical description of which differs from the plane electromagnetic wave model under consideration (1.53) and (1.54). In [31], the amplitude of gravitational waves induced by a similar short laser radiation pulse was estimated as $h_0 \sim 10^{-37}$.

When operating in continuous mode to generate a system of standing electromagnetic waves with components (1.53) and (1.54) in an electromagnetic resonator, the maximum laser radiation power is significantly lower than the required value $P_L \sim 10^3$ W [32]. For a given radiation power, Eq. (1.66) yields the amplitude of induced gravitational waves of the order $h_0 \sim 10^{-27}$ (at the frequency $f_{gw} \simeq 100$ Hz) and $h_0 \sim 10^{-29}$ (at the frequency $f_{gw} \simeq 10^3$ Hz).

Therefore, gravitational waves induced by a system of electromagnetic waves at a differential frequency cannot be detected using contemporary detectors since the difference between the amplitude of these gravitational waves and the sensitivity limit of modern detectors is 4–5 orders of magnitude, while for prospective detectors the difference is 2–3 orders of magnitude.

This result may be interpreted as pointing towards technical, rather than fundamental, problems in the implementation of the Hertz experiment using the approach proposed. It should be noted that although realistic estimates of the amplitude of low-frequency gravitational waves that can be obtained within the scope of the approach proposed $h_0 \sim 10^{-27} - 10^{-29}$ do not imply any potential for detecting them directly via existing experimental facilities, this value is much higher than the amplitude of high-frequency gravitational waves induced by perturbations of matter $h_0 \sim 10^{-40}$ or a powerful pulse of laser radiation $h_0 \sim 10^{-37}$.

We should also note that the effect of gravitational wave amplification at the difference frequency of electromagnetic waves $h_0 \sim \tilde{\omega}^{-2}$ may be regarded as the basis for further research in the field of methods for generating gravitational waves in laboratory conditions via modifying the electromagnetic field configuration in the resonator and increasing the detector sensitivity.

References

1. Maggiore, M.: Gravitational Waves. Volume 1: Theory and Experiments, p. 576. Oxford University Press (2007). https://doi.org/10.1093/acprof:oso/9780198570745.001.0001
2. Fomin, I.V., Chervon, S.V., Morozov, A.N.: Gravitatsionnye Volny Ranney Vselennoy (Gravitational Waves of the Early Universe), 154 p. BMSTU Press (2018)
3. Abbott, B.P., Abbott, P., Abbott, T.D., et al.: (LIGO Scientiéc Collab., Virgo Collab.): Observation of gravitational waves from a binary black hole merger. Phys. Rev. Lett. **116**(6), 061102 (2016). https://doi.org/10.1103/PhysRevLett.116.061102
4. Abbott, B.P., Abbott, P., Abbott, T.D., et al.: (LIGO Scientific Collaboration, Virgo Collaboration, Fermi Gamma-ray Burst Monitor, and INTEGRAL): Gravitational waves and gamma-rays from a binary neutron star merger: GW170817 and GRB 170817A. Astrophys. J. Lett. **848**, L13 (2017). https://doi.org/10.3847/2041-8213/aa920c
5. Gertsenshtein, M.E., Pustovoit, V.I.: On the detection of gravitational waves of low frequencies. JETP. **43**, 605–607 (1962) [Translate Gertsenshtein, M.E., Pustovoit, V.I.: Sov. Phys. JETP **16**, 433 (1963)]
6. Kadlecova, H., Klimo, O., Weber, S., Korn, G.: Gravitational wave generation by interaction of high power lasers with matter using shock waves. Eur. Phys. J. D. **71**(4), 89 (2017). https://doi.org/10.1140/epjd/e2017-70586-y
7. Fomin, I.V., Gladyshev, V.O., Gorelik, V.S., Kauts, V.L., Kayutenko, A.V., Sharandin, E.A.: Laboratory sources of gravitational waves. J. Phys. Conf. Ser. **1705**(1), 012004 (2020). https://doi.org/10.1088/1742-6596/1705/1/012004
8. Grishchuk, L.P., Sazhin, M.V.: Radiation of gravitational waves by an electromagnetic resonator. JETP. **65**(2), 441–454 (1973)
9. Grischuk, L.P., Sazhin, M.V.: Generation and detection of standing gravitational waves. JETP. **68**, 1569–1582 (1975)
10. Grishchuk, L.P.: Gravitational waves in space and in the laboratory. Sov. Phys. Usp. **20**, 319 (1977). https://doi.org/10.1070/PU1977v020n04ABEH005327
11. Denisov, V.I.: The gravitational field inside a spherical electromagnetic resonator. Moscow Univ. Phys. **32**(5), 41–45 (1977)
12. Gertsenshteyn, M.E.: Wave resonance of light and gravitational waves. JETP. **14**(1), 84–85 (1962)
13. Zeldovich Ya, B.: Electromagnetic and gravitational waves in a stationary magnetic field. JETP. **38**(4), 652–654 (1974)
14. Kolosnitsyn, N.I., Rudenko, V.N.: Gravitational Hertz experiment with electromagnetic radiation in a strong magnetic field. Phys Scr. **90**(7), 074059 (2015). https://doi.org/10.1088/0031-8949/90/7/074059

15. Gorelik, V.S., Pustovoit, V.I., Gladyshev, V.O., et al.: Generation and detection of high frequency gravitational waves at intensive electromagnetic excitation. J. Phys. Conf. Ser. **1051**, 012001 (2018). https://doi.org/10.1088/1742-6596/1051/1/012001
16. Pustovoit, V.I., Gladyshev, V.O., Gorelik, V.S., Kauts, V.L., Kayutenko, A.V., Morozov, A.N., Fomin, I.V., Sharandin, E.A., Gladysheva, T.M., Portnov, D.I.: Electromagnetic and gravitational waves conversion in a nonlinear dielectric medium by intense light source irradiation. J. Phys. Conf. Ser. **1348**(1), 012008 (2019). https://doi.org/10.1088/1742-6596/1348/1/012008
17. Pustovoit, V.I., Gladyshev, V.O., Kauts, V.L., Morozov, A.N., Gorelik, V.S., Fomin, I.V., Portnov, D.I., Sharandin, E.A., Kayutenko, A.V.: High frequency gravitational waves generation by optical methods. J. Phys. Conf. Ser. **1557**, 012034 (2020). https://doi.org/10.1088/1742-6596/1557/1/012034
18. Morozov, A.N., Pustovoyt, V.I.: Generation and registration of high-frequency coupled gravitational waves. Herald of the Bauman Moscow State Technical University. Ser. Nat. Sci. **1**(88), 46–60 (in Russ.) (2020). https://doi.org/10.18698/1812-3368-2020-1-46-60
19. Morozov, A.N., Pustovoit, V.I., Fomin, I.V.: On the gravitational waves coupled with electromagnetic waves. Prostranstvo, vremya i fundamentalnye vzaimodeystviya [Space Time Fundam. Interact.]. **2**, 53–63 (in Russ.) (2020). https://doi.org/10.17238/issn2226-8812.2020.2.53-63
20. Morozov, A.N., Pustovoit, V.I., Fomin, I.V.: Bound gravitational waves in a dielectric medium and a constant magnetic field. Eur. Phys. J. Plus. **135**(12), 950 (2020). https://doi.org/10.1140/epjp/s13360-020-00961-0
21. Morozov, A.N., Pustovoit, V.I., Fomin, I.V.: Generation of gravitational waves by a standing electromagnetic wave. Gravit. Cosmol. **27**(1), 24–29 (2021). https://doi.org/10.1134/S020228932101014X
22. Aggarwal, N., et al.: Challenges and opportunities of gravitational-wave searches at MHz to GHz frequencies. Living Rev. Rel. **24**(1), 4 (2021). https://doi.org/10.1007/s41114-021-00032-5
23. Akutsu, T., Ando, M., Arai, K., et al.: KAGRA: 2.5 generation interferometric gravitational wave detector. Nat. Astron. **3**(1), 35–40 (2019). https://doi.org/10.1038/s41550-018-0658-y
24. Maggiore, M., Van den Broeck, C., Bartolo, N., et al.: Science case for the Einstein telescope. JCAP. **2020**, 050 (2020). https://doi.org/10.1088/1475-7516/2020/03/050
25. Ackley, K., Adya, V.B., Agrawal, P., et al.: Neutron star extreme matter observatory: a kilohertz-band gravitational-wave detector in the global network. Publ. Astron. Soc. Austral. **37**, 047 (2020). https://doi.org/10.1017/pasa.2020.39
26. Peng, H., Torr, D.G.: The electric field induced by a gravitational wave in a super-conductor: a principle for a new gravitational wave antenna. Gen. Relat. Gravit. **22**(1), 53–59 (1990). https://doi.org/10.1007/BF00769245
27. Vinet, J.Y.: Elasto-optical detection of gravitational waves. Annales de l'l.H.P. Physique Theorique. **30**(3), 251–262 (1979)
28. Sabin, C., Bruschi, D.E., Ahmadi, M., et al.: Phonon creation by gravitational waves. New J. Phys. **16**, 085003 (2014). https://doi.org/10.1088/1367-2630/16/8/085003
29. Bagnoud, V., Blazevic, A., Borneis, S.: PHELIX: a petawatt-class laser recently com-missioned for experiments in plasma and atomic physics. J. Phys. Conf. Ser. **194**, 152028 (2009). https://doi.org/10.1088/1742-6596/194/15/152028
30. Korzhimanov, A.V., Gonoskov, A.A., Khazanov, E.A., et al.: Horizons of petawatt laser technology. Physics-Uspekhi. **54**(1), 9–28 (2011). https://doi.org/10.3367/UFNe.0181.20n01c.0009
31. Lageyre, P., D'Humieres, E., Ribeyre, X.: Gravitational influence of high power laser pulses. Phys. Rev. D. **105**(10), 104052 (2021). https://doi.org/10.1103/PhysRevD.105.104052
32. Sprangle, P., Hafizi, B., Ting, A., et al.: High-power fiber lasers for directed-energy applications. Appl. Opt. **54**(31), F201–F209 (2015). https://doi.org/10.1364/AO.54.00F201

Chapter 2
Color Theory: Basic Color Models 10.0 Basic Music Theory 5.0: Harmonics

Carl Swopes

Abstract Vibrating strings and Chladni Plates (Cymatics) have been used to demonstrate or explain harmonic vibrations. Per these examples, Jacob's Ladder may or may not define a valid concept for a harmonic, sub-harmonic, or a harmonic distribution. Vibrating strings and Chladni Plates define harmonic vibrations based on the concept of a standing wave with the associated antinodes and nodes. Jacobs Ladder is based on sequential, exponential values of golden ratio proportions. It is independent of the system it is being used to evaluate. Per this color research, Jacob's Ladder has been applied to vibrations in free space (midair). Music theory allows Jacob's Ladder to be applied to the actively vibrating strings of a piano.

This color research believes that there are differences between things that vibrate, things that are vibrated, and things that ring when impacted. The electromagnetic spectrum is a constant vibration. Vibrating string and Chladni Plate experiments are examples of things being vibrated. The mechanical vibrations of sound and piano strings are examples of things that ring when impacted. Something that is ringing could also be or produce pressure waves. At the atomic level, everything vibrates, with few exceptions.

The color theory of this research developed while this researcher was attempting to define color within the visible light spectrum, by using numbers and geometry. A numerical and geometric color pattern was discovered within the visible light spectrum. If the entire electromagnetic spectrum is light, then the color pattern should be applicable to the entire electromagnetic spectrum. A repeating color pattern was discovered within the electromagnetic spectrum, and it was named Jacob's Ladder.

The frequencies at the lower end of the electromagnetic spectrum are also found within the range of audible sound (20 Hz to 20,000 Hz). If frequencies are frequencies, cycles/second, then Jacob's Ladder would be applicable to audible

https://sites.google.com/view/color-theory-basic-mechanics/home

C. Swopes (✉)
Swopes Inc./Color Research, Las Cruces, NM, USA

A. Adhikari et al. (eds.), *Quantitative Frontiers*, Industrial and Applied Mathematics, https://doi.org/10.1007/978-981-96-9435-8_2

sound (mechanical vibrations). To test this assumption, the color pattern was applied to basic music theory.

Pythagoras doubled or halved Primary Tone "A" at 440 Hz to obtain four octaves above and below the Primary Tone "A" reference. Nine music octaves are defined which partially overlap their adjacent music octaves. As a group, the nine music octaves, A_0 to A_8, have little correlation to the 13 color octaves, C_0 to C_{12} (A_0 to A_{12}) for the range of audible sound. Individually, a single music octave and a single color octave can approximate each other. Color octaves divide the range of audible sound into 13 color octaves, compared to the nine octaves of basic music theory. Color octaves are color harmonics and color harmonics are color octaves.

Each music octave contains 12 semitones, even temperament, and arranged per the counting of the Perfect Fifth. It results in the chromatic scale of the Circle of Fifths. The Circle of Fifths is like a spiral staircase. Ascending 360° clockwise defines a rising color octave or color harmonic. Descending 360° counterclockwise defines a lowering color octave or color sub-harmonic.

Jacob's Ladder promotes the concept of color harmonic vibrations with a correlation to color, radiation, sound, and basic music theory. It is based on the music octave and chromatic semitones as described by Pythagoras' Circle of Fifths. We are in Primary Tone "F". Sunlight is in Primary Tome "F".

Jacob's Ladder is an exponential, 12th-root, "golden ratio-based" distribution of harmonic frequencies (vibrations) along the x-axis. Jacob's Ladder is not linear. It is on an exponential, logarithmic, and golden ratio curve. There are 12 color references (Six Primary Colors and Six Primary Color Interfaces). Within Jacob's Ladder, adjacent color harmonics or color octaves meet at the same point, but they do not overlap.

Vibrating string and Chladni Plate experiments, with their associated standing waves, antinodes, and nodes are accepted methods of defining a harmonic vibration. Jacob's Ladder as, a harmonic distribution, color harmonics, and color octaves, calls the definition of a harmonic into question. An effort will be made to contrast vibrating string and Chladni Plates experiments with the mechanics of Jacob's Ladder.

Keywords Color theory · Vibrating strings · Cymatics · Chladni plates · Standing waves · Jacob's Ladder · Color harmonics · Color sub-harmonics · Basic music theory · Color octaves · Semitones · Rising octave · Lowering octave · Golden ratio · Nodes · Antinodes · Circle of fifths · Reflected waves · Primary red [Anchor] · Frequency · Resonant frequency · Harmonic frequency · Mechanical vibrations · Electromagnetic radiation · Pressure waves · Noise · Kuiper belt · Oort cloud · RGB (red · green · blue) · CMY (cyan · magenta · yellow) · Optimization · Composite color wave · Fibonacci sequence · Metatron's cube

2.1 Harmonics, Standing Waves, and Jacob's Ladder

Figure 2.1 shows the pattern of a vibrating string, **standing wave** per the first, second, and third harmonic.

The first harmonic is the antinode or the maximum **y-axis displacement** of the standing wave. The standing wave is created by a reflected wave. Nodes are points where vibrations are zero or effectively zero.

The second, third, etc. harmonics are identified experimentally. Many current vibrating string experiments may choose to vary the tension on the string while using the same frequency to create additional standing waves.

Jacob's Ladder identifies a numerical and geometry correlation between the frequencies associated with color, radiation, sound, and music per the color hexagram, Circle of Fifths and A440 method of tuning. In effect it adheres to a set of standards. Jacob's Ladder is independent of the media.

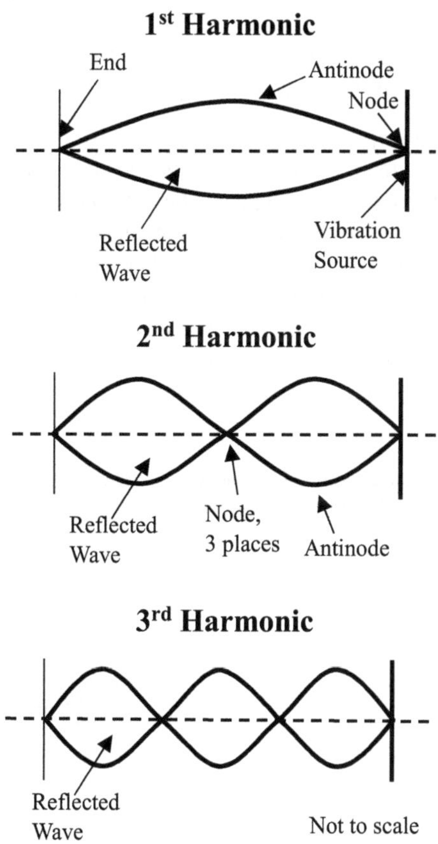

Fig. 2.1 Vibrating string standing wave & harmonic

Chladni Plates and vibrating strings have standing waves, reflected waves, antinodes, nodes, and vibration source. The Chladni Plate reflected waves are caused by the edge of the plate.

Note 2.1: Standing waves, reflected waves, antinodes, and nodes are not found in Jacob's Ladder.

Per Fig. 2.1 the formation of a standing wave with antinodes, and nodes in a vibrating string can be influenced by various conditions:

- Length of the vibrating string
- String thickness
- Material of construction (fibers, metal, plastic, etc.)
- String density
- String tension
- End point attachments or pulley
- Vibration mechanism
- System resonance—synchronous reinforcement by a reflected vibration
- Figure 2.2 does not detail the setup of the experiment
- The experiments in Figs. 2.1 and 2.2 were not detailed to where they are repeatable. Similar experiments are possible but not the same one.

As reported, for a vibrating string per Fig. 2.1, increasing the frequency by sequential whole numbers (positive integers or whole number multiples) causes more nodes to appear. The antinodes (sub-harmonics) below the first harmonic are not discoverable. Only the higher harmonics are discoverable. Per Fig. 2.2, the frequency of a vibrating plate is adjusted upwards until the next clear pattern is observed. It may not be a whole number multiple.

If the vibrations of Figs. 2.1 and 2.2 were audible or radiation, then those vibrations would relate more to Jacob's Ladder. Otherwise, the vibrations are harmonic mechanical vibrations of a given vibrating string or plate. As low

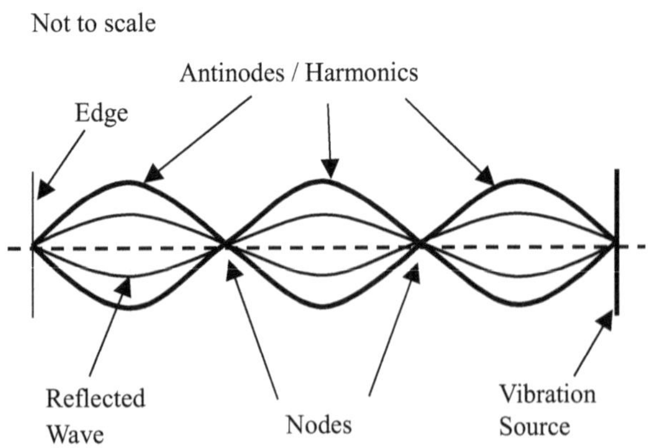

Fig. 2.2 Antinodes & nodes of a Chladni Plate—Example

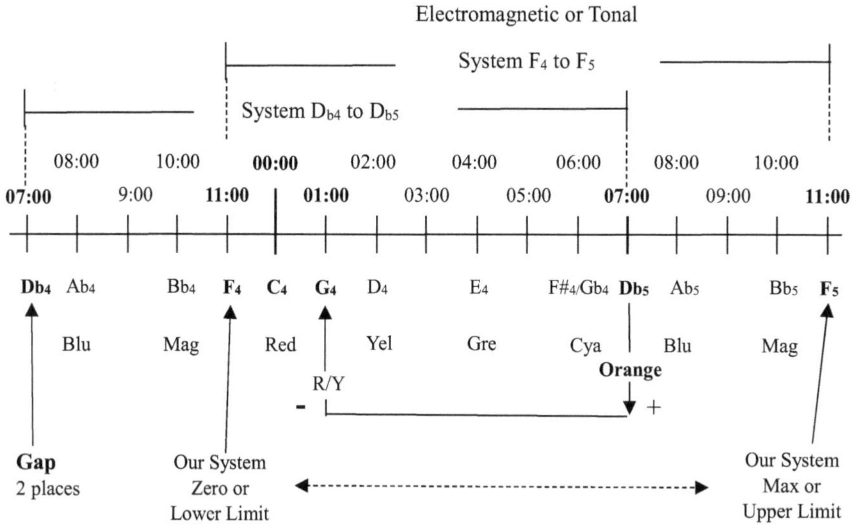

Fig. 2.3 Electromagnetic harmonic or tonal color octave expressed linearly in primary tone F

frequency mechanical vibrations, they may not register within the range of audible sound. If not audible sound, the vibrations could register as pressure waves.

The string or Chladni Plate is being vibrated versus the string or Chladni Plate being the source of the vibration. Per the vibrating string and vibrating Chladni Plate experiments, the string or Chladni Plate does not continue to ring and maintain the standing wave once the vibrational source is removed.

Figure 2.3 is an electromagnetic color harmonic or tonal color octave expressed linearly in Primary Tone "F" per the Circle of Fifths and A440 method of tuning. Per Color Theory and Basic Music Theory, nodes would exist for Primary Tone "F_4–F_5" at the 11-o-clock position. The beginning of the color harmonic or color octave is at the 7-o-clock position. Primary Tone "F" is 1 of 12 "Primary Tones" within a chromatic music scale, or it relates to 1 of 12 "Primary Color References."

Figure 2.3 identifies the basic structure of the color harmonic and color octave. Figure 2.3 is a linear expression of the Circle of Fifths which may help in visualizing the Jacob's Ladder correlation to the electromagnetic spectrum and the range of sound.

Note 2.2:

Figure 2.3 is a linear expression of the color octave or color harmonic. System F_4 to F_5 does not wrap around to pick up A_{b4} and B_{b4}. It picks up the A_{b5} and B_{b5} of the next higher color octave or color harmonic. "The Gap" happens at the 7-o-clock (D_b) position. The Gap causes Orange to happen at the 1-o-clock position with an energy release when Primary Red and Primary Yellow are not interfacing. Jacob's Ladder also addresses the concept of a vibrating string in terms of wavelengths. But the vibrations exist in free space (midair). Per this color research, it is currently unknown if the vibrating string or vibrating plate in Figs. 2.1 and 2.2 could address the concepts of music or Orange.

Note 2.3:
The lower frequency Color Octave's North Pole interfaces with the higher frequency Color Octave's South Pole at $D_\flat$ or 7-o-clock position. At the $D_\flat$ interface, the Color Octaves meet at a common point, but they do not overlap. A small gap separates the adjacent Color Octaves. These adjacent Color Octave interfaces are moving in relatively opposite directions. The system interface response to a vibration moving past the 7-o-clock ($D_\flat$) position is displayed as Orange at the 1-o-clock position. Primary Red and Primary Yellow are not interfacing to create Orange in the spectral displays of a prism or rainbow. If Primary Red and Primary Yellow were interfacing (coupling), there should be a color gradient between Primary Red and Primary Yellow. The spectral displays of a prism or rainbow contain Orange as one of the seven bands of color. The color bands could imply that each color band represents a very narrow range of frequencies or a single frequency.

Note 2.4:
Possibly, a static magnetic field equates/relates to a node, standing wave or the Gap between adjacent harmonics or color octaves, see Figs. 2.1, 2.2, and 2.3. A node is a point of zero movement and zero frequency. A static magnetic field has zero movement and zero frequency.

Table 2.1 considers the harmonic correlation of the F_4 music octave at $\approx$349.23 Hz versus the comparable F_6 color octave at $\approx$340.3212604 Hz. Music octave F_4 to octave F_5 has a Total Range value of $\approx$349.23 Hz. F_5 is the Upper Limit. The comparable color octave F_6 to F_7 would have a Total Range value of $\approx$210.3309869 Hz. F_7 is the Upper Limit. Basic Music Theory is not just whole number multiples.

The music octave is based on the concept of "doubling" or "halving" the given reference frequency, Primary Tone "A_4" = 440 Hz. The initial construction of the Circle of Fifths is in Primary Tone "F". Once the Circle of Fifths has been constructed, any of the 12 Primary Tones can be selected to identify the order of the 12 Primary Tones for that Primary Tone. The 12 Primary Tones (semitones) of music theory are identified numerically by $^{12}(\surd 2)^{x}$ for doubling (rising octave) or $^{12}(\surd 1/2)^{x}$ for halving (lowering octave) per the A_4 = 440 Hz reference. See Table 2.1.

Rising music octaves are whole number multiples of the A440 Hz reference. Lowering music octaves are the inverse of whole number multiples of the A440 Hz reference (1/whole number multiple). Starting at music octave F_0, all higher music octaves are obtained by doubling the F_0 value, see Table 2.1.

The music octaves of Table 2.1 do not identify standing waves, antinodes, nodes, or any points of no vibrations (zeros). From the A440 method of tuning, half of the "music octaves" are identified by whole number (n) multiples (doubling). The other half, from the A440 reference, is identified by $1/n$ (halving).

Table 2.1 Music octave at $\mathbf{F_4}$ versus color octave at $\mathbf{F_6}$

F_4 349.23 Hz	Music Octaves	Frequency × 1/2 or 2	Frequency × phi or PHI	Color Octaves	Range
	Halving	Values	Values	(#) = 450 THz	Lower Limit
					18.21979281
			18.96527178	= (#) × phi^{64}	$Color_0$ ↑
			30.68650344	= (#) × phi^{63}	$Color_1$
F_0	(*) × 1/16	21.825 Hz	49.65188499	= (#) × phi^{62}	$Color_2$
F_1	(*) × 1/8	43.65 Hz	80.33856605	= (#) × phi^{61}	$Color_3$
F_2	(*) × 1/4	87.31 Hz	129.9907384	= (#) × phi^{60}	$Color_4$
F_3	(*) × 1/2	174.61 Hz	210.3297695	= (#) × phi^{59}	$Color_5$
$\mathbf{F_4}$	**× 1**	**349.23 Hz (*)**	340.3212604	= (#) × phi^{58}	$\mathbf{Color_6}$
F_5	(*) × 2	698.46 Hz	550.6522473	= (#) × phi^{57}	$Color_7$
F_6	(*) × 4	1396.91 Hz	890.9754776	= (#) × phi^{56}	$Color_8$
F_7	(*) × 8	2793.83 Hz	1441.630912	= (#) × phi^{55}	$Color_9$
F_8	(*) × 16	5587.65 Hz	2332.611547	= (#) × phi^{54}	$Color_{10}$
			3774.250804	= (#) × phi^{53}	$Color_{11}$
			6106.875853	= (#) × phi^{52}	$Color_{12}$ ↓
					9492.729267
Tone	Doubling	Tonal value	Color diversity	Exponential	Upper Limit

As a group, the 9 music octaves, A_0 to A_8, have little correlation to the 13 color octaves, $Color_0$ to $Color_{12}$. Individually, a single music octave and a single, color octave can approximate each other

The frequencies in the "Frequency × phi or PHI" column equate to a Primary Red [Anchor] value. Primary Tone "F" relates to the M/R color Interface at the 11-o-clock position

The concept behind the color octave is the harmonic repeating of the tonal pattern. The 12 Primary Tones (semitones) are identified by $^{12}(\sqrt{PHI})^x$ for doubling (rising octave) or $^{12}(\sqrt{phi})^x$ for halving (lowering octave). See Table 2.1. The color octaves are exponential and do not actually double or halve.

Color octaves are Jacob's Ladder harmonics and Jacob's Ladder harmonics are color octaves. This concept applies to the entire range of sound and the entire range of the electromagnetic spectrum. The color octaves of Table 2.1 do not identify standing waves, antinodes, nodes, or any points of no vibrations (zeros).

Jacob's Ladder extends beyond the classical end of the electromagnetic spectrum or range of sound because its scale is exponential. Jacob's Ladder accommodates time base changes. Frequency can be cycles/sec, cycles/min, or cycles/time for mechanical vibrations, pressure waves, or electromagnetic radiation, within reason.

2.2 Cymatics and Chladni Plates

Cymatics is a science, and a vibrating Chladni Plate is one of the devices used to demonstrate various three-dimensional (3D) vibrational patterns. The resulting patterns are influenced by frequency, size, and shape of the vibrating plate.

The original experiments were conducted with a vibrating plate that was lightly covered with sand. Subsequent experiments have replaced sand with oil. Other experiments have replaced the vibrating plate with a vibrating pan containing oil or water. The implication is that the 3D Chladni Plate patterns exist in free space as sound waves (mechanical vibrations) and radiation (electromagnetic vibrations).

Antinodes and nodes are associated with Chladni Plates. The concepts are the same as the antinodes and nodes of a vibrating string. See Figs. 2.1 and 2.2. But the nodes on a Chladni Plate are three-dimensional patterns on both sides of the plate. The Chladni Plate patterns form along the lines of no vibrations (nodes).

It is generally believed in science that everything vibrates and that everything has a resonant frequency. When all system vibrations are combined, the net is the natural or resonant frequency. The resonant frequency itself or if conditions cause the resonant frequency to be amplified, it could result in a system shaking itself to death. This is not the same resonant or harmonic frequency as found in electronics. In electronics, a resonant or harmonic frequency equates to noise.

Note 2.5 (Fig. 2.4):
Standing wave: a wave that does not appear to oscillate.

- Standing waves form because of wave interaction. If waves reflect off a boundary they go in the opposite direction. In the vibrating string and Chladni Plate experiments, outgoing waves interact with incoming waves.

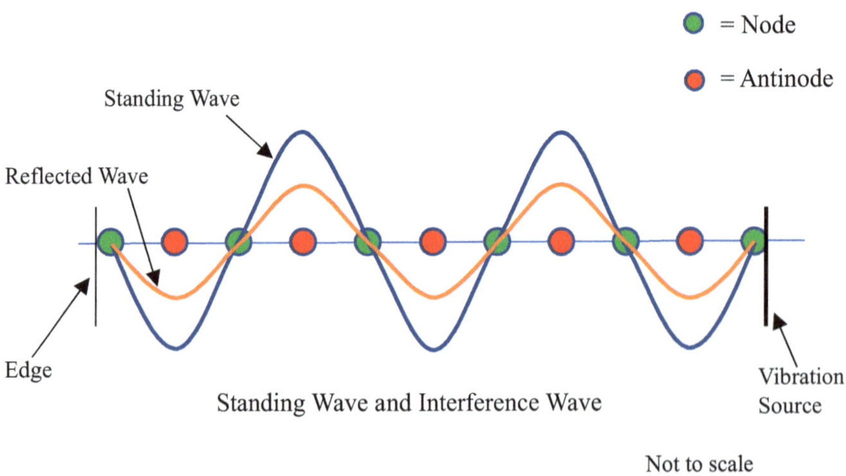

Fig. 2.4 Chladni plate standing waves

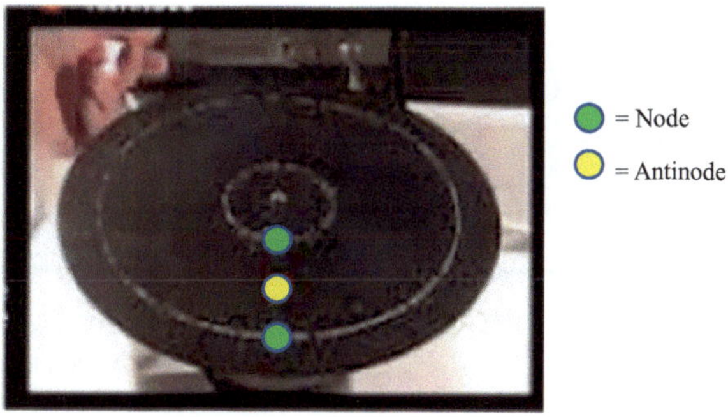

Chladni Plates and Standing Waves – Example of 3rd Harmonic

Fig. 2.5 Chladni plate standing waves

Standing waves have nodes and antinodes.

- Nodes—the points with the smallest amplitude (zero).
  - The points that do not move.
- Antinodes—these are the points with the largest amplitude.
  - The points that have the maximum movement.

Note 2.6 (Fig. 2.5):
Chladni Plates:

- Chladni Plate patterns result from standing waves.
  - Waves of a set frequency are propagated through a plate.
  - As the waves reach the edge of the plate they are reflected.
  - The reflected waves interact with the propagated waves resulting in interference which forms a standing wave when properly synched.
- Wherever reflected waves cause the plate to stop moving is where the sand collects and represents a node (zero motion).
- The areas on the plate without sand indicate an area of movement and represent an antinode (above and below the plate, y-axis deflection).

In contrast, Jacob's Ladder (the electromagnetic spectrum, range of sound, or basic music theory) does not have the edge of a vibrating plate to create reflected waves that result in the creation of standing waves, antinodes, or nodes. The theoretical zeros in Jacob's Ladder are the upper and lower limits of adjacent color harmonics or color octaves. These "zeros" have an assigned value, tone, or frequency. The nodes of a vibrating string or Chladni Plate have no frequency. For Jacob's Ladder, zero is relative. For a vibrating string or Chladni Plate, zero is absolute. If the Chladni Plate vibration is audible, then a correlation to Jacob's Ladder is possible. The single vibration would fit within a specific color harmonic

2nd Harmonic Nodes = 1 | 3rd Harmonic Nodes = 2 | 6th Harmonic Nodes = 5

Chladni Plates and Standing Waves – y-axis amplitude

Fig. 2.6 Chladni plates

or color octave of Jacob's Ladder. A single vibration or whole number multiple of the single vibration would identify with a color harmonic or color octave in Jacob's Ladder.

Could the asteroid belt, Kuiper belt, and/or Oort cloud act like the edge of a Chladni Plate with the solar wind being a vibrational source? Could this concept create reflected waves that develop into standing waves, antinodes, and nodes? Could it design and define a solar system?

Note 2.7 (Fig. 2.6):
Chladni Plates:

- Research suggests that the use of Chladni Plates can allow the visualization of different harmonics or resonant frequencies in three dimensions.
- The harmonic or resonant frequencies demonstrated by a Chladni Plate are represented by the maximum y-axis amplitude of the antinode of the resultant standing wave pattern.

The creation and position of the nodes per the second, third, and sixth harmonic are not harmonic per Jacob's Ladder. This is not to say Chladni Plate harmonics are not harmonic. It says that this color research did not use standing waves to identify the harmonics or color octaves of Jacob's Ladder. Jacob's Ladder uses the Exponential Equation Formula: $y = a \times b^x$—for proportional Total Range values.

$$y = 2.671832554 \times 10^{14} \times \text{phi}^x$$

$$y = 267.1832554\ \text{THz} \times \text{phi}^x$$

267.1832554 THz is the new Baseline Total Range value identified by the 12th-root [$({}^{12}\sqrt{\text{phi}})^x$] calculations.

Above the Baseline value, the calculation is:

$$y = 267.1832554 \text{ THz} \times \text{PHI}^x, \left(\text{phi}^{-x}\right)$$

As stated in Note 2.5 and shown in Fig. 2.4, Standing waves form because of wave interaction. If waves reflect off a boundary they go in the opposite direction. In the vibrating string and Chladni Plate experiments, outgoing waves interact with incoming waves.

It is also interesting to note that Chladni Plate patterns are related to the shape of the vibrated plate. Square vibrated plates tend to make square patterns. Round vibrated plates tend to make circular patterns.

2.2.1 *Technology Gap: What Is a Harmonic?*

Is a harmonic the antinode of a standing wave and the whole number multiples of its originally identified frequency, see Figs. 2.1, 2.2, 2.4 and Note 2.7? This defines the harmonic as being expressed in terms of the y-axis and a media. Demonstrations show that a higher frequency or greater string tension creates more standing waves.

The harmonics of Jacob's Ladder are defined along the x-axis. The color harmonic or color octave has a defined lower limit, upper limit, and Total Range value. For Jacob's Ladder, color potential is defined per the *y-axis*.

The Composite Color Wave shows the six Primary Colors as potentials in the ±*z-axis*. RGB or CMY are harmonics of each other. These potentials are associated with the 12 Primary Color References or 12 Primary Tones per the Circle of Fifths. The relationship is with individual frequencies within the lower limit and upper limit of a color harmonic or color octave.

Standing waves are also associated with electromagnetic waves (transmission lines, electronics, radio transmitters, etc.). In these applications, a standing wave is undesirable because there is no signal output. A harmonic within a signal is seen as noise. The concepts of Jacob's Ladder should be applicable to the original signals in transmission lines, electronics, radio transmitters, etc. as a means of optimization.

A vibrating string, Chladni Plate, and Jacob's Ladder demonstrate different methods of defining a harmonic vibration. The vibrating string and Chladni Plate concepts have been around for a very long time. Jacob's Ladder is a new concept. Are they all saying the same thing? How would the concept of a harmonic be applied to Biology, Chemistry, our solar system, space-time, or universe? Therefore, this is a technology gap.

Note 2.8:

Chladni Plates:

- Chladni Plate patterns were initially discovered by Robert Hooke and subsequently studied by Ernst Chladni.

- Chladni Plate patterns are the result of relatively low frequency vibrations being used to vibrate a plate that is lightly covered with a fine grain sand.
- Chladni Plate patterns can be explained by standing waves.
- Variations to Chladni Plate patterns are made by:
  - Mounting the vibrating plate off-center
  - Using oil instead of sand
  - Vibrating a shallow pan of water or oil
  - Using different shaped plates
  - Varying the frequency of the vibration

2.3 Jacob's Ladder vs. Vibrating String and Chladni Plate

As a contrast to a vibrating string or vibrating plate experiment, Figs. 2.7, 2.8, 2.9 are associated with Jacob's Ladder. The lines or points in the referenced Figures and Tables are color boundaries versus standing waves, antinodes, and nodes. Multiple frequencies are involved in the creation of the color harmonic or color octave patterns. The identified patterns are applicable to color, the electromagnetic spectrum (radiation), sound (mechanical vibrations), and music (color octaves per the Circle of Fifths and the A440 method of tuning). They have a defined order.

In Jacob's Ladder, single frequencies are associated with a specific color harmonic or color octave. The Total Range value of a color harmonic or color octave

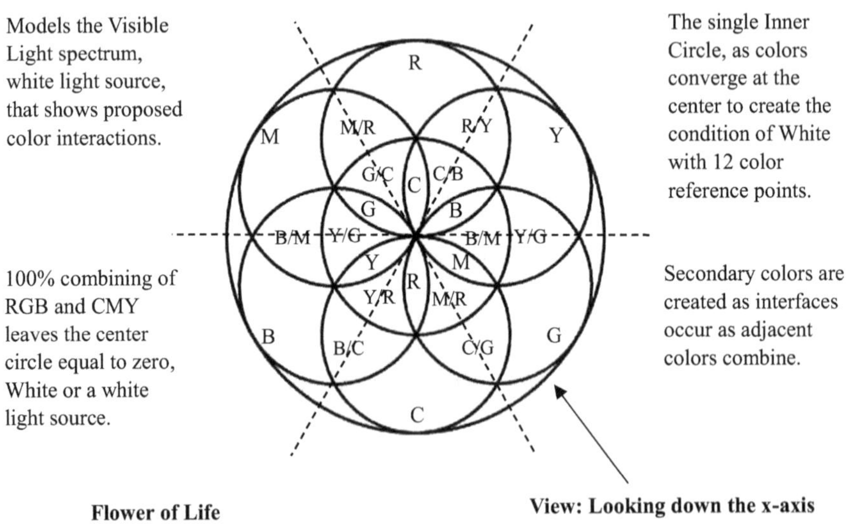

Fig. 2.7 Color combining model (Aka Metatron's Cube)

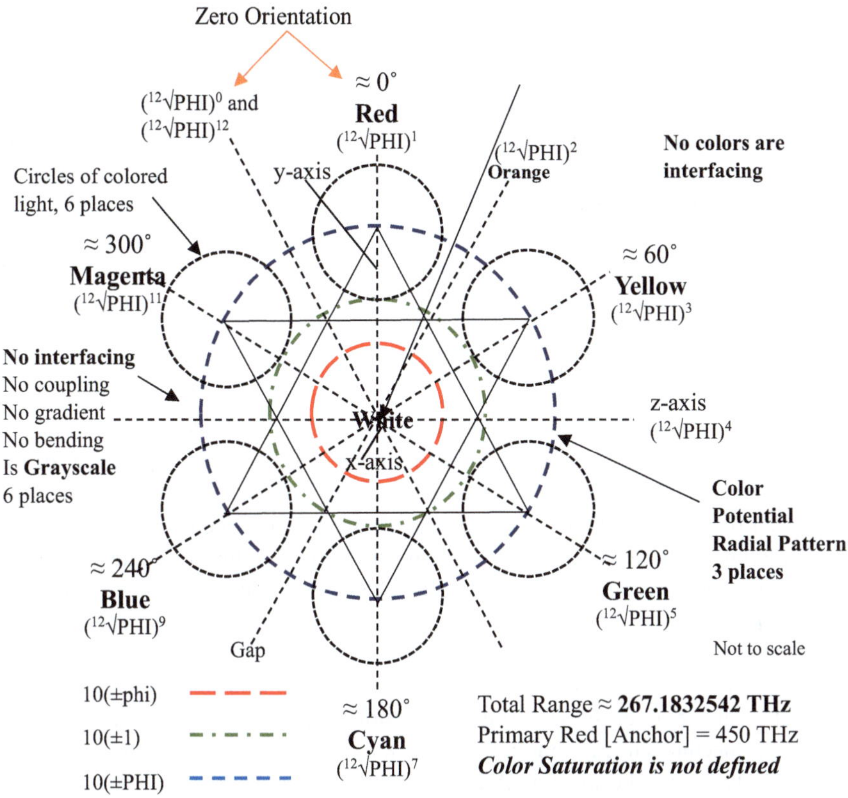

Fig. 2.8 Circled hexagram showing 10(±phi), 10(±1), and 10(±PHI) unit circles

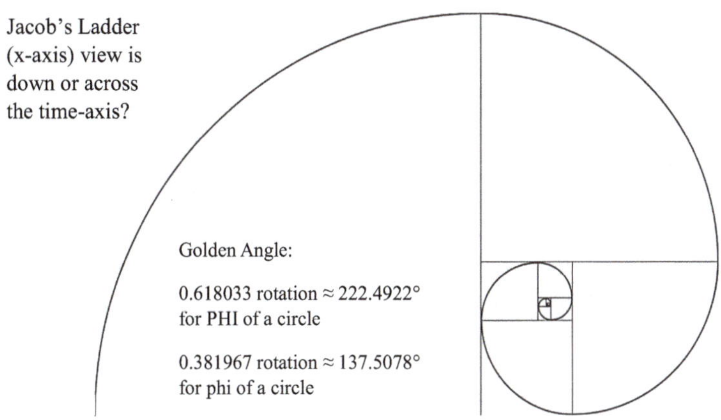

Fig. 2.9 Golden ratio spiral

is equally divided into the 12 Primary Color references or 12 Primary Tones. There are no reflected waves.

The color harmonics or color octaves identified by Jacob's Ladder, per this color research, are in Primary Tone "F" according to the Circle of Fifths and the A440 method of tuning. The harmonics or color octaves identified by Jacob's Ladder, per this color research, are keyed to the Primary Red [Anchor] and Total Range values.

Jacob's Ladder is a harmonic distribution that is independent of the frequencies (vibrations) being evaluated. Using a cause-and-effect approach, the frequencies of the electromagnetic spectrum or the range of sound did not create Jacob's Ladder. Just as Jacob's Ladder did not create the frequencies of the electromagnetic spectrum or the range of sound.

Figure 2.7 "Color Combining Model" resembles patterns displayed by a Chladni Plate (not shown). The patterns are color boundaries versus nodes. The similarities in the patterns demonstrate the relationship between Jacob's Ladder, Chladni Plates, and vibrating strings. Differences could be related more to the application of the technology and the focus of the experiment.

When the A440 method of tuning was developed, the focus was on doubling and halving the pitch to define a "rising" music octave or "lowering" music octave. This color research approached the color octave as a pattern to be repeated per exponential values of the 10(phi) circular constant. The music octave and the color octave are expressions of the same thing. But they tell different stories.

The standing wave, antinodes, and nodes of a vibrating string or a vibrating plate are not independent of the vibrating string, vibrating plate, associated hardware, or vibrating mechanism. In combining two *adjacent* Primary Colors, the results are like the interference between a vibrating string/plate and the reflected wave.

Per the linear 1-2-3 color pattern and Color Math, Red = 1, Green = 2, Blue = 3 (RGB) and Cyan = −1, Magenta = −2, Yellow = −3 (CMY):

$$\text{Red } (1) + \text{Yellow } (-3) = -2 \text{ or Magenta}$$

$$\text{Green } (2) + \text{Cyan } (-1) = 1 \text{ or Red}$$

The concentric circles in Fig. 2.8 are proportional, harmonic unit circles, at right angles (orthogonal) to the x-axis, representing 10(±phi), 10(±1), and 10(±PHI). The color harmonics and color octaves of Jacob's Ladder are along an exponential and golden ratio curved x-axis. The "phi" golden ratio proportion relates to a sub-harmonic or a lowering color octave. The "1" golden ratio proportion relates to the Baseline of Jacob's Ladder. The "PHI" golden ratio proportion relates to a harmonic or a rising color octave.

Per the concept of 10(x) circular constants, "x" is not equal to zero and x goes to positive and negative infinity. Deflection is radial to the x-axis. Propagation is along the x-axis. Negative is regarding orientation. Per the Circle of Fifths, increases in frequency are clockwise. Decreases in frequency are counterclockwise.

A single frequency within Jacob's Ladder could appear as a diverging golden ratio spiral as frequently increases, see Fig. 2.9. As frequency decreased, the golden ratio spiral could converge, see Fig. 2.9. Full spectrum electromagnetic radiation or full spectrum sound could appear as a diverging or converging golden ratio spiral (vortex). The 12 color references or the 12 Primary Tones could appear as 12 concentric, proportionally spaced, stranded cables of a diverging or converging, golden ratio spiral, as modified by an exponential curve. Its deflection is radial and orthogonal to the x-axis, as it travels along the x-axis (Jacob's Ladder). The curve of Jacob's Ladder and the Composite Color Wave is not shown, see Note 2.9.

The theoretical divergence or convergence of the golden ratio spiral regarding Jacob's Ladder is problematic. A coordinate system has not been developed.

List 2.1 Golden Ratio Spiral and Vibrations Examples

- Jacob's Ladder
  - Visible light color models
  - Electromagnetic spectrum
  - Sound (ultrasound, audible sound, infrasound)
  - Color Octaves (Circle of Fifths and A440 method of tuning) where we are in, color octave Primary Tone F
- Golden Ratio—phi: 1: PHI (the color pattern of Color Theory)
- Golden Ratio Circular Constants (GRCC)—10(±phi), 10(±1), 10(±PHI)
- Fibonacci Sequence
- Crystal structures/patterns
- Fractal patterns
- Sunflower seed patterns
- Galaxy and Nebula patterns
- Hurricane/Typhoon patterns
- Nautilus Shells
- Snail shells
- Human body and body parts
- The pattern of the leaves on plants and trees
- Pinecones
- Water going down a drain
- Etc.

Note 2.9:

Figure 2.9 "Golden Ratio Spiral" is a *converging or diverging* spiral. If Fig. 2.11 were *cylindrical* helixes and the view was down the x-axis, the helixes would appear as three concentric and proportional circles. Red and Cyan have the same unit circle diameter. Green and Magenta have the same unit circle diameter. Blue and Yellow have the same unit circle diameter.

The color hexagram is used to define the geometry of the color pattern. The color hexagram correlates with the Circle of Fifths and the A440 method of tuning. Zero for the color model and Primary Tone F (per the Circle of Fifths) is at the 11-o-clock position. The color model and the Circle of Fifths use an arbitrary sized circle that is divided into 12 equal sections to identify the 12 color references or the 12 Primary Tones. This arbitrary sized circle is at a right angle (orthogonal) to the x-axis as

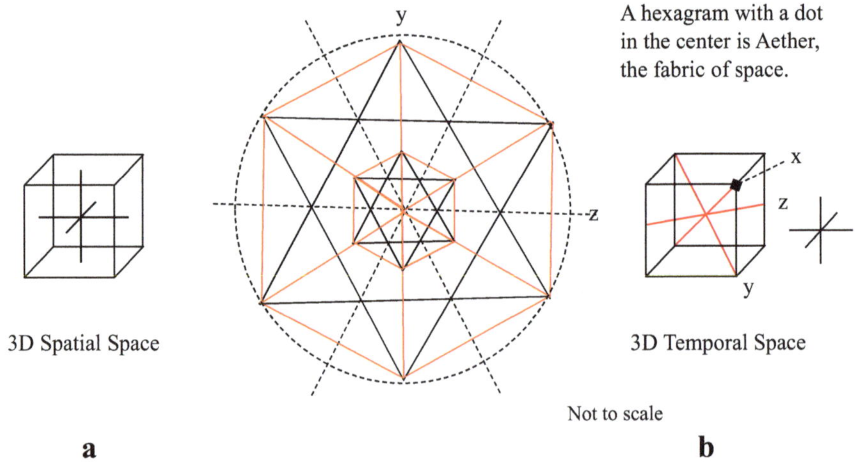

Fig. 2.10 3D Metatron's Cube (temporal space)

defined by the curve of Jacob's Ladder. The arbitrary sized circle is represented by Fig. 2.10.

The uniquely oriented cube space of Fig. 2.10a represents the Circle of Fifths with its clocking and the color hexagram (color model) in the configuration of Metatron's Cube (three-dimensional vibrations). The cartesian coordinate system (orientation) of Fig. 2.10a is not same as Fig. 2.10b. Figure 2.10a *represents spatial space and* Fig. 2.10b *represents vibrational/temporal space.*

Figure 2.11 is a proposed waveform for Jacob's Ladder and RGB and CMY Color Potential (aka Composite Color Wave) which is orthogonal to the x-axis. *Jacob's Ladder is on an exponential and golden ratio curve with unresolved characteristics.* Jacob's Ladder has no standing waves, reflected waves, antinodes, or nodes.

The potentials (waveforms) are in the yz-plane (magnetic field), not the xy-plane (electric field). This color research promotes the concept of the associated waveforms being helixes versus sinusoidal waves. As demonstrated in Fig. 2.10b, the orientation of Fig. 2.11 causes the x-axis and y-axis to appear as a single line, per cycle. The xy-axis is in the shape of a "V" or the top half of an "X".

If everything vibrates, then everything is contributing to the big and little concerts of existence. Some contributions could be constructive. Some contributions could be destructive. A sustainable system would probably need a balance of the constructive and destructive elements to achieve harmony.

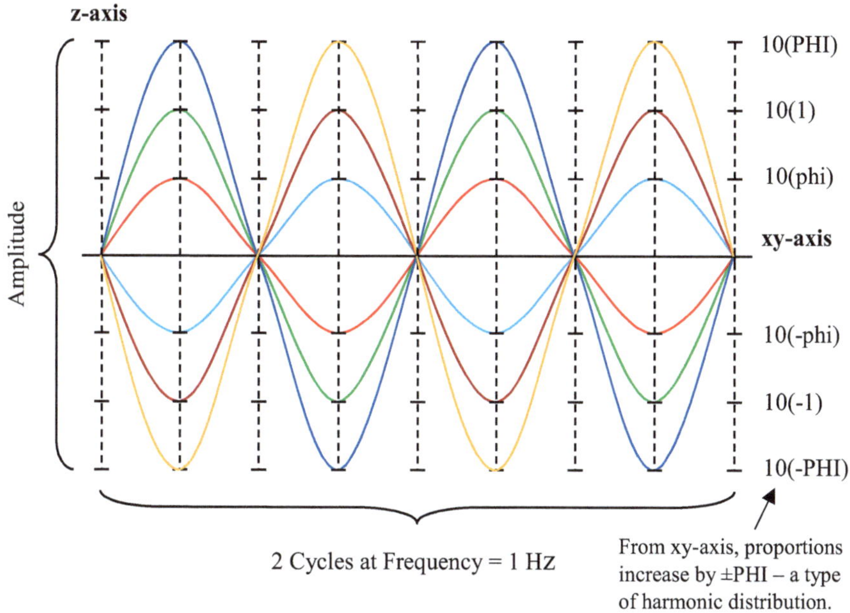

Fig. 2.11 Drawing of Proposed Waveforms for Color Potential

2.4 Conclusion and Open Issues

Table 2.2 compares 16 features of vibrating strings, Chladni Plates, and Jacob's Ladder regarding a harmonic. Features are not listed in any order. A mechanical vibration could be pressure waves. Pressure waves may or may not register as audible sound.

The Jacob's Ladder harmonic was contrasted with the harmonic of the vibrating string experiment and the harmonic of the Chladni Plate experiment. The vibrating string experiment and the Chladni Plate experiment are based on standing waves. A standing wave is the physical interaction of a vibration with its reflected wave to create a standing wave (antinode). The standing wave has antinodes and nodes. The antinode (harmonic) is the maximum y-axis deflection of the vibrating string or vibrating plate. The node is a point of no movement.

The harmonic of the vibrating string experiment and the harmonic of the Chladni Plate experiment are classical examples of a harmonic per the introduction of a mechanical vibration into a string or plate. The string or plate does not continue to vibrate or ring like a bell if the vibrating stimulus is removed.

Jacob's Ladder is a harmonic, exponential distribution based on the proportions of the golden ratio. Jacob's Ladder correlates to color, the entire range of sound (mechanical vibrations), the entire range of the electromagnetic spectrum (radiation), and basic music theory per the Circle of Fifths and A440 method of tuning.

Table 2.2 Harmonic—Vibrating string, Chladni Plate, and Jacob's Ladder

Vibrating string	Chladni Plates	Jacob's Ladder
Standing wave	Standing wave	No standing wave
Reflected wave	Reflected wave	No reflected wave
Nodes and antinodes	Nodes and antinodes	No nodes, and no antinodes
Physically vibrated media by a mechanical device	Physically vibrated media by a mechanical device	Physical mechanical vibration propagations in a medium
No radiation or sound waves	No radiation or sound waves	Radiation and sound waves
Harmonics—Antinodes	Harmonics—Antinodes	Harmonics—Jacob's ladder
Range—limited	Range—limited	Range—All radiation and sound
No hardware info.	Little hardware info.	n/a
No materials info.	No materials info.	n/a
Results—y-axis antinodes	Results—y-axis antinodes	yz-plane harmonics and octaves
Media—string	Media—Metal plate	Media—Radiation and sound
Media dependent	Media dependent	Media independent
Performance oriented	Performance oriented	Mathematically defined
Standards—none	Standards—none	Standards—Circle of Fifths, A440 method of tuning, golden ratio, exponential equation formula, color model
Venue—2D Spatial Space	Venue—3D Spatial Space	Venue—3D Temporal Space
Accepted method	Accepted method	New color theory

Chapter 3
Possible Extension of the Linguistics Semiology and Geometric Lie Algebra Foundation of Atomic Structure and Periodic System to Superconductivity

Godfrey E. Akpojotor, Alexander O. Animalu, Eric Trell, Samuel Edeagu, and O'Paul Isikaku-Ironkwe

Abstract In earlier papers, we have shown how the semiology, here broadly understood as the logic, language, and physics of Nature and culture, comes into play in the Igbo as well as Urhobo linguistic geometry based on the cube-hexagon/torus step pyramid alphabet/number system that has led us to the "geno-dual cuBits" representation of the role of symmetries in atomic structure and the periodic table of elements. The recent link of the periodic table to superconductivity requires us to revisit this natural formulation to suggest it as a natural path to the mechanism of superconductivity. By considering the semiology of linguistics in terms of an extension of a correlation of the usual 26 English alphabets with the corresponding numbers, (a = 1, b = 2, ..., z = 26) and attaching the numbers to

G. E. Akpojotor (✉)
Department of Physics, Delta State University, Abraka, Nigeria

Centre for Research and International Programmes, Delta State University, Abraka, Nigeria

International Centre for Basic Research, Abuja, Nigeria
e-mail: akpogea@delsu.edu.ng

A. O. Animalu
International Centre for Basic Research, Abuja, Nigeria

Department of Physics and Astronomy, University of Nigeria, Nsukka, Enugu State, Nigeria

E. Trell
Faculty of Health Sciences, University of Linköping, Linköping, Sweden
e-mail: erik.trell@liu.se

S. Edeagu
International Centre for Basic Research, Abuja, Nigeria

Department of Electronic Engineering, University of Nigeria, Nsukka, Enugu State, Nigeria

Communications Research Group, University of Kent, Canterbury, UK
e-mail: samuel.edeagu@unn.edu.ng

O. Isikaku-Ironkwe
Bernd T. Matthias Superconductivity Lab., RTSD Technologies, San Diego, CA, USA

A. Adhikari et al. (eds.), *Quantitative Frontiers*, Industrial and Applied Mathematics, https://doi.org/10.1007/978-981-96-9435-8_3

the alphabets so as to form a cube-hexagon/torus step pyramid, we show that this methodology permits corresponding Sanskrit/Devanagari alphabets to be arranged in a 6×6 matrix of numbers 1,2,...,36 correlated with gravitational lensing and arranged geometrically in a similar cube-hexagon/torus step pyramid. The linguistic/structural view that emerges is shown to be very much in line with Mendeleev original molding of his two-dimensional rendering of the periodic table on the dual Sanskrit grammar/phonetics and leads to a striking Lie geometric/structural correspondence with observed atomic structure and periodic system of the elements. The implications for digital signal processing with computers and possible extension for superconductivity are discussed.

Keywords Atomic structure · Digital signal processing · Geometric Lie algebra · Periodic system · Semiology · Superconductivity

3.1 Introduction

The Periodic Table is a database of all known elements. It is a predictive tool used by scientists in the search for novel properties when different elements are combined. Matthias and others have actively used the periodic table in the search for novel superconductors. By treating superconductors as "super-atoms," we can trace them in periods like normal elements, to which end it is helpful to understand how the periodic table and its mechanisms are naturally formulated in the archaic Igbo as well as Urhobo semiology, broadly understood as the logic, language, and physics of Nature and culture [1–3].

Just some perspectives: The unique evolution of language in the Homo Sapiens speciation demanded extensive modifications of all tracts of the vocal apparatus and parallel with the anatomical alterations of a syntactic ordering in sounds along the y axis with the easiest/earliest coming first as this depends on time and their further maturation into pressure waves along the x axis, i.e., according to the Panini grammar which was the re-actualized model for Mendeleev's periodic system.

This process has been estimated to be in culminative stage some 100,000 years ago. And it all took place in Africa before the emigration waves started some 30,000 years later and the paradoxical but explainable evidence is that the later and farther the exodus the less the retention of phonemes and contexts with resulting need to renew the cultivation, while in the stayers it could be carried on by its sayers and sages through all rises and falls; mighty empires; invasions; periods of bloom and renaissance and of suppression, and persecutions and cities once gleaming like "Heliopolis, the celestial city of ancient Egypt and India"; and the "lost Testament of the ancestors of Adam"; and the stone monoliths of Alok Ikom bearing a linear form of writing and a complex system of codified information and geometric images showing an advanced knowledge of mathematics several centuries B.C.; and the Igbo-Ukwu metal working craft from the ninth century A.C. with bronze castings that are among the most inventive and technically advanced ever made of ritual vessels, of sculptures forming modern art [4] and of sophisticated geometrical objects including a hexagonally engraved torus of high transformation virtue... [1–3].

However, reading and reciting Nature's language remained analphabetic till the second millennium B.C., when phonemic letters developed in the middle East from simplified Egyptian hieroglyphs, in turn often of central African origin. They are arranged in Panini hierarchy order, which is also that of language and hence of Nature in the Igbo—Urhobo semiology reverberating in all expressions and signs of the world and consequently also between the late-coming letters and the ageless numerical harmonies and distributions. Thus numbers and their digital Gram Code of African Adam [1–3] numeral system were brought to light in Africa, too, first as random notches on animal bones some 80,000 years ago, and 40,000 years later arranged in linear sequences and in the Ishangobone [4] year 20,000 BC in arrays with angular and prime number ordering and with uses from tally sticks and simple counting tools in the beginning to more qualified mathematical, astronomical, prediction, and ceremonial instruments and practices living on in many media in close social interaction.

What is then the morale in this hasty exposé? That the data and their processing here is as real and rational as of the material fossils of the Rift Valley. They are not the haphazard jumble of meaningless totems and magic formulas they are often held for in the post-colonial Western apprehension; they are the genuine da Vinci source code from the birth of man and the birth of number and the birth of language alike persevered in *those who sustained and managed in the mother land the whole way to the present era of reconciliation.*

So we may go on with our study within the purview of the quest in the world of physicists for a unified field theory that would be independent of change of *scale and conformal deformation* of space and time, so that in all the historical sequence from the African prehistoric times to the European Renaissance and onto the twentieth and twenty-first century dichotomy and trichotomy of logic, the geometry and numbers conserved in linguistics provide a useful correlation of the logic and physics of culture [5–7]. There we reported the development of the *science of semiolog*y of linguistics that has permitted not only the English 26 alphabets but also the corresponding Sanskrit/Devanagari alphabets [8] to be arranged in square matrix array of numbers 1,2,...,36 to form a cube-hexagon/torus step pyramid correlated with gravitational lensing. The interest of the linguistic/structural view that emerged is that it is very much in line with Mendeleev's original development of his two-dimensional rendering of the Periodic Table which we have reconstructed by employing Lie's algebraic method in his 1871 Norwegian Ph.D. Thesis [5–9] for resolving the dichotomy between cubic and spherical geometries from correspondence with observed atomic structure.

3.2 Lie-Geometric Method of Alphabets-Numerals Transformation

In the past, Lie groups, Lie algebras, and some of their applications [8] were mainly concerned with the space-time of special relativity with metric (+++−) with respect to the real contravariant bases (x,y,z,ct) and also metric (++++) with regard to the bases (x,y,z,ict). Even though Einstein's relativity theory was developed after Lie's work, we have recently shown how in a Minksowski space for which space has four dimensions, the inclusion of ellipsoidal and toroidal deformations of a sphere in Sophus Lie's original group of continuous transformation of a sphere lead, through the application of the *principle of duality* [9] between points, lines, and planes in three-dimensional projective geometry to explicit realization of generalizations of the conventional Lie algebra as deformation of the conventional Minkwoski space. This is readily achieved via the complementary processes of inscribing unit sphere in 2^3 (=8) cubical Lie neighborhood, i.e., a cubical box defined by its tangent planes in Fig. 3.1a and cube-hexagon 4^3 (=8×8) representation of SU(3) ~ SO(3)×O(5) Lie transformation diagram [10] in Fig. 3.1e leading altogether to the (cube-hexagon) system of lines in Fig. 3.1b, c and the cube-hexagon/torus in Fig. 3.1d.

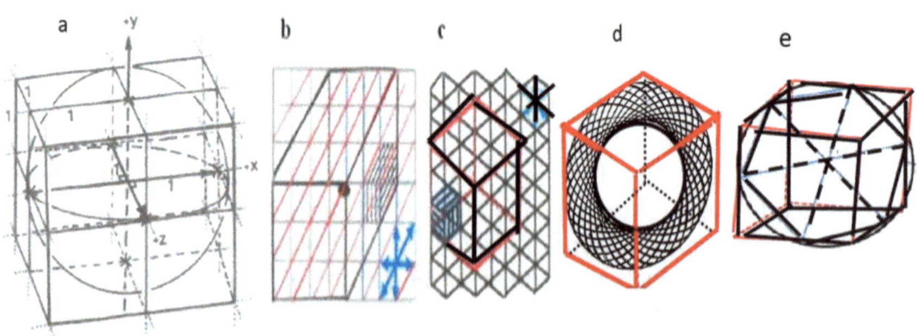

Fig. 3.1 (**a**) Unit sphere in 2^3 (=8) cubical Lie neighbourhood; (**b**) crisscrossing of three-dimensional real space generating its cuBits that enable variable parallelepiped packing in identical cube-hexagon hyperspace perspective with; (**c**) holographic grid for hexagonal projection in diagonal A2 hyperplane across midpoints of cuBits of space and correlated with; (**d**) Cube-hexagon/torus (geno-dual cuBits) equivalent to; (**e**) cube-hexagon 4^3 (=8×8) representation of SU(3)~SO(3)×O(5) Lie transformation diagram

To fix the idea, consider in geometric algebra terms, the following Lax pairing transformation [11]:

$$\left(g^0_{\mu\nu}\right) = \begin{pmatrix} 1 & 0 & 0 & 0 \\ 0 & -1 & 0 & 0 \\ 0 & 0 & -1 & 0 \\ 0 & 0 & 0 & -1 \end{pmatrix} \to \left(\tilde{g}_{\mu\nu}\right) = \left(1 + \overset{\sim}{\beta}\right)\left(g^0_{\mu\nu}\right) = \begin{pmatrix} 1 & 0 & -1 & 0 \\ 0 & -1 & 0 & -1 \\ -1 & 0 & -1 & 0 \\ 0 & 1 & 0 & -1 \end{pmatrix};$$
$$\text{where } \overset{\sim}{\beta} \equiv \begin{pmatrix} 0 & 0 & 1 & 0 \\ 0 & 0 & 0 & 1 \\ -1 & 0 & 0 & 0 \\ 0 & -1 & 0 & 0 \end{pmatrix}; \tag{3.1}$$

This characterizes the transformation of the quadratic form in four-dimensional space
$s^2 = x^\mu g^0_{\mu\nu} x^\nu \equiv (ct)^2 - x^2 - y^2 - z^2$ into a torus $s^2 = x^\mu \tilde{g}_{\mu\nu} x^\nu \equiv (ct)^2 - x^2 - y^2 - z^2 - 2cty$, with parametric equations,

$$\begin{cases} x = (is - ct\sin\theta)\cos\phi,\ y = (is - ct\sin\theta)\sin\phi \\ z = ct\cos\theta, \text{ where } y/s = \pm i\sin\theta \end{cases} \tag{3.2}$$

The insight from the Lax pairing of the *symmetric* tensor $g^0_{\mu\nu}$ and the *antisymmetric* $\overset{\sim}{\beta}$ is that it has yielded a "unified field" tensor $\tilde{g}_{\mu\nu}$, with a torus as the "generic" geometric object. Furthermore, whereas the equation of the conventional light cone given by $(ct)^2 - x^2 - y^2 - z^2 = 0$ is the Minkowski space-time characterized by the O(3,1) Lorentz group, the equation of the light cone given by $(ct)^2 + x^2 + y^2 + z^2 = 0$ is an Euclidean space characterized by an O(4) Lorentz group whose generalization to O(5) is referred to in Fig. 3.1e. Two important features of the O(3,1) and O(4) Lorentz groups are worth noting.

On the one hand, one needs to overcome the restriction of scale and conformal invariance to massless fermion fields in Lorentz-invariant O(3,1) theories especially Einstein's tensor field of general relativity theory $G_{\mu\nu} \equiv R_{\mu\nu} - \frac{1}{2}Rg_{\mu\nu} = -\kappa\Theta_{\mu\nu}$, $R_{\mu\nu\rho\sigma}$ being the curvature tensor defined by a metric tensor $g_{\mu\nu}$ and $R_{\mu\nu} = g^{\rho\sigma}R_{\mu\nu\rho\sigma}$ is the Ricci tensor with $R = g^{\rho\sigma}R_{\rho\sigma}$ and $\kappa = 8\pi G/c^4$ and $\Theta_{\mu\nu}$ is a *symmetric* energy-momentum tensor whose conservation is given by $\partial^\mu D_\mu = \Theta_{\mu\mu} \Rightarrow G_{\mu\nu,\,\nu} = 0$. This is achieved by replacing the Dirac current J_μ for fermions by a tensor current $J_{\mu\nu} = \gamma_\mu\gamma_\nu + M^{-1}(\gamma_\mu p_\nu + p_\mu\gamma_\nu)$ where $J_{0\mu}$ is the most general parity-conserving current in the O(4,2) algebra of Dirac's γ-matrices and M is a 3×3 mass matrix characterizing an SU(3) triplet, so that, for coupling to U(1) gauge field, A_μ, in the minimal form $J_{0\mu}A^\mu$, a generalization of Einstein's field theory [12] is achieved as a cube-hexagon lattice O(4, 2) × SU(3) × U(1) gauge

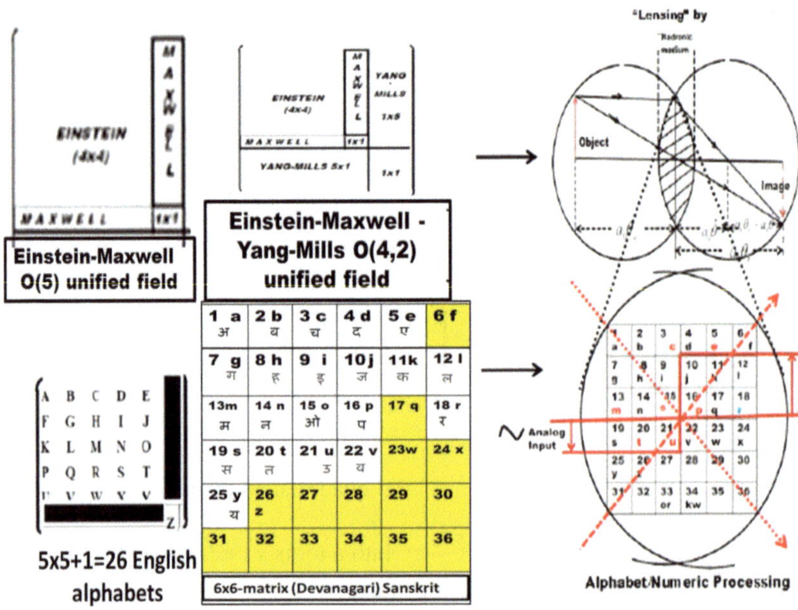

Fig. 3.2 Correlation between the progressive unification of Einstein (4×4) & Maxwell fields (top left) in O(5) to 6×6 metric tensor of Einstein-Maxwell-Yang-Mills unified gauge field in O(4,2) in "lensing" by matter (hadronic medium) (top right) and a generalization of 26 = 5×5+1 English alphabet system (bottom left) to corresponding Sanskrit alphabet system in 6×6 magic square alphabet word (analog) input processing into numeric (digital) output (bottom right)

theory, for which scale invariance ($\partial^{\mu} D_{\mu} = \Theta_{\mu\mu} = 0$) and conformal invariance $\left(\partial^{\mu} J_{\mu\nu} = 2x_{\lambda}\Theta^{\lambda}_{\nu} = 0\right)$ are non-trivial.

On the other hand, Lie algebra of the spin group, SO(4) consists of real antisymmetric matrices, (to which $\tilde{\beta}$ introduced in Eq. (3.1) belongs) whose two-sheeted covering group Spin(4) introduced by Brauer and Weyl [13] governs Mendeleev's Periodic System of chemical elements [14].

Altogether, the progressive generalization of O(4) to O(5) in Einstein-Maxwell unified to O(4,2) in Einstein-Maxwell-Yang-Mills lattice gauge field theoretic explanation of gravitational "lensing" by matter (see Fig. 3.2) provides an apparent geometric perspective (also indicated in Fig. 3.2) for a correlation of a corresponding generalization of the 5×5+1 = 26 English Alphabet system to corresponding Sanskrit alphabet system in 6×6 magic square matrix of numbers 1,2,…,36, for processing word (analog) input into numerical (digital) out.

To show how the correlation in Fig. 3.2 is very much in line with Mendeleev's original use of his two-dimensional rendering of the periodic table on the dual Sanskrit grammar/phonetics correspondence with atomic structure and periodic system of elements, we show in Fig. 3.3a, b how the correlation attaching numbers to English alphabets (a = 1, b = 2,… z = 26) so as to form the cube-hexagon/torus step pyramid system permit the 26 English alphabets and the Sanskrit alphabets

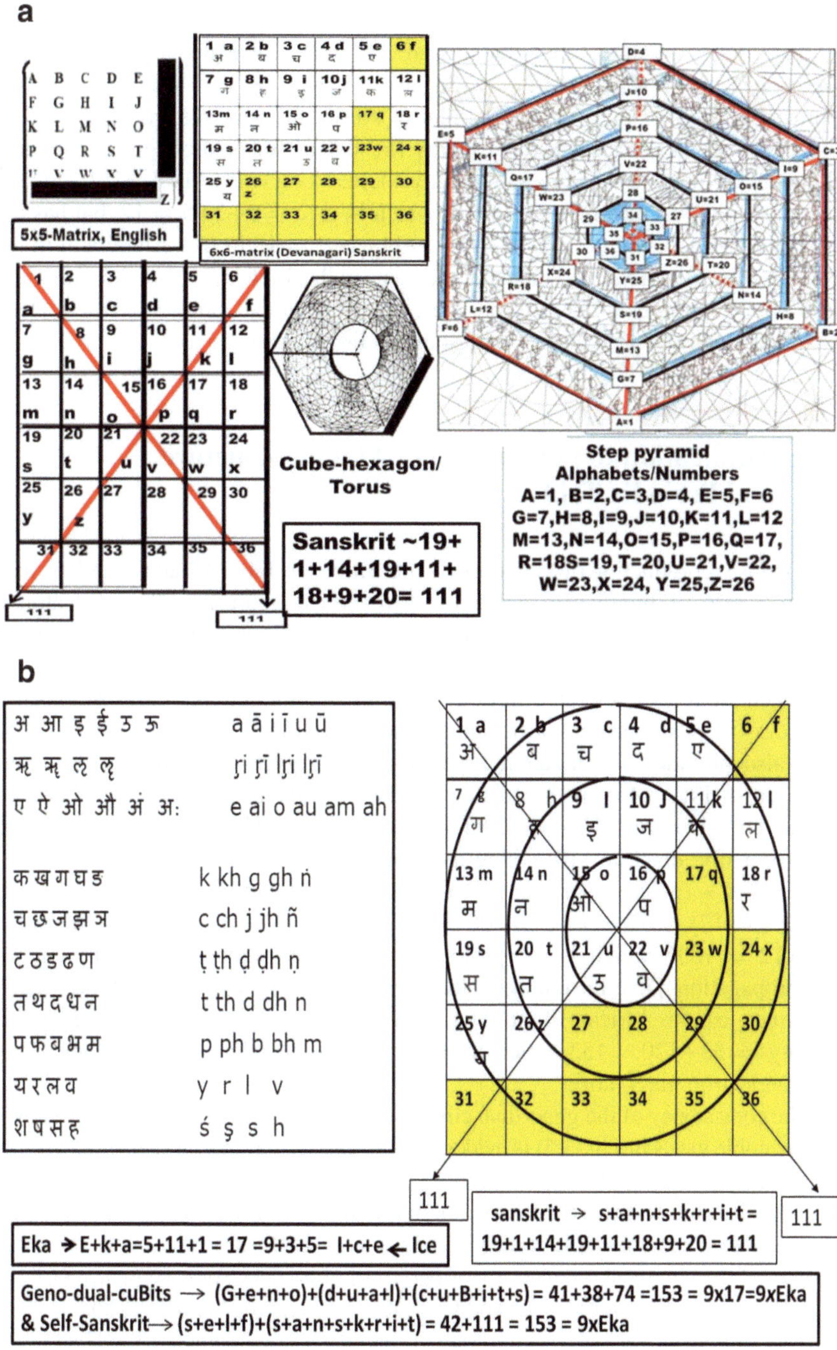

Fig. 3.3 (**a**) Arrangement Of numbers attached to alphabets as a cube-hexagon/torus Step pyramid System. (**b**) The Sanskrit alphabet in the Devanagari script/atomic structure (cuBits) correlation

in Devanagari script[3] to be correlated in a 6×6 matrix arranged geometrically to form a step pyramid. In order to find the numerical equivalent of a "word" one makes the substitution of w = 23, o = 15, r = 18, d = 4 in the algebraic expression, w+o+r+d = 23+15+18+4 = 60. In this way, we have indicated in Fig. 3.3b that the Sanskrit "Eka" used by Mendeleev for predicting new chemical elements in Table 3.1 has the numerical value, E+k+a = 5+11+1 = 17 which is cognate with the English word, Ice for which I+c+e = 9+3+5 = 17, but has the phonetic/linguistic value E+k+a+a = 5+11+1+1 = 18 in *Eka*-Aluminum in the modern name, Gallium, and the other elements in the Table 3.1. Note the ambiguity with phonetic values of *Eka*-Boron, the correct numerical values of *T(ri)* = 20 in Tri-manganese *Dvi* = 32 in Dvi-tellurium and Dvi-Caesium.

3.3 Application to Foundation of Atomic Structure and Periodic Table

The precise reproduction of Mendeleev's predictions in Table 3.1 above underscores the relevance of linguistics in the Lie geometric algebraic method for atomic structure. Firstly, it leads to a principle of cube-hexagon hyperlink reflecting a holographic pattern of growth in a toroidal universe [defined by model of the hydrogen atom (as elaborated in Fig. 3.4a)], predicting how the universe expands according to a fixed proton-electron mass ratio of 1836.

It also provides a realization (in terms of linguistic geno-dual cuBits interpretation in Fig. 3.4b) of the relation of the 2^3 (=8) cubical Lie neighborhood to the cube-hexagon 4^3 (=8×8) representation of SU(3) ~ SO(3)×O(5) Lie transformation diagram [10] where the numerical equivalent of the name "geno-dual cuBits" ~ (g+e+n+o) + (d+u+a+l) + (c+u+b+i+t+s) = (7+5+14+15) + (4+21+1+12) + (3+21+2+9+20+19) = 41+38+74 = 153 gives the value 153×12 = 1836 which is the proton/electron mass ratio predicted from module propagation [15] Intriguingly, the identification in Fig. 3.4b of the "ichi" craftsman as an African (Igbo~i+g+b+o = 9+7+2+15 = 33) gives Igbo-Ptolmys = 33+120 = 153 = Self-Sanskrit which is correlated in semiology to the proton/electron mass ratio.

The consistency of the geno-dual cuBits with both the atomic structure (as shown for the noble gases in Fig. 3.5) is evident. By noting further that "cuBits" can also be translated as "Energy" (~e+n+e+r+g+y = 5+14+5+18+7+25 = 74), we arrive at an application to the Mendeleev's periodic table of elements shown in the plot of the atomic spectroscopic term [16, 17] Energy levels (in Rydbergs) versus z^2 (square of the chemical valence, z) for two rows of iso-electronic sequence of ions, Li^+, Be^{2+}, …, F^{7+} in Fig. 3.6a and Cu^+, Zn^{2+}, …., Ni^{10+} in Fig. 3.6b.

This completes our use of semiology of linguistics and geometric Lie algebra for the foundation of the Atomic Structure and Periodic Table which is the result we are after.

Table 3.1 The full list of Mendeleev's predictions with their Sanskrit names

Mendeleev's given name	Modern name
Eka-aluminum	Gallium
Eka-boron	Scandium
Eka-silicon	Germanium
Eka-manganese	Technetium
Tri-manganese	Rhenium
Dvi-tellurium	Polonium
Dvi-caesium	Francium
Eka-tantalum	Protactinium
Atomic structure correlation with Mendeleev Sanskrit name (Ekaa ~ E+k+a+a = 5+11+1+1 = 18)	Atomic structure correlation with modern name
$Al^{13}:(1s^2)(2s^2 2p^6)(3s^2 3p)$ $2+8+3=13$; *Ekaa-Aluminum* $= 18+13=31$	$Ga^{31}:(1s^2)(2s^2 2p^6)(3s^2 3p^6 3d^{10})(4s^2 4p)$ $2+8+18+3=31=$ Gallium
$B^5:(1s^2)\ (2s^2 2p)$ $2+3=5$; *Ekaa-Boron* $= 18+5=23$	$Sc^{21}:(1s^2)(2s^2 2p^6)\ (3s^2 3p^6 3d)(4s^2)$ $2+8+9+2=21=$ Scandium $V^{23}:(1s^2)(2s^2 2p^6)\ (3s^2 3p^6 3d^3)(4s^2)$ $2+8+11+2=23=$ Vanadium
$SI^{14}:(1s^2)(2s^2 2p^6)(3s^2 3p^2)$ $2+8+4=14$; *Ekaa-Silicon* $= 18+14=32$	$Ga^{32}:(1s^2)(2s^2 2p^6)(3s^2 3p^6 3d^{10})(4s^2 4p^2)$ $2+8+18+4=32=$ Germanium
$Mn^{25}:(1s^2)(2s^2 2p^6)(3s^2 3p^6 3d^5)(4s^2)$ $2+8+13+2=25$; *Ekaa-Manganese* $= 18+25=43$	$Tc^{43}:(1s^2)(2s^2 2p^6)(3s^2 3p^6 3d^{10})x$ $(4s^2 4p^6 4d^6)(5\ s)$; $2+8+18+14+1=43=$ Technitium
$Ta^{73}:(1s^2)(2s^2 2p^6)(3s^2 3p^6 3d^{10})x$ $(4s^2 4p^6 4d^{10} f^{14})(5s^2 5p^6 5d^3)(6s^2)$; $2+8+18+32+11+2=73$; *Ekaa-Tantaluum* $= 18+73=91$	$Pa^{91}:(1s^2)(2s^2 2p^6)(3s^2 3p^6 3d^{10})(4s^2 4p^6 4d^{10} 4f^{14})x$ $(5s^2 5p^6 5d^{10} f^2)(6s^2 6p^6 6d)(7s^2)$; $2+8+18+32+20+39+2=91=$ Pro- tactinium
$Mn^{25}:(1s^2)(2s^2 2p^6)(3s^2 3p^6 3d^5)(4s^2)$ $2+8+13+2=25$; Tri-Mn $= 20+25=45$	$Rh^{45}:(1s^2)(2s^2 2p^6)(3s^2 3p^6 3d^{10})x$ $(4s^2 4p^6 4d^8)(5\ s)$; $2+8+18+16+1=45$ Rhenium
$Te^{52}:(1s^2)(2s^2 2p^6)(3s^2 3p^6 3d^{10})(4s^2 4p^6 3d^{10})x$ $(5s^2 5p^4)$; $2+8+18+18+6=52$ DviTe $= 32+52=84$	$Po^{84}:(1s^2)(2s^2 2p^6)(3s^2 3p^6 3d^{10})(4s^2 4p^6 4d^{10} 4f^{14})x$ $(5s^2 5p^6 5d^{10})(6s^2 6p^4)$; $2+8+18+32+18+6=84$ Polonium
$Cs^{55}:(1s^2)(2s^2 2p^6)(3s^2 3p^6 3d^{10})(4s^2 4p^6 4d^{10})x$ $(5s^2 5p^6)(6\ s)$; $2+8+18+18+8+1=55$ DviCe $= 32+55=87$	$Fr^{87}:(1s^2)(2s^2 2p^6)(3s^2 3p^6 3d^{10})(4s^2 4p^6 4d^{10} 4f^{14})x$ $(5s^2 5p^6 5d^{10})(6s^2 6p^6)(7\ s)$; $2+8+18+32+18+8+1=87$ Francium

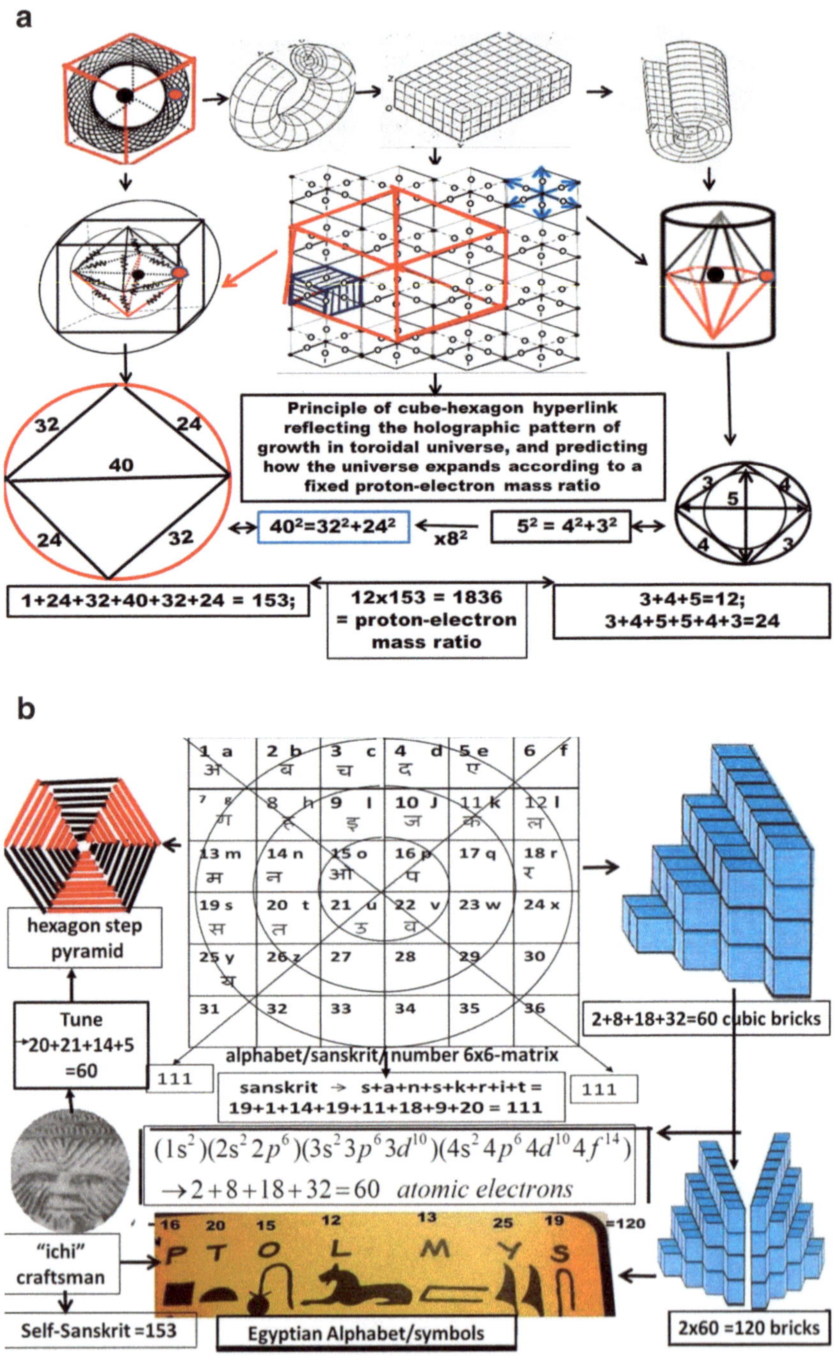

Fig. 3.4 (**a**) Model of H-atom (Top-left). (**b**) Linguistic & Lie geometric correlation with atomic structure

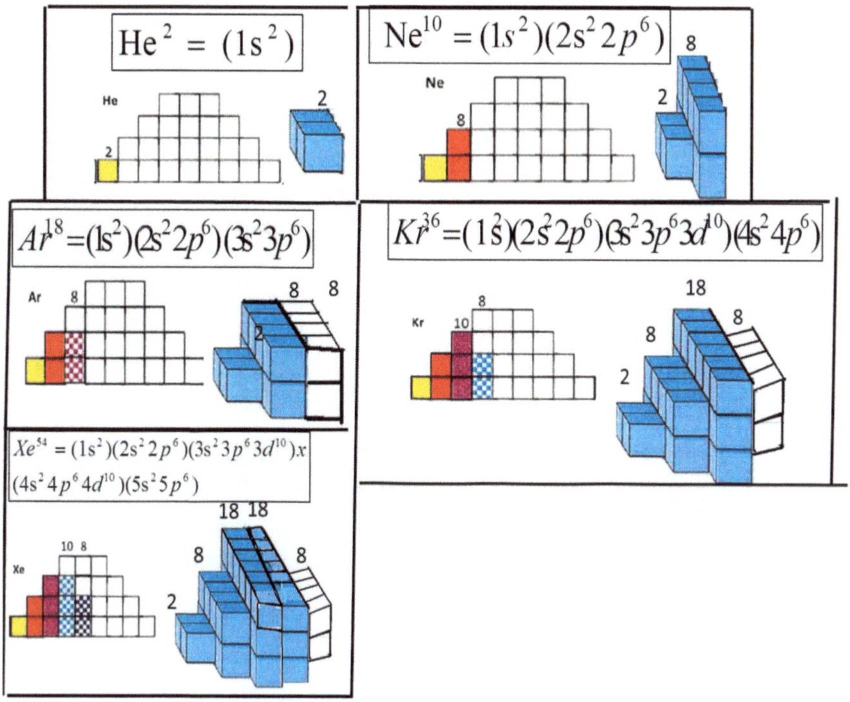

Fig. 3.5 According to the notation used to describe the electronic configuration of atoms as discussed in all text-books of introductory atomic physics, the letters, s, p, d, signify electrons having orbital angular momentum 0, 1, 2,.... in Planck constant units; the number to the left of the letter denotes the principal quantum number of one orbit, and the to the right denotes the number of atomic electron orbit

3.4 Possible Extension of the Application to the Foundation of Superconductivity

Since the discovery of superconductivity in 1911, there has been search for its understanding in order to predict room temperature superconductors [18]. This has actuated tens of theoretical attempts to account for superconductivity, but there is still no consensus theory to predict the transition temperatures of both the conventional and non-conventional superconductors [19, 20]. Instead, it is the use of periodic table to discover superconductors that has been more successful. For example, Matthias discovered over 100 low temperature superconductors using his understanding of the periodic table [21, 22]. Also, Bednorz and Muller discovery of the relatively high-T_C superconducting cuprates as well as the quick increment by chemical pressure were from their understanding of the periodic table [23, 24]. There are reports of regularity of the transition temperatures in certain periods of the periodic table [25, 26]. Then recently, a general principle for calculating the

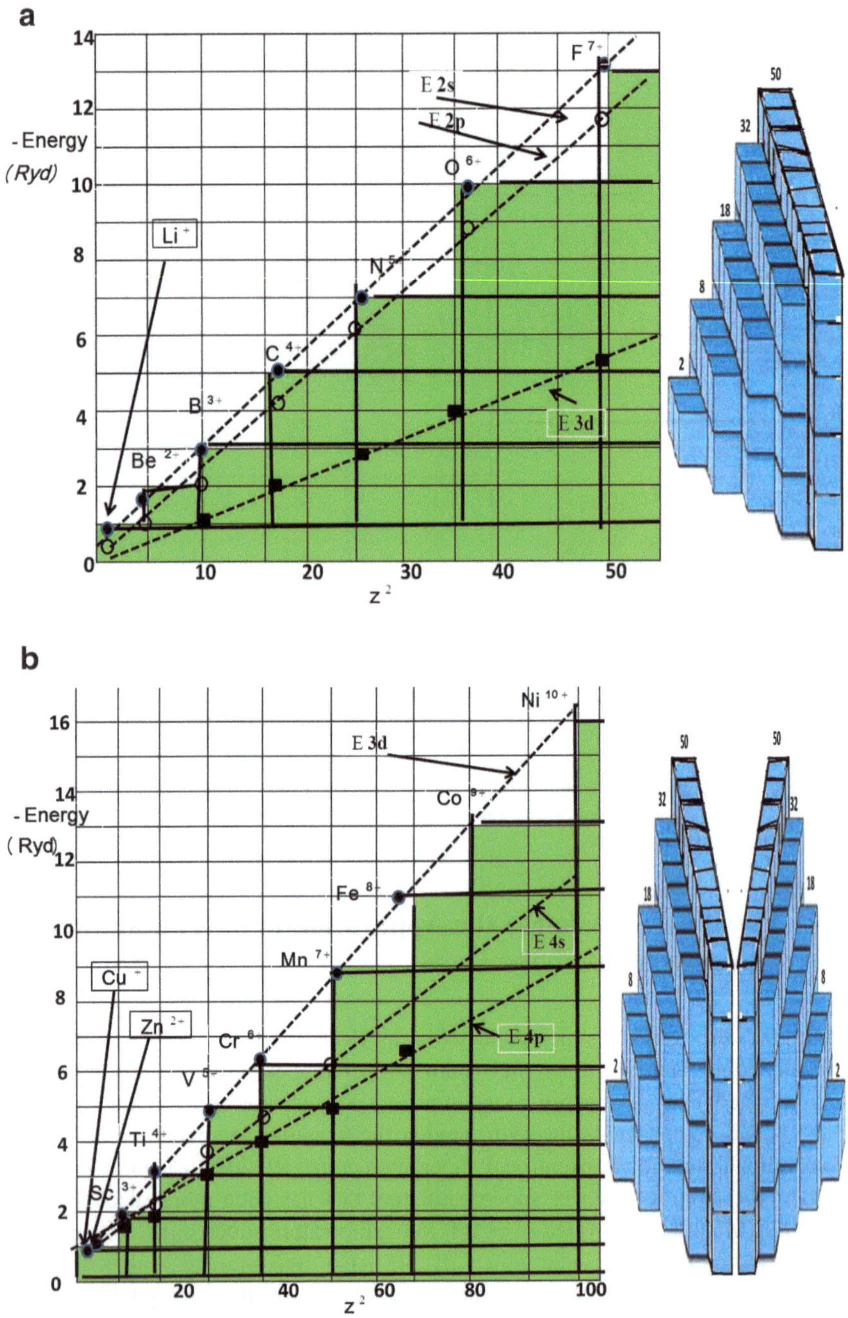

Fig. 3.6 (**a**) Atomic spectroscopic energy levels versus square of the chemical valence and correlated Cubits structure. (**b**) Atomic spectroscopic energy levels versus square of the chemical valence and correlated Cubits structure

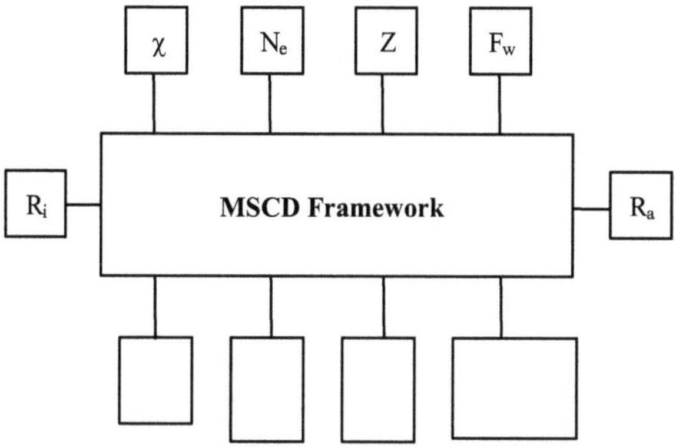

Fig. 3.7 Material specific characterization dataset (MSCD) framework

superconductivity of hydrides based on the periodic table alone was reported [27]. The authors claimed that their calculation could be used to possibly predict periods to discover certain superconducting materials with the periodic table. Therefore, their calculation is in the same spirit with the Material Specific Characterization Dataset (MSCD) developed by one of us (P. Ironkwe) [22, 28].

In general, the MSCD is Material Specific Characterization Dataset (MSCD) Framework is an electronic structure system based on Periodic Table parameters. A key theoretical foundation of MSCD framework is that combining elements yield a new "element" that has the average properties of the elements and their combining ratios. Unlike the Matthias rules [29] which are a set of guidelines to find low temperature superconductors that depends on the valence electron, the key parameters of MSCD framework are electronegativity (χ) valence electros (N_e), atomic number (Z), formula weight (F_w), ionic and atomic radius (R_i and R_a), and first ionization energy ($\sum 1iE$). These set of data and more for a material constitute its MSCD framework (Fig. 3.7).

The guidelines for a material to be superconductive in the MSCD framework are as follows:

1. Electronegativity χ, falls between 1.7 and 2.1 on Pauling scale.
2. Valence electrons, N_e, lies between 2.0 and 3.7.
3. Atomic number Z, is <25, >7.
4. $\frac{N_e}{\sqrt{Z}}$ is >0.8 but <1.
5. The radius ratio is >0.7 but <1.
6. $\frac{F_w}{Z}$ is high; >45 and up to 100 (the higher, the better for high T_c).

Further, taking these aforementioned parameters into account, the transition temperature (T_c) is obtained as

$$T_c = k_1 \left[\left(\frac{N_e}{\sqrt{z}} \right)^{\pi}, \frac{F_w}{A_n}, \frac{F_w}{Z}, \left(\frac{X \cdot N_e}{Z} \right)^2 \right] \tag{3.3}$$

where k_1 is a factor that depends on the period of material in the Periodic Table.

MSCD Framework provides the conditions for high-T_c superconductivity and also the formula for computing the T_c of the proposed superconductors.

3.5 Concluding Remarks

We have in this paper shown how semiology (i.e., physics and the logic of culture) comes into play in the "linguistic" geometry based on the cube-hexagon/torus step pyramid alphabet/number system that has led us to the "geno-dual cuBits" representation of the role of symmetries in atomic structure and the periodic table of elements. The limitation of the generalization of Lie geometric algebra method is only on the correlation of the number (26) of letters of the English alphabets and the 6×6 magic square of numbers 1,2,…,36 in the cube-hexagon geometry. To see the implications for the physical interpretation of relativity theory, we require a corresponding generalization of the conventional (static) de-Sitter space metric to a screw (dynamic) geno-de Sitter space metric by adding a *deformation* term $2\varepsilon drdt$ (ε being an arbitrary parameter measure of the sine of the angle between $d\vec{r}$ and $d\left(\vec{c}\,t\right)$) to the conventional Schwarzschild line element of Einstein's general relativity theory in spherical polar coordinates (θ, ϕ, r, ct) (with $c = 1$)):

$$(ds)^2 = r^2\left[(d\theta)^2 + \sin^2\theta(d\phi)^2\right] + \frac{1}{\left(1 - \frac{2M}{r}\right)}(dr)^2 - \left(1 - \frac{2M}{r}\right)(dt)^2 + 2\varepsilon drdt \tag{3.4}$$

In geometric terms, the envisaged progressive generalization from the conventional (static) to screw (dynamic) de-Sitter space that is characterized as illustrated in Fig. 3.8:

In previous work [9], we have found an application in Digital Signal Processing [9] via a correspondence $(r,ct,1-M/r) \leftrightarrow (I,V,R)$ between de-Sitter space-time (r,ct) lensing geometry for an object of mass M in general relativity theory of gravity and Ohm's law current (I), voltage (V) and resistance (R) for nanostructures. In the current study, we have proposed another application: the possibility of extending the semiology of linguistics and geometric Lie algebra for the foundation of the Atomic Structure and Periodic Table to accounting for superconductivity.

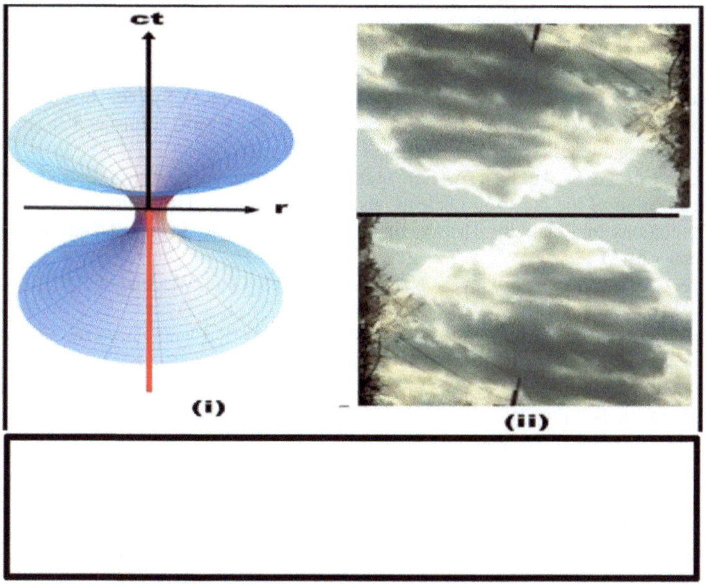

Fig. 3.8 Representative of (i) conventional (static) de-setter and (ii) its screw (dynamic) view as usual cloud formation (sourced from AOL)

References

1. Animalu, A.O., Edeagu, S., Akpojotor, G., Trell, E.: Semiology of linguistics and geometric Lie algebra foundation of atomic structure and periodic system. Afr. J. Phys. **10**, 1–16 (2017)
2. Trell, E., Akpojotor, G., Edeagu, S., Animalu, A.: Structural wave-packet tessellation of the periodic table and atomic constitution in real $R^3 \times SO(3)$. J. Phys.: Conf. Ser. **1251**, 012047 (2019)
3. Animalu, A., Akpojotor, G., Edeagu, S., Trell, E.: Generalized Lie algebraic geometry in $R^3 \times SO(3)$ configuration space for SU(3) of elementary particles and for wave-packing of Atomic Structure. J. Phys.: Conf. Ser. **1557**, 012009 (2020)
4. Kak, S.: How Sanskrit led to the creation of Mendeleev's periodic table. http://swarajyamag.com/books/sanskrit-and-mendeleevs-periodic-table-of-elements (2015)
5. Lie, M.S.: Over en Classe geometriske Transformationer. Ph.D. thesis, Kristiania (now Oslo) University (1871)
6. Trell, E., Santilli, R.M.: Marius Sophus Lie's doctoral thesis Over en Classe Geometriske Transformationer. Algebras Groups Geom. **15**, 395–445 (1998)
7. Trell, E.: Lie, Santilli, and nanotechnology: From the elementary particles to the periodic table of the elements. AIP Conf. Proc. **1637**, 1100–1109 (2014)
8. Gilmore, R.: Lie Groups, Lie Algebras and Some of Their Applications. John Wiley, New York (1974)
9. Animalu, A., Edeagu, S., Johansen, S., Strand, J., Trell, E.: Application of Lie Algebra and its generalizations to Physics and Digital Signal Processing. Afr. J. Phys. **7**, 62–81 (2014)
10. Trell, E.: The eightfold eightfold way: application of Lie's true geometriske transformationer to elementary particles. Algebra Groups Geom. **15**, 447–471 (1998)
11. Animalu, A.O.E.: Dirac equation in five dimensions and consequences. Hadron. J. **28**, 119–138 (2005)

12. Animalu, A.O.E.: Geno-Bragg's Law and 10*×10 representation of SU(3) symmetry for quasicrystal structures. J. Comput. Methods Sci. Eng. **13**, 1–17 (2013)
13. Brauer, R., Weyl, H.: Spinors in n dimensions. Am. J. Math. **57**, 425 (1935)
14. Rumer, Y.B., Fet, A.I.: The Group Spin(4) and the Mendeleev System. Institute of Nuclear Physics, Siberian Branch, Academy of Sciences of the USSR, Translated from Teoreticheskaya/MatematicheskayaFizika. **9**, 203–210 (1971)
15. Trell, E., Edeagu, S., Animalu, A.: Self-organized isotropic vector matrix translation apparatus for realization of the electron, nucleon, and periodic system. International Meeting Physical Interpretations of Relativity Theory, Bauman Moscow State Technical University July 3–7 Moscow (2017)
16. Moore, C.E.: Atomic energy levels. Nat. Bur. Stds (U.S. GPO, Washington, DC, Vols. I, II and III (1949)
17. Animalu, A.O.E.: Electronic structure of transition metals III d-band resonance and Regge-pole theory. Phys. Rev. **B10**(12), 4964–4972 (1974)
18. Lilia, B., et al.: The 2021 room-temperature superconductivity roadmap. J. Phys. Condens. Matter. **34**, 183002 (2022)
19. Schmalian, J.: Failed theories of superconductivity. Modern Phys. Lett. B. **24**, 2679–2691 (2010)
20. Akpojotor, G.E., Animalu, A.O.E.: A Possible Fundamental Theory of Cooper Pair Formation in Unconventional Superconductivity. J. Adv. Phys. **2**, 200–207 (2013)
21. Matthias, B.T.: Chapter V. Superconductivity in the periodic system. In: Gorter, C.J. (ed.) Progress in Low Temperature Physics, vol. 2, pp. 138–150. Elsevier (1957)
22. Ironkwe, P.I.: Transition temperatures of superconductors estimated from periodic table properties. https://arxiv.org/ftp/arxiv/papers/1204/1204.0233.pdf (2012)
23. Bednorz, J.G., Müller, K.A.: Possible high T_c superconductivity in the Ba-La-Cu-O system. Zeitschrift für Physik B. **64**(2), 189–193 (1986)
24. Cao, Z.Y., et al.: Spectroscopic evidence for the superconductivity of elemental metal Y under pressure. NPG Asia Mater. **15**, 5 (2023)
25. Simon, A.: Superconductivity and the periodic table: from elements to materials. Phil. Trans. R. Soc. A. **373**, 20140192 (2015)
26. Chapnik, I.M.: Superconductivity and the periodic table. Phys. Status Solidi. **B155**, K39–K44 (1989)
27. Semenok, D.V., Kvashnin, A.G., Kruglov, I.A., Oganov, A.R.: Actinium hydrides AcH10, AcH12, and AcH16 as high-temperature conventional superconductors. J. Phys. Chem. Lett. **9**(8), 1920–1926 (2018)
28. Animalu, A.O.E., Akpojotor, G.E., Ironkwe, P.I.: Generalization of conventional BCS model to iso-superconductivity model of high-T_c superconductivity in the cuprates and pnictides. Afr. J. Phys. **2**, 46–62 (2009)
29. Matthias, B.T.: Empirical relation between superconductivity and the number of valence electrons per atom. Phys. Rev. **97**, 74–76 (1955)

Chapter 4
Color Theory: Basic Color Models 12.0 Basic Music Theory 7.0: Basic Time Hypothesis

Carl Swopes

Abstract The primary focus of this paper is the basic conceptualization of "time". Time will be approached as "clock time (*t*)" and as "temporal frequency." A sustainable concept of time will provide a solution to Einstein's twin paradox and time dilation as a black hole is approached. Einstein's twin paradox and time dilation as a black hole is approached are assumed to be opposite extremes of each other.

Keywords Jacob's Ladder · Color harmonics · Color octaves · Temporal density · Velocity · Magnetic field · Clocking · Clock time (*t*) · Time dilation · Space-time · Spatial dilation · Topography · Temporal resistance · Temporal potential · Temporal current · Temporal slope · Temporal wavelength · Temporal frequency · Cartesian coordinates · Ohm's Law · Velocity Formula · Sierpinski Triangle · Circle of Fifths · Golden ratio · Spatial space · Temporal space · Electromagnetic (EM) space · Fabric of space · Composite color wave · Orange hypothesis

4.1 Clock Time (*t*) vs. Temporal Frequency Spectrum

"Clock time (*t*)" will be defined per the Ohm's Law Triangle and the Velocity Formula Triangle. The components of the Ohm's Law Triangle and the Velocity Formula Triangle were equated to each other. Clock time (t) is equated to resistance in Ohm's Law. Similar equations, similar results. Expressing clock time (*t*) as resistance helps to break any preconceived concepts of clock time (*t*).

This paper also considers "time" to be a "temporal frequency spectrum" as defined by the changes in the time vector of Einstein's gravity manifold. Changes in the time vector magnitude are relative to the change in curvature of the gravity

https://sites.google.com/view/color-theory-basic-mechanics/home

C. Swopes (✉)
Swopes Inc./Color Research, Las Cruces, NM, USA

A. Adhikari et al. (eds.), *Quantitative Frontiers*, Industrial and Applied Mathematics, https://doi.org/10.1007/978-981-96-9435-8_4

manifold. A temporal frequency spectrum would have a direct correlation to the curvature of the gravity manifold. The concept of temporal slope is used to define changes in the x-axis of Temporal Space.

Temporal frequencies correlate to Jacob's Ladder. Jacob's Ladder is independent of what it is used to evaluate.

Einstein's space-time (special relativity) identifies three dimensions of space (xyz-axis) + one dimension of time (time-axis). General relativity uses time [clock time (t)] as a one-dimensional vector (magnitude) in the gravity manifold. Einstein acknowledged the connection between gravity and time. He did not consider time as a temporal frequency spectrum.

Clock time (t) is linear, one-dimensional, and based on Earth's orbital period. The International Standard for clock time (t) is 1-s (particle). We take our clock time (t) with us wherever we go. Temporal frequency is based on gravity where Earth = 1 (*one*). Temporal frequency values above or below Earth = 1 (*one*) is time dilation per the temporal density value (temporal frequency_2/temporal frequency_1). Temporal density is the *rate that time cycles* relative to Earth = 1 (one).

Velocity flattens the curved fabric of space and results in a temporal frequency that is relatively lower than Earth = 1 (one). The concept of temporal frequency resolves Einstein's twin paradox. It also suggests that traveling at a relativistic speed causes a traveler to age slower. Relativistic speed is traveling close to the speed of light or at a percentage of the speed of light. Relativistic speed could also refer to the speed of one object being relative to another object (GPS satellite).

References made to "space," "space-time," and "fabric of space" are in most cases interchangeable with each other. In the conceptualization of "time," this paper will introduce the concept of the balanced, system of systems, multi-dimensional (MD) Fabric of Space = 4D Spatial Space + 4D Temporal Space + 4D Electromagnetic (EM) Space per a Sierpinski Tetrahedron geometry. The balanced, system of systems, multi-dimensional (MD) Fabric of Space = *MD Fabric of Space or Fabric of Space*. The fourth dimension is a sub-space specific version of Primary Tone "F" per Jacob's Ladder. The assumption is that everything vibrates. If valid, specific characteristics of the MD Fabric of Space can be associated with a specific sub-space (4D Spatial Space, 4D Temporal Space, or 4D EM Space). Primary Tone "F" is equated to the concept of the "current moment" or "Now."

Per this color research, Jacob's Ladder is the x-axis for 4D Temporal Space and 4D EM Space. The x-axis for cartesian coordinates (4D Spatial Space) is a spatial axis (coordinate). Jacob's Ladder is curved by its use of exponents, golden ratio proportions, and $\sqrt[12]{\text{PHI}^x}$ or $\sqrt[12]{\text{phi}^x}$ logarithmic calculations. It identifies color harmonics or color octaves for the entire range of the electromagnetic spectrum and range of sound.

If everything vibrates, then Jacob's Ladder in Primary Tone "F" is a harmonic distribution (rhythm, order) within everything (Spatial Space, Temporal Space, EM Space, MD Fabric of Space). Primary Tone "F" is used in this paper as the common factor that connects Spatial Space, Temporal Space, EM Space, and MD Fabric of Space.

Jacob's Ladder correlates with the Circle of Fifths. This makes the color harmonics of Jacob's Ladder a variation of the chromatic scale developed by Pythagoras. Therefore, the Jacob's Ladder defined color octaves (mechanical vibrations) of basic music theory = the color harmonics (EM radiation) of Jacob's Ladder. And they are in Primary Tone "F". Jacob's Ladder is assigned to Temporal Space because it is time related, but it is independent of whatever it is used to evaluate.

If the entire viewable universe is in Primary Tone "F", the timing and clocking of basic music theory could be related to the enigmatic concept of "the current moment" or "Now". The concept of Temporal Space introduces temporal slope, temporal wavelength, temporal frequency, and temporal density to complement Einstein's theories on flat fabric of space and curved fabric of space per the time vector of the gravity manifold (funnel). Per Einstein's twin paradox, the twins became aware of their clock time (t) time dilation after they were reunited. This paper proposes that the time dilation occurred because the twins existed at different temporal frequencies within the balanced system of systems, curved MD Fabric of Space. If valid, it implies that 4D Spatial Space, 4D Temporal Space, and 4D EM Space are continuous within the MD Fabric of Space.

Per the current general understanding of the universe, mass (Spatial Space) tells space-time (MD Fabric of Space) how to curve. Space-time tells mass how to move. Einsteinian physics tells us about depressions (curvature, topography) in the MD Fabric of Space (space-time) resulting in the presence of gravity.

Gravity is not a force. Per general relativity, gravity is equated to going downhill. Correctly directed velocity allows a spaceship to climb uphill. Regarding Einstein's gravity manifold (funnel), correctly directed velocity allows a spaceship to move to a flatter or less steep area of the gravity manifold. This paper addresses the curvature of the gravity manifold as the temporal slope of the x-axis of 4D Temporal Space. The curvature of the gravity manifold corresponds to gravity. The temporal slope of the x-axis corresponds to temporal frequency. The assumption is that the length of the gravity manifold's time vector equates to temporal wavelength. As the temporal slope may increase or decrease, we can keep our clock time (t) in our pocket and take it wherever we go.

Einstein and other great scientists have made incredible contributions to our knowledge of "space" and "time." Yet, we still do not completely understand the mechanics of the fabric of space, gravity, or time. This paper attempts to identify some of the temporal mechanics and how those temporal mechanics interface with the MD Fabric of Space. Instead of accepting "space" and "fabric of space" as a 3D Cartesian coordinate system, and space-time as a 3D Cartesian coordinate system + time-axis, this paper constructed the balance, system of systems, MD Fabric of Space for temporal mechanics to have a venue.

Time is typically approached as something that is measured and/or perceived linearly. If Jacob's Ladder is valid, then "time as a frequency" (Temporal Space) is curved. If the combination of Jacob's Ladder with the Composite Color Wave is valid, time is four-dimensional per the Temporal Space orientation. The associated coordinate system could be circumferential versus rectangular.

Currently, "time" (Temporal Space) is not defined regarding a set of mechanics or characteristics. This paper uses the Ohm's Law Triangle and Velocity Formula Triangle to make assumptions about the mechanics and characteristics of clock time (*t*), as a type of temporal resistance. Clock time (*t*) would decrease as the temporal slope angle increases. And clock time (*t*) would increase as the temporal slope angle decreases. This would make clock time (*t*) problematic. Whatever the x-axis is doing, it should not affect the clock time (*t*) that is in your pocket.

Temporal Space and EM Space have the same orientation. Per the Composite Color Wave, color potentials are harmonic golden ratio proportions and are orthogonal to Jacob's Ladder (Temporal Space x-axis).

Jacob's Ladder correlates to color, the electromagnetic spectrum, the range of sound, and basic music theory per the Circle of Fifths. Per the Circle of Fifths, the color harmonics and color octaves of Jacob's Ladder are in Primary Tone "F" (the 11-o-clock to 11-o-clock position). The Orange Hypothesis demonstrates that color harmonics or color octaves begin and end at the 7-o-clock position. The Orange Hypothesis explains the presence of Orange in the spectral display of a prism or rainbow. And that Orange is not a Primary Color.

Einstein addressed some of the characteristics of time in special relativity and time dilation in general relativity per his "twin paradox" thought experiment and his expansion of the *George Lemett "black hole" theory*. In mathematics, dilation means to make something larger or smaller.

The interactions between 4D Spatial Space, 4D Temporal Space, and 4D EM Space influence what is realized in the balanced system of systems of the MD Fabric of Space. Conditions in 4D EM Space could influence 4D Temporal Space resulting in *time dilation* or *spatial dilation*. Spatial dilation is another way of saying teleportation. If EM Space causes time (Temporal Space) to become zero, distance (Spatial Space) goes to zero (bypassed) within the balanced, system of systems of the MD Fabric of Space.

The basic time and time dilation hypothesis of this color research will be addressed by the concepts of temporal wavelength and the Composite Color Wave. The intent is to apply the concept of temporal frequency to characteristics that will provide a better perspective of ***temporal mechanics***. Approaching time as a temporal frequency spectrum should provide a better perspective on Jacob's Ladder concepts and/or characteristics (Figs. 4.1, 4.2, and 4.3).

This paper initially conceptualized the Fabric of Space as a 2D Sierpinski Triangle. *Form follows function.* Next, the Fabric of Space was conceptualized as a 3D Sierpinski Tetrahedron. This configuration requires three (3) tetrahedron for the base and one (1) tetrahedron on top. The additional tetrahedron at the base is a sub-space-specific (system of systems) version of Primary Tone "F". If valid, the sub-spaces of the balanced system of systems, multi-dimensional (MD) Fabric of Space = 4D Spatial Space, 4D Temporal Space, and 4D EM Space. Primary Tone "F" is the fourth dimension and synchronizing element (clocking, orientation, rhythm, current moment, Now, etc.). Everything vibrates.

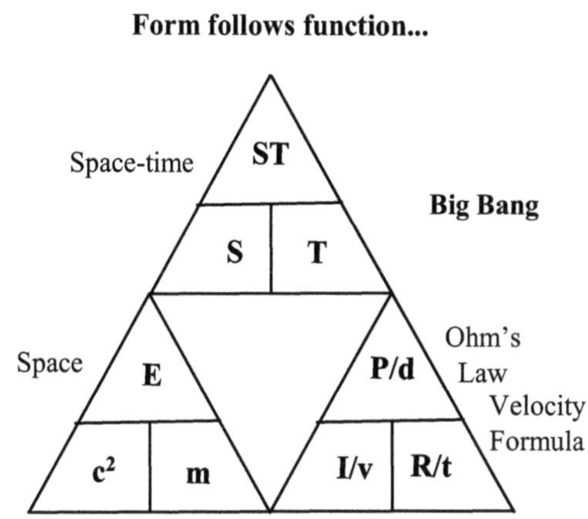

Fig. 4.1 Sierpinski triangle

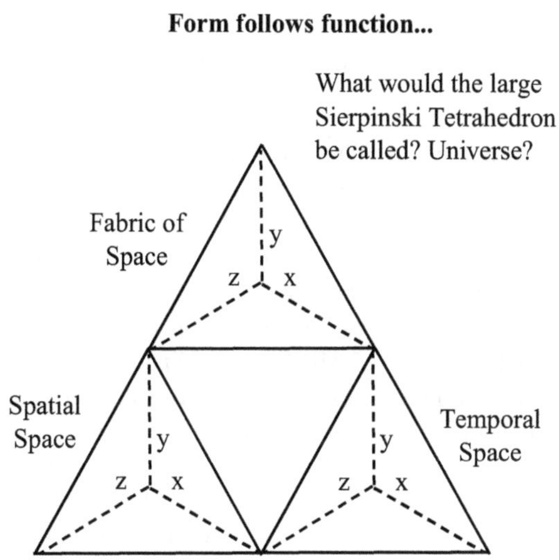

Fig. 4.2 Sierpinski tetrahedron

Color Math. The combining of Red, Green, and Blue (RGB) Primary Colors (EM Space radiation) as measured as frequencies (Temporal Space) within Jacob's Ladder can be expressed as a balance equation where:

Eq. 1: Red (1) + Green (2) = Blue (3)
Eq. 2: Red (1) + Green (2) − Blue (3) = 0 or White, white light source

−Blue = Yellow and −Yellow = Blue, the difference being their orientation. Temporal Space and EM Space have the same orientation.

Eq. 3 Proposed: Temporal Space (1) + EM Space (2) = Spatial Space (3)
Eq. 4 Then: Temporal Space (1) + EM Space (2) − Spatial Space (3) = 0 or balanced, system of systems, multi-dimensional (MD) Fabric of Space

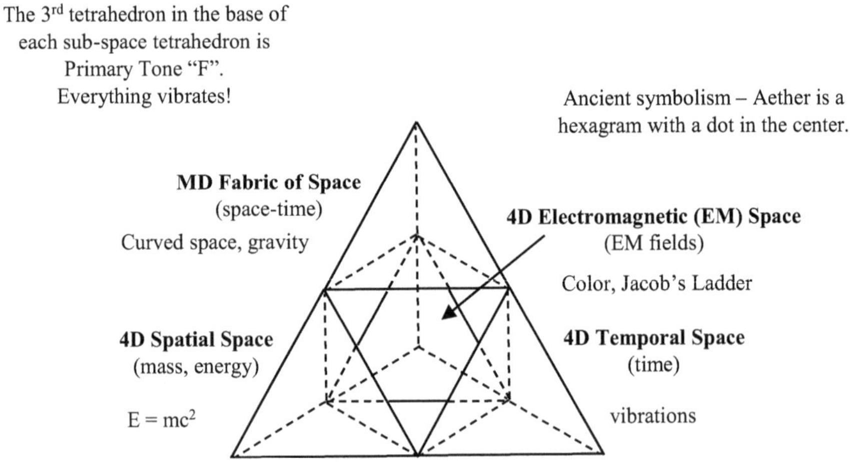

Fig. 4.3 Sierpinski tetrahedron with third tetrahedron in the base

Mass belongs to 4D Spatial Space. If the MD Fabric of Space includes 4D Spatial Space as a sub-space (system of systems), then 4D Spatial Space cannot sit on top of the MD Fabric of Space to form a depression. The presence of 4D Spatial Space could only exclude the MD Fabric of Space to form an inclusion, bubble, or void within the MD Fabric of Space. The resulting *curved MD Fabric of Space* is re-balanced with proportional changes in 4D Temporal Space and 4D EM Space. The re-balanced curved MD Fabric of Space is not equal to flat MD Fabric of Space.

The normal cartesian coordinate system orientation of 4D Spatial Space does not match the orientation of 4D Temporal Space and 4D EM Space. The orientation of the color hexagram, color models, a uniquely oriented cube, Circle of Fifths, Jacob's Ladder, Flower of Life, or Metatron's Cube may be better explained with polar coordinates, cylindrical coordinates, or spherical coordinates which are per the orientation of 4D Temporal Space and 4D EM Space, see Fig. 4.4a, b.

Knowing where and how "time" fits into the general configuration of the MD Fabric of Space helps tell the story of the Universe. If the MD Fabric of Space has structure and sub-spaces, then it may not be an inert elastic sheet that deforms when a bowling ball is placed on it. The MD Fabric of Space could be a structured and reactive environment.

This research has used the color hexagram as a color model that correlates to the Circle of Fifths. In Fig. 4.4, there is nothing to imply that the color model is expandable or how Fig. 4.4a relates to Fig. 4.4b. When Fig. 4.1 was converted to a Sierpinski Tetrahedron per the geometry of Fig. 4.3, it became obvious that the modeling was expandable, see Figs. 4.5, 4.6, 4.16, and 4.17.

Note that Figs. 4.5 and 4.6 are at the MD Fabric of Space level. Figures 4.16 and 4.17 are the same as Figs. 4.5 and 4.6. But Figs. 4.16 and 4.17 are identified as a particle of the MD Fabric of Space.

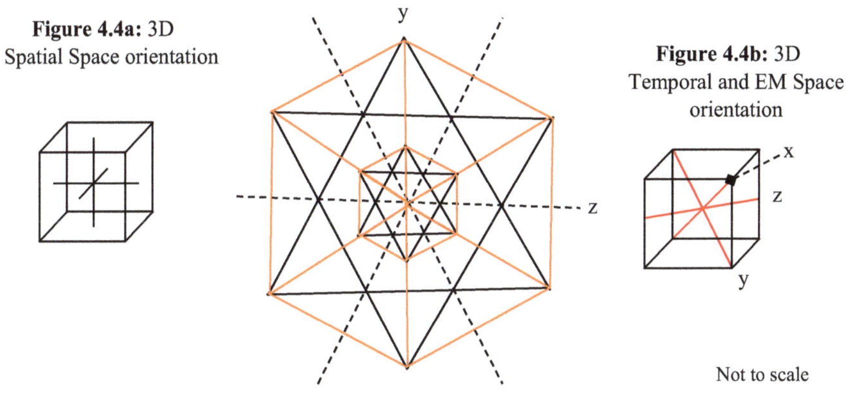

Fig. 4.4 Metatron's Cube (temporal space orientation)

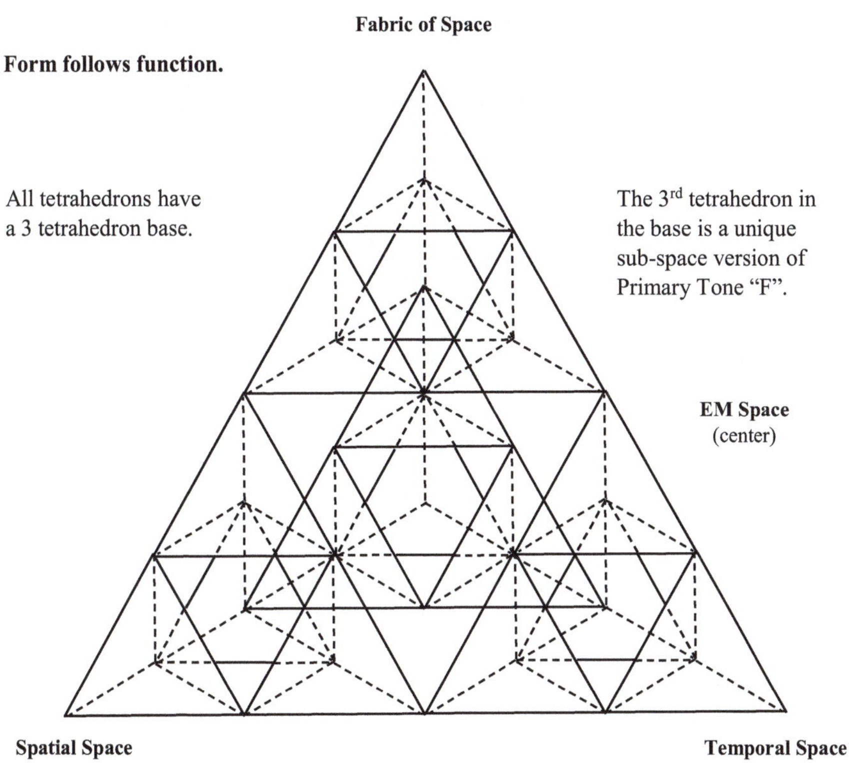

Fig. 4.5 Sierpinski Tetrahedron (crystal, fractal) with technical errors

Per the Ohm's Law Triangle and the Velocity Formula Triangle correlation, Clock Time (t) = Temporal Resistance or a form of Temporal Density along the x-axis (Jacob's Ladder) within 4D Temporal Space. Clock Time (t) should not

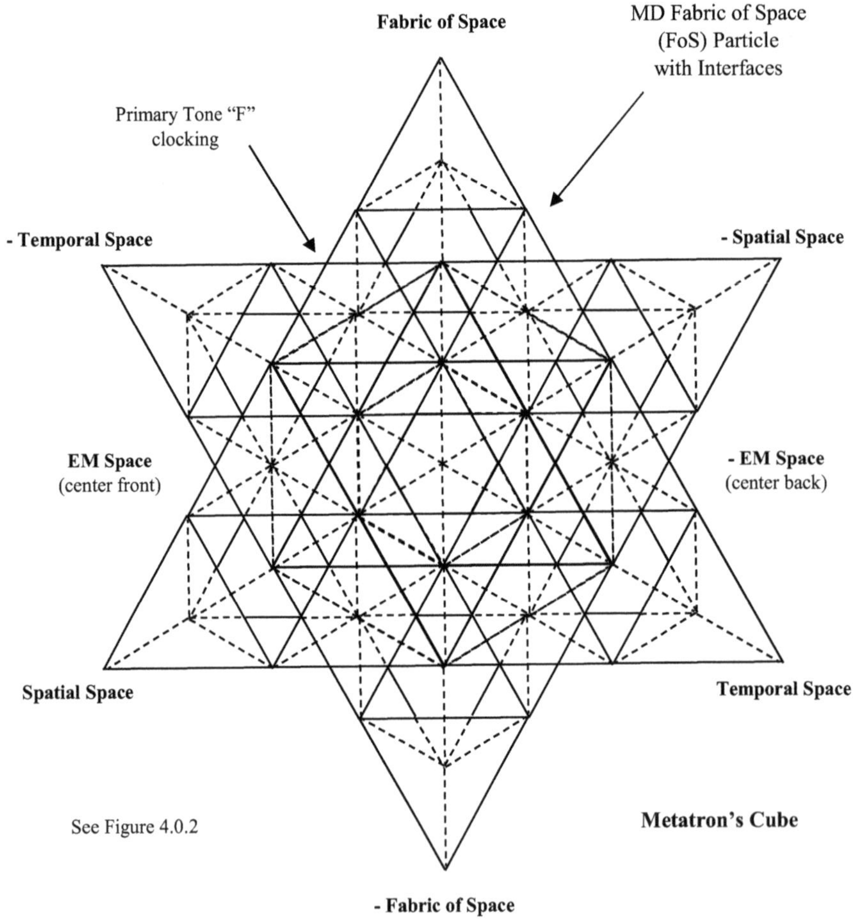

Fig. 4.6 Sierpinski Star Tetrahedron (crystal, fractal)

be affected by changes in temporal slope. Temporal Frequency correlates to the time vector of Einstein's gravity manifold per general relativity. Einstein addresses "time" more directly in special relativity as space-time.

The elements of the Ohm's Law Triangle and the Velocity Formula Triangle [clock time (*t*)] cannot effectively characterize time dilation per "Einstein's twin paradox" and time dilation when "approaching a black hole" theories. The temporal slope value of the x-axis is used to define temporal wavelength, temporal frequency, temporal density, and time dilation of the curved MD Fabric of Space. They are characteristics of temporal mechanics.

The time vector magnitudes of Einstein's gravity manifold (funnel) are equated to temporal wavelengths and temporal slope values. The time vector, temporal wavelength, and temporal slope values are tied directly to changes in the x-axis of 4D Temporal Space (time, Jacob's Ladder) due to the curved MD Fabric of Space.

Earh = 1 (one). Let Earth's temporal density value (rate that time cycles) be Earth = 1 (*one*). When another temporal density value is divided by Earth's reference value of Earth = 1 (*one*), the quotient is the "temporal density value" (time dilation) between two MD Fabric of Space locations. As the temporal slope angle (x-axis) goes to the vertical (y-axis), the temporal slope value becomes smaller. A shorter temporal slope value (time vector magnitude) equates to a higher temporal frequency.

The Earth = 1 (*one*) value correlates to the time vector, temporal wavelength, temporal frequency, temporal density, and temporal slope value. Temporal frequency correlates to Jacob's Ladder and provides an interesting insight into the basic nature of temporal mechanics. Understanding how things work is the basis for science, mathematics, technology, etc. Time as a frequency does not work well with many of the equations used to describe how things work, Ex. $d = v \times t$.

Time as a frequency corresponds to "things" being cyclic and harmonic in Nature, history, politics, plagues, etc. The cyclic happening of an event is called its period. This is a reference to temporal frequency but identified as clock time (t). Temporal frequency is a spectrum, but clock time (t) is linear. We take our clock time with us wherever we go and apply it to most of the things we encounter. Einstein's twin paradox is time dilation due to differences in temporal frequency (pitch).

If time approaches zero, distance approaches zero. Time is space and space is time. Therefore, time must also have an elasticity or weighted characteristic that facilitates time dilation—the temporal slope value. The temporal density (rate that time cycles) of relatively flat MD Fabric of Space has a longer temporal wavelength (time vector, lower temporal frequency) than the temporal density of curved MD Fabric of Space that is near a black hole. Areas of curved MD Fabric of Space should be older than flatter areas of the MD Fabric of Space.

There are four fundamental forces of nature. The weak atomic force and strong atomic force are associated with 4D Spatial Space. Gravity is associated with the MD Fabric of Space. The fourth fundamental force of nature is the electromagnetic (EM) force per 4D EM Space. Time is not considered a fundamental force of nature.

4.2 Topography of the MD Fabric of Space

Overtones and Undertones are the higher and lower color harmonics or color octaves per Jacob's Ladder in Primary Tone "F" of a base tone (frequency) that are present with the base tone. A base tone is the tone (frequency) that is displayed, played, sounded, etc. In electronics, these additional harmonics are seen as noise. For a musical instrument, overtones and undertones are due to resonance and give a musical instrument its unique tonal quality or characteristics (timbre). Timbre is also known as "tone color" or "voice." Overtones and undertones cannot be demonstrated with the classical vibrating string experiment or Chladni plate. The harmonics of a vibrating string or Chladni plate are based on standing waves.

Figure 4.5 is the Sierpinski Tetrahedron. The current configuration identifies the MD Fabric of Space and its sub-spaces (Spatial Space, Temporal Space, and EM Space). Going down 1-tier, the same structure would apply to a 4D Spatial Space, 4D Temporal Space, or 4D EM Space tetrahedron. This is the fractal nature of the Sierpinski Tetrahedron. If a force is applied to a crystal, it will produce an electric charge. If an electric charge is applied to a crystal, it will vibrate (quartz watch). Interesting...

If the universe has some sort of order and/or shape, what would that order or shape look like? This paper is the researcher's first effort at creating the universe.

Figure 4.3 was used to construct Fig. 4.5. Dotted lines are hidden edges and corners. Some of the visible lines in Fig. 4.3 became hidden edges in Fig. 4.5 and were not corrected—technical drawing errors.

Figure 4.6 is the star tetrahedron version of the Sierpinski Tetrahedron. It equates to the color hexagram, Jacob's Ladder, Circle of Fifths, color harmonics, color octaves, and the Composite Color Wave. It is in the 4D Temporal Space orientation.

There is a potential for polarization and/or orientation in the MD Fabric of Space. There is polarization and/or orientation in 4D Spatial Space, 4D Temporal Space, and 4D EM Space which are the sub-spaces of the MD Fabric of Space.

Curved MD Fabric of Space could introduce changes within the sub-spaces (Spatial Space, Temporal Space, and EM Space) that result in changes to the characteristics of the curved MD Fabric of Space. Or changes in the sub-spaces (Spatial Space, Temporal Space, and EM Space) could introduce changes in the curved MD Fabric of Space. Gravity is at the MD Fabric of Space level.

4.2.1 Slope vs. Curvature

Temporal slope is regarding the slope of the x-axis of 4D Temporal Space as it correlates to the time vector of Einstein's gravity manifold. Gravity is explained as the curvature of the fabric of space due to the presence of mass (Spatial Space).

Note 4.1: Slope Definition
Slope is the change in one set of coordinates versus another set of coordinates—rise/run. This could be two points on a straight line or curved line. The slope at a point on a curved line is equal to the slope of the tangent line that best approximates the curved line at that point. The slope at a point on a circle is the tangent line that is perpendicular to the radius (a straight line from the center (focus) of the circle to the point on the circle).

Note 4.2: Curvature Definition
Curvature is the inverse of the radius of the circle that is the better approximation of the curvature at that point. *Curvature* measures how fast a *curve* changes direction at a given point.

The slope is read from left to right. First the rise (y-axis) then the run (x-axis).

- Slope is positive if it is increasing from left to right.
- Slope is negative if it is decreasing from left to right.
- The slope is zero if the line is horizontal—no change in x-axis value.
- Slope is undefined if the line is vertical—no change in y-axis value.

Note 4.3: The Concept of Temporal Slope
The concept of temporal slope addresses the slope of the x-axis of 4D Temporal Space. If the x-axis is experiencing a change in slope, it would imply that the MD Fabric of Space itself is curving.

The focus is on the x-axis of 4D Temporal Space versus a $\Delta y/\Delta x$ within the coordinate system of Spatial Space, EM Space, or the curving surface of the gravity manifold. If the "fabric of space" equates to the MD Fabric of Space, then it could have a crystalline structure, see Fig. 4.6. As a crystalline structure, the MD Fabric of Space would be reactive.

Per this paper, as a balanced, system of systems, flat MD Fabric of Space is assumed to be balanced to zero. Curved MD Fabric of Space is assumed to react with piezoelectric like characteristics. When a force is applied to a crystal, the crystal produces a charge. When a charge is applied to a crystal, the crystal vibrates. The presence of mass causes the MD Fabric of Space to curve which results in a Temporal Space charge and an EM Space charge. Respective vibrations should also be produced.

4.2.1.1 Calculus: The Derivative of the Equation of the Curve

The derivative of the function (quadratic equation, parabola) identifies the slope (rate of change) at any point along the line (Fig. 4.7).

The derivative of the equation for a curved line allows the slope to be calculated at any point along the curve. The derivative of the function x^2 is $2x$. Therefore, the rate of change at any point along the curve is $y = 2x$.

Figure 4.8 identifies the slope of a point on a circle. Coordinates are per the center (focus) of the circle which define the cosine (adjacent leg, x-axis) of the angle and the sine (opposite leg, y-axis) of the angle. Therefore, the x-coordinate of a point on a circle is the cosine of the angle, and the y-coordinate is the sine of the angle. Tangent is the basic concept of "rise/run" or slope (y/x). The change in the value of "y", due to the change in the value of "x", is a basic concept in calculus. The resulting formula in Fig. 4.8 is the derivative of a trigonometric function. Slope is between two points. In Fig. 4.8, the first point is the origin at coordinate 0,0. The second point is a point on the circle (Table 4.1).

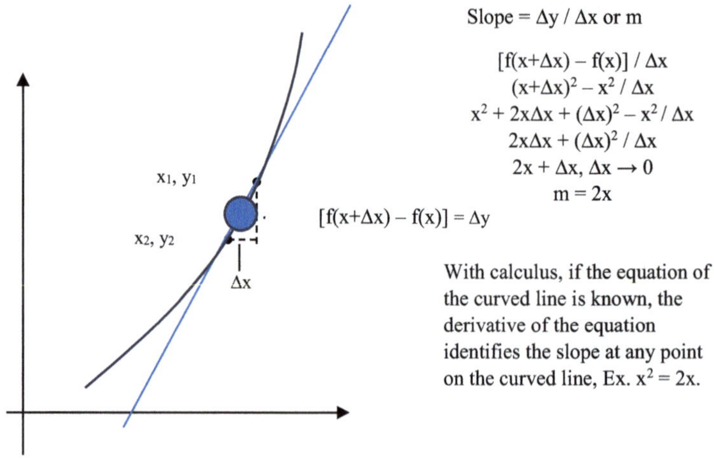

Fig. 4.7 Slope at a point on a curved line—Calculus

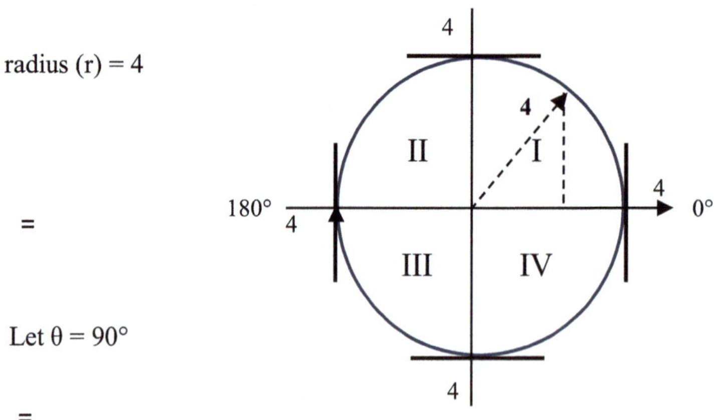

Fig. 4.8 Slope of a circle (temporal space)

Table 4.1 Basic Trig Functions

Function	Equation
sine	$y = \text{sine}\ \theta$
cosine	$x = \text{cosine}\ \theta$
tangent	$y/x = \text{tangent}\ \theta$

- The slope of a circle is the radius of the circle connecting to a point on the circle, as measured in degrees or radians (θ).
- The slope at 90° or 270° is undefined. Slope is an increasingly small decimal value just before the slope becomes undefined at 90° or 270°. Significant changes occur 1° to either side of 90° or 270° (*temporal wavelength, x-axis*).

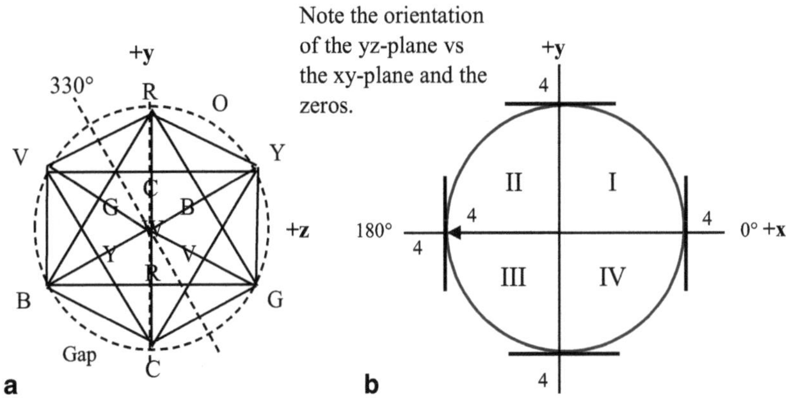

Fig. 4.9 (**a**) Color hexagram ordered information temporal space orientation. (**b**) Slope of a point on a circle spatial space orientation

- The slope at 0° or 180° is zero. The value is a very large number before the slope becomes zero at 0° or 180°. Significant changes occur 1° to either side of 0° or 180° (*temporal potential, y-axis*).
- Quadrant I or III slope values are negative. Quadrant II or IV slope values are positive (Figs. 4.9 and 4.10).
- Gravity (curved MD Fabric of Space due to the presence of mass).
- Time (curved MD Fabric of Space due to the presence of mass).
- Jacob's Ladder (x-axis, Temporal Space, etc.)
- Clocking (yx-plane, Temporal Space, EM Space, Jacob's Ladder, etc.)
- Color Harmonics and Color Octaves (Jacob's Ladder).
- Temporal Slope (time vector, temporal wavelength).
- Temporal Density [rate that time cycles with Earth = 1 (*one*)].
- Temporal frequency or temporal wavelength → Temporal Density.
- Time vector value equates to temporal wavelength.
- Temporal wavelength equates to temporal frequency.
- Time dilation: deltas in temporal frequency between two locations.
- Does matter exist at higher/lower values of temporal frequency? Per Einstein's twin paradox, where in "space" was the twin that was traveling near the speed of light?

4.2.2 Temporal Density

Assumption 4.1 Temporal Density Summation
Space-time is four-dimensional.

- Three spatial dimensions (xyz-axis)

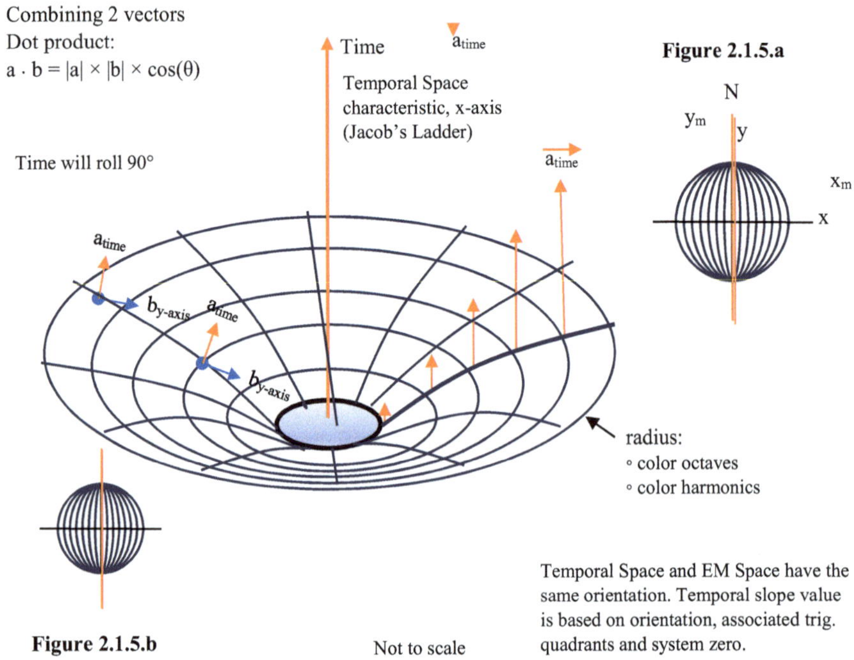

Fig. 4.10 Curved fabric of space–temporal slope

- One dimension of time (time-axis, time vector)

In 4D Temporal Space, the time vector magnitude equates to temporal wavelength (temporal spectrum).

- Temporal wavelength is the temporal slope value.
- Temporal wavelength correlates to temporal frequency.

Let Earth's values for temporal slope, gravity, time vector magnitude, temporal frequency, and temporal density be equal to **Earth** = **1** ***(one)***. Where temporal wavelength_1, temporal frequency_1, and temporal slope value_1 represent Earth's values for 1 *(one)*. A "Δ" from Earth = 1 *(one)* temporal density is time dilation.

- temporal wavelength_1/temporal wavelength_2 = Δ temporal density
- temporal frequency_2/temporal frequency_1 = Δ temporal density
- temporal slope value_1/temporal slope value_2 = Δ temporal density

General relativity predicts that time passes slower near a black hole because the time vector magnitude decreases. In general relativity, time is just a vector or a vector's magnitude. "Time" as represented as 4D Temporal Space, a shorter time vector magnitude implies a theoretically shorter "temporal wavelength." As temporal wavelengths become shorter, the Temporal Space is at a higher vibrational rate.

Per this paper, as temporal frequency increases above the reference value of Earth = 1 (*one*), the Temporal Space vibrates (cycles) at a faster rate, not slower. As temporal frequency decreases below the reference value of Earth = 1 (*one*), the Temporal Space vibrates (cycles) at a slower rate.

$$\text{Clock time } (t) \times \text{temporal density value} \approx \text{rate of clock time } (t) \text{ passage}$$

Note: Replacing clock time (*t*) with a temporal frequency does not work well in an equation unless the temporal frequency is Earth = 1 (*one*).

In Einstein's twin paradox, velocity flattens the curved MD Fabric of Space. The time vector magnitude increased. Temporal frequency decreased. The temporal slope value increased. Temporal density decreased. The effect is that time cycled at a slower rate for the twin aboard the spaceship traveling close to the speed of light. The twins did not share a common "current moment" (Now) while the one twin was traveling close to the speed of light. When the velocity of the spaceship returned to "normal" and the spaceship returned to Earth, the Earthbound twin and the Earth were much older. The twin aboard the spaceship returned to the future Earth. *Could it be assumed that traveling at a relativistic speed will cause someone to arrive later than sooner?*

What if..., the "big bang" happened at relativistic speed?

A spaceship approaching a black hole would experience an increase in the temporal slope angle of the curved MD Fabric of Space. Gravity is seen as going downhill. The time vector magnitude decreases. The temporal slope value decreases. Temporal density increases. Temporal frequency increases. Gravity would add potential to a mass per the applicable gravity constant (acceleration). An increase in potential causes an increase in current when resistance is constant—Ohm's Law. An increase in displacement causes an increase in velocity when clock time (*t*) is constant—Velocity Formula.

Therefore, as a spaceship approaches a black hole, the going downhill effect (gravity) increases. The time vector (temporal wavelength) is becoming shorter causing the temporal frequency (temporal density) to increase.

As the time vector magnitude decreases, clock time (*t*) should not change. *If clock time (t) decreased* as the time vector magnitude decreased, then potential (displacement) decreases. Current (velocity) would increase towards infinity until it eventually becomes undefined at 90°. Either way, temporal density (the rate that time cycles) would propel mass into the future. With the Earth's current moment being at a lower temporal density, the Earth is in the relative past for anything that gets close to a black hole. This is Einstein's twin paradox in reverse. The twin travelling close to the speed of light was at a lower temporal density than Earth = 1 (*one*). When that twin returned to Earth, the Earthbound twin and the Earth were older. For a spaceship that traveled close to a black hole, the occupants and the spaceship would be much older than those that were Earthbound. There would be a point of no return, if returning was possible.

For "light" to escape the MD Fabric of Space of a black hole, it would have to travel backwards in time due to temporal density (temporal frequency, time dilation). The "speed of light" is a constant (in a vacuum), and it is only expressed as a positive value per our "current moment" (Now) and current understanding of "light" [electromagnetic (EM) radiation].

Per general relativity, gravity is caused by the curved MD Fabric of Space due to the presence of mass. If the concept of temporal density is valid for a black hole, then what *mass* in time is causing the MD Fabric of Space to curve per the "current moment"? Per this paper, a black hole would have a current moment that is constantly moving into the future. Is the mass of a black hole ***cumulative*** based on its position in the future due to its temporal density (rate that time cycles)? The measurement of mass is a "current moment" measurement, not a temporal density measurement.

What happens to the MD Fabric of Space and its sub-spaces when "time" is applied as a temporal frequency (temporal density, spectrum, 4D Temporal Space) with specific characteristics versus clock time (t)—a sequence of discrete moments in "1D time"? One example is that it makes the measurement of the mass of a black hole problematic. Per this paper, a black hole does not share the same current moment with Earth = 1 (*one*). For a black hole and possibly the universe, "time" may be more of a spectrum that is better represented by temporal frequencies in Primary Tone "F" of Jacob's Ladder.

Figure 4.11 is a two-dimensional representation of a heavy ball rolling around on a suspended rubber sheet, with a few modifications. The modeling was used by Einstein to demonstrate the displacement (curvature) of the fabric of space due to the presence of mass. A *vertical line* (y-axis) has an undefined slope. To say that it has no slope is ambiguous. All lines have a slope, even if the slope cannot be defined.

4.3 4D Temporal Space and Orbital Mechanics

Per the Ohm's Law Triangle, and the Velocity Formula Triangle, "time" is addressed as clock time (t). Figure 4.1 equates current, potential, resistance, with displacement, velocity, and time. Time is equated to temporal resistance. Time as a frequency (spectrum) per 4D Temporal Space has different characteristics. In this theoretical characterization, time equates to the time vector magnitude of Einstein's gravity manifold, temporal wavelength, temporal frequency, temporal density, temporal slope, etc.

Per Fig. 4.1 and Table 4.2, time is the x-axis for 4D Temporal Space. The 4D Temporal Space characteristic that relates to the value of time is temporal resistance, when Earth = 1 (one), see Fig. 4.11. Any value above or below Earth = 1 (one) involves time dilation due to differences in temporal frequency.

Due to velocity and/or the curved MD Fabric of Space, the rate that time cycles could be different for two independent observers (Einstein's twin paradox

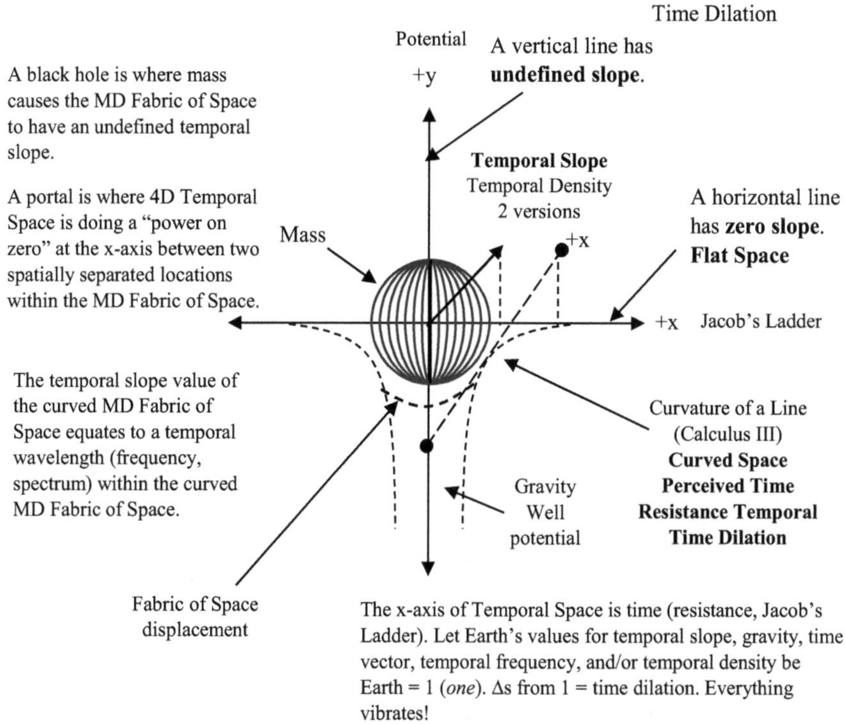

Fig. 4.11 Slope of a line, curved vs. flat space, per Einstein's twin paradox

Table 4.2 Definitions: Spatial Space, Temporal Space, Electromagnetic (EM) Space, and Fabric of Space

4D Spatial Space definition—Relating to, occupying, or having the character of space, relating to the position, area, and size of things, etc. Spatial Space may have Temporal Space and/or EM Space characteristics. Characteristics are regarding the Ohm's Law Triangle, the Velocity Formula Triangle, and time vector per Primary Tone "F".
4D Temporal Space definition—Of or pertaining to time (vibrations, harmonics, octaves, temporal slope, temporal wavelength, and/or temporal frequency, etc.). Temporal *Space* may have Spatial Space and/or EM Space characteristics. Characteristics are regarding the Ohm's Law Triangle, the Velocity Formula Triangle, and time vector per Primary Tone "F".
4D Electromagnetic (EM) Space definition—The concept was evaluated as basic color theory (color harmonics), basic music theory (color octaves), and Jacob's Ladder. EM Space may have Spatial Space and/or Temporal Space characteristics. Characteristics are regarding the Ohm's Law Triangle, the Velocity Formula Triangle and time vector per Primary Tone "F".
Multi-dimensional (MD) Fabric of Space definition—This is where the whole is greater than the sum of its parts. The Fabric of Space includes the sub-spaces of 4D Spatial Space, 4D Temporal Space, 4D EM Space, Jacob's Ladder, Primary Tone "F", clocking, etc. Within MD Fabric of Space, one sub-space can manipulate characteristics in another sub-space.

or approaching a black hole). The difference in clock time (t) or time as a frequency as experienced by two observers due to velocity and/or the curved MD Fabric of Space is called *time dilation*.

Orbital mechanics is another method this paper uses to conceptualize "temporal mechanics." A spacecraft in a stable circular orbit around a mass is in freefall (microgravity). Its velocity is offsetting the downward pull (acceleration) of gravity.

Newton's first Law of Motion: an object will not change its motion unless a force acts on it.

For an orbiting spacecraft to slow down, it must accelerate. Acceleration pushes the spacecraft into a higher stable orbit, but its orbital period is slower. Acceleration pushes the spacecraft away from the gravitational pull of the mass it is orbiting. The orbital period (time vector, temporal wavelength) for the spacecraft becomes longer. Acceleration caused the spacecraft to relocate to a less curved area of the curved MD Fabric of Space of the mass it orbits. Its time vector became longer. Therefore, its temporal frequency (temporal density) becomes lower.

Note that the orbital period or unit value for clock time (t) has not been identified. On Earth, the orbital period of the planet is the basis for the International Standard (SI) for clock time (t) being 1 s. The calculation of orbital velocity is different from the Velocity Formula Triangle. It is the gravitational constant that introduces the element of clock time (t) into the calculation for orbital velocity.

The calculation for orbital velocity is:

$$\mathbf{V\ orbit} = \sqrt{}\,(\boldsymbol{GM})\,/\boldsymbol{R}\ \text{[is measured in m/s]}$$

G = gravitational constant
M = mass of the body from the center
R = radius of the orbit from the center of the mass plus the distance above the mass

At a high enough acceleration, a spacecraft will obtain its escape velocity and break free of the gravitational pull (curved MD Fabric of Space) of the mass it is orbiting. Its outwards velocity exceeds the downwards velocity of gravity.

Time is associated with the gravitational constant versus radius or period of the orbit. As "R" increases "V orbit" decreases. As "R" decreases "V orbit" increases.

For an orbiting spacecraft to speed up, it must reduce its "R" value by falling deeper into the curved MD Fabric of Space caused by the mass it is orbiting. This is accomplished by slowing down (decelerating)—Einstein's twin paradox.

Calculating orbital velocity helps confirm the association between time and gravity. Just as it helps confirm the disassociation between clock time (t) and orbital period as it relates to "time." The weight of an orbiting mass is zero because gravity is zero. This implies that gravity is separate from time, with both having a relationship to the curved MD Fabric of Space they are associated with.

Figure 4.12 demonstrates that 4D Spatial Space, 4D Temporal Space, and 4D EM Space have similar characteristics related to frequency. Color harmonics and color octaves may equal *temporal dimensions* or *the rate that time cycles (vibrates) at a point along the x-axis*. Figure 4.12 is the visible light and Circle of Fifths model.

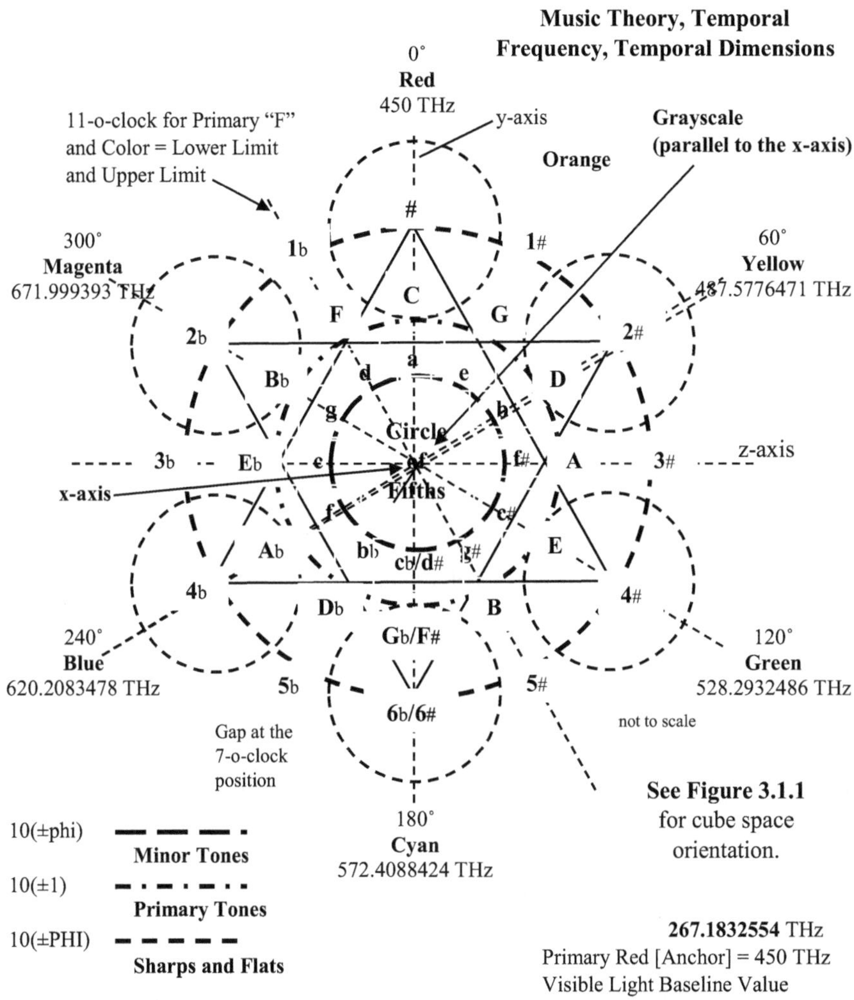

Fig. 4.12 Circled hexagram showing unit circles and pythagorean circle of fifths color wheel

Characteristics—Color Circle of Fifths Encoding (Fig. 4.12):

- Chromatic scale—12 Primary Tones and/or 12 Primary Color references
- Equal temperament—equally spaced around a circle
- Counting per the Perfect Fifth—tonal arrangement/order
- Clocking per the Perfect Fifth (orientation)
- Harmonic and clocked colors per Jacob's Ladder
- A direct numerical and geometric correlation between color octaves and color harmonics. 4D Temporal Space and 4D EM Space have similar characteristics and the same orientation

- Each color octave, semitone, color harmonic, Primary Color, Primary Color Interface, or frequency represents a unique 10(x) circular constant
- Orange Hypothesis per the lower frequency North Pole vs. the higher frequency South Pole Gap (interface) at the 7-o-clock position
- Geometries that relate to the Platonic Solids (regular solids), Flower of Life, color hexagram, golden ratio proportions, Metatron's Cube, etc.

In 4D Temporal Space, the yz plane orientation is per Fig. 4.12. The system zero is at 330°. *This is not the yz-plane of 4D Spatial Space.* For 4D Spatial Space, the system zero is at 0° (12-o-clock). Currently, the Earth's static magnetic field North Pole is approximately 11° from the geographic North Pole. The Sun's static magnetic field North Pole is in alignment with the Sun's geographic North Pole.

Therefore, our solar system has a North Pole, South Pole, and the "Characteristic Color Circle of Fifths Encoding" of the Sun. The prism experiments of Sir Isaac Newton converted sunlight (a white light source) into a spectral display (the six Primary Colors + Orange). ***Sunlight is tuned to Primary Tone "F".***

Per the Orange Hypothesis of this color research, Orange is created when a signal jumps the Gap between the north pole of the lower frequency color octave and the south pole of the higher frequency color octave at the 7-o-clock position, see Fig. 4.12. The same would be true for a signal traveling in the opposite direction. A pole reversal would cause the signal to be 180° out of phase. Since the 7-o-clock position is at a higher energy level than the 1-o-clock position, there is an energy release.

4.3.1 MD Fabric of Space Phenomena: Portals

When time equals zero, distance equals zero. Therefore, when temporal frequency (Temporal Space) equals zero, distance (Spatial Space) equals zero per the concepts of temporal slope, temporal wavelength, temporal frequency, temporal density, and time dilation.

A magnetic portal (EM Space) causing time (Temporal Space) to equal zero at the x-axis, could be a naturally occurring phenomenon of bypassing the physical distance (Spatial Space) between spatially separated locations in the MD Fabric of Space. If Fig. 4.12 is capable of a "power on" zero at the x-axis, then a magnetic field (EM Space) could cause temporal frequency (Temporal Space) to equal zero at the x-axis and the MD Fabric of Space is balanced with distance (Spatial Space) equaling zero (bypassed) for the duration of the phenomenon. The average value of the Composite Color Wave is zero at the x-axis.

When time equals zero, distance equals zero. With teleportation, the "teleported" does not pass or move through the 4D Spatial Space between the spatially separated locations in the MD Fabric of Space. The teleported would move from a "here and now" to a "there and now."

The concept of teleportation seems to be a variation of "time dilation." Instead of having a delta in time (time vector, temporal density, temporal slope value) between

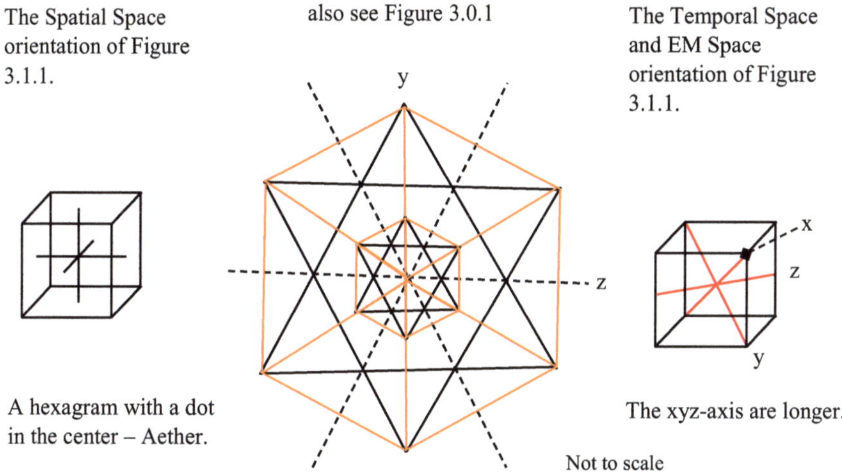

Fig. 4.13 Metatron's Cube (temporal space)

two spatially separated locations in the MD Fabric of Space, there is a "spatial dilation" at the interface (portal) between two spatially separated locations in the MD Fabric of Space. But the teleported going from a lower/higher temporal density (time vector, temporal density, temporal slope value, temporal frequency) would be subject to the lower/higher temporal density (rate that time cycles) on arrival at the "there and now."

Consideration must be given to any delta in environmental conditions or change in position between the two spatially separated locations in the MD Fabric of Space. Any moving portal gate location could have a limited window of opportunity to establish a successful portal connection.

Figure 4.13 identifies the orientation of 4D Spatial Space versus the orientation of 4D Temporal Space and 4D EM Space. Metatron's Cube, a uniquely oriented cube, hexagram, star tetrahedron, Flower of Life, and "hexagram with a dot in the center" (Aether) share the orientation of 4D Temporal Space.

Prof. PhD Jack Scudder, NASA researcher, plasma physicist, and astronomer at the University of Iowa, has identified what he calls X-points in Earth's upper atmosphere. X-points are places of electromagnetic reconnection between the Earth's magnetosphere (magnetic field) and the Sun's magnetosphere. X-points are believed to be temporary magnetic portals that open between the Earth and the Sun. X-points are naturally occurring events that require no hardware.

Per this paper, the interfacing magnetospheres (EM Space) are causing time (Temporal Space) to do a "power on" zero at the x-axis. When time is zero, distance is zero. Distance is a 4D Spatial Space characteristic within the curved MD Fabric of Space, see Figs. 4.14 and 4.15.

Metatron's Cube is believed to be a portal or a symbol for a portal. As a symbol for a portal, Metatron's Cube would be an ancient schematic. Per Fig. 4.13,

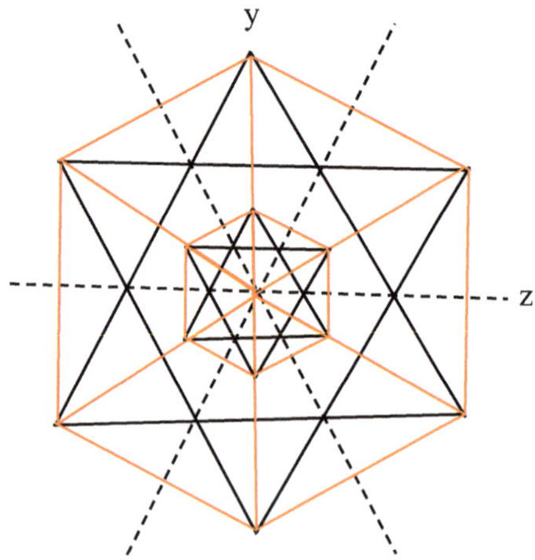

Fig. 4.14 Metatron's Cube (fabric of space)

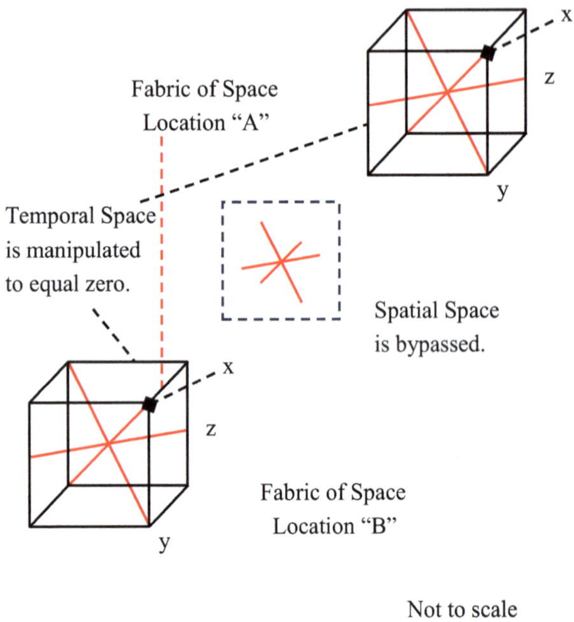

Fig. 4.15 Spatial space dilation model

Metatron's Cube is associated with 4D Temporal Space per its orientation. If true, then a tuned, clocked, and directed magnetic field (EM Space) in the configuration of Metatron's Cube could create a power on zero at the x-axis in 4D Temporal Space. The concept is that of a unique configuration of tuned and oriented magnetic fields (EM Space) facilitating a power on zero at the x-axis and 4D Temporal Space between two spatially separated locations in the MD Fabric of Space. Within the balanced system of systems, MD Fabric of Space, 4D Spatial Space is bypassed. In this case, temporal potential (time, Temporal Space) is reduced to zero at the x-axis, therefore distance is also zero (*spatial dilation*).

Time dilation relates to Einstein's twin paradox. Per this thought experiment, traveling close to the speed of light (velocity) flattens curved MD Fabric of Space causing a decreased temporal slope value (longer time vector, lower temporal frequency). A lower temporal frequency for the twin traveling close to the speed of light causes time to cycle at a slower rate, see Figs. 4.10 and 4.11.

The twin in a spaceship traveling close to the speed of light did not share a common "current moment" or "Now" with the twin that remained on Earth. At the end, the twin that had traveled close to the speed of light (a lower temporal frequency) had effectively traveled to the future. Following this line of thought, the future was accessed by traveling at a lower temporal frequency (temporal density) than the Earth = 1 (*one*) reference value. Per Einstein's twin paradox (a thought experiment) could be interpreted as implying that "time" and therefore "space" is a type of "velocity or frequency polarized spectrum." We see and experience the Universe based on our velocity and/or temporal frequency. The concept could help define the "current moment," "Now," Jacob's Ladder, Primary Tone "F". Also, Einstein's twin paradox thought experiment implies that "time" and "space" are continuous. Otherwise, the twin traveling near the speed of light existed outside of "time" and/or "space" (space-time). Clock time (t) was used to confirm the time dilation.

Hypothesis 4.1 Spatial Dilation
Instead of velocity (speed of light, current) creating a "temporal dilation" within the MD Fabric of Space, electromagnetic radiation (magnetic current, EM Space) creates a "spatial dilation" between two spatially separated locations in the MD Fabric of Space. 4D EM Space radiation (color harmonics, color octaves) combines to create a power on zero at the x-axis, causing Earth = 1 (*one*) "temporal frequency" (4D Temporal Space) to equal zero at the x-axis for the duration of the phenomenon. *In math, dilation means to make larger or smaller.* If time goes to zero, distance goes to zero. If the MD Fabric of Space location "A" is separate from MD Fabric of Space location "B", then a 4D EM Space *power on* zero at the x-axis between locations A&B makes them adjacent to each other. The MD Fabric of Space is a balanced system of systems. A temporal zero at the x-axis between locations A&B is balanced with the intervening distance being equal to zero (bypassed). The threshold of a theoretical stable portal is believed to be two-dimensional and obstruction free. There could be temporal density differences between the two spatially separated locations, see Figs. 4.7 and 4.13, Figs. 4.14 and 4.15.

Per the color equations [Asin(10(±phi)), Bsin(10(±1)), and Csin(10(±PHI))] used to produce the "voltage as a function of time [$V(t)$]" *Composite Color Wave*, resulted in the waveforms assuming the 4D Temporal Space orientation. A, B, and C are Alice Scaling Factors. Time base correction is required. The color equations allow the creation of RGB waveforms and CMY waveforms that are 180° out of phase with each other. The waveforms (EM Space potentials) are orthogonal to Jacob's Ladder (Temporal Space, x-axis). Figure 4.12 is orthogonal to the x-axis.

4.3.2 Research Objectives Review

This paper has made correlations or contrasted the characteristics of "clock time (t)" and those of "temporal frequency." Clock time (t) is associated with the concept of resistance temporal. Temporal frequency is associated with the time vector magnitude of Einstein's gravity manifold.

Per this paper, gravity must belong to 4D EM Space. It is like magnetism. Gravity increases as the temporal slope *angle* (time vector magnitude, temporal wavelength) of the curved MD Fabric of Space increases, see Figs. 4.10 and 4.11. Note, as the temporal slope angle increases, the temporal slope value (temporal wavelength) decreases and temporal frequency increases.

Gravity is an MD Fabric of Space phenomenon that is addressed by the General Theory of Relativity. It is generally accepted that gravity is due to the presence of mass within the MD Fabric of Space. In terms of cause-and-effect, gravity must have a flat MD Fabric of Space (potential) version of itself. Just as, time must have a flat MD Fabric of Space (potential) version of itself. If the cause is the presence of mass within the MD Fabric of Space, then what characteristics must the MD Fabric of Space possess?

This paper proposes that the presence of mass in the MD Fabric of Space creates a depression (inclusion) and that is all it does. The interaction of 4D Temporal Space with the depression results in temporal characteristics. The interaction of 4D EM Space with the depression results in EM characteristics. Most EM characteristics are addressed by the Maxwell-Heaviside Eqs. A basic conceptualization of 4D EM Space was first addressed by Jacob's Ladder. The conceptualization of the characteristics of 4D Temporal Space has been proposed in this paper.

This paper has used concepts from the *Ohm's Law Triangle* and the *Velocity Formula Triangle* to approach the concept of time and time dilation from a nontraditional perspective. The component parts of the Ohm's Law Triangle and the Velocity Formula Triangle express similar numerical relationships between their associated equations—a pattern.

The Velocity Formula Triangle introduced the time component, where $t = d/v$. The time component was equated to the resistance component in the Ohm's Law Triangle, where $R = P/I$. Similar equations, similar results. Equating time to resistance allows the characteristics of resistance, per the Ohm's Law Triangle, to be applied to time. The equated components uniformly increased and decreased in value as component values were manipulated. The component associations, as defined, were found to be valid.

The Ohm's Law Triangle and Velocity Formula Triangle components were associated with the concept of 4D Temporal Space. The component associations were used to evaluate Einstein's twin paradox (time dilation) and Einstein's version of "approaching a black hole" regarding time dilation. Einstein's twin paradox and approaching a black hole were assumed to be opposite extremes of each other.

A valid time dilation theory would explain Einstein's twin paradox and time dilation as a black hole is approached. *While attempting to make a clock time (t) correlation to the time vector magnitude of Einstein's gravity manifold, the concept of temporal wavelength was discovered (time vector magnitude = temporal slope value = temporal wavelength = temporal frequency).*

There is *clock time (t)* [a particle] and there is *temporal frequency (a spectrum).* If temporal density = 1, then clock time (t) would apply. Temporal density is *the rate that time cycles (vibrates).* With clock time (t), "time passes" per the period of Earth's orbit. Temporal frequency is defined by temporal wavelength. A clock time (t) year is still a clock time (t) year, regardless of its temporal density. If the temporal density is higher or lower than the Earth = 1 (*one*) reference, then time dilation is at work. A temporal density value lower than the Earth = 1 (*one*) reference is time dilation where time cycles slower. A temporal density value higher than the Earth = 1 (*one*) reference is time dilation where time cycles faster. Again, a clock time (t) year is still a clock time (t) year. Time cycling faster or slower relates to Temporal Space being at a higher or lower vibrational rate due to curved MD Fabric of Space.

The *spatial-related characteristics* of "space" were defined as $E = mc^2$ and conceptualized as 4D Spatial Space. The *temporal related characteristics* of "clock time (t)" (a particle) were defined as the components of the Ohm's Law Triangle and the Velocity Formula Triangle. The conceptualization of 4D Temporal Space identifies "time" as temporal frequencies (a spectrum) that have similar characteristics as EM radiation. The *electromagnetic (EM) characteristics* of "space" were defined as electromagnetic radiation and conceptualized as Jacob's Ladder and 4D EM Space. The fourth (4th) dimension of 4D Spatial Space, 4D Temporal Space, and 4D EM Space is a unique sub-space version of Primary Tone "F". *Multi-dimensional (MD) Fabric of Space* is a balanced system of systems of its sub-space components, 4D Spatial Space, 4D Temporal Space, and 4D EM Space, see Figs. 4.5 and 4.6.

A single and consistent explanation of *time* and *time dilation* regarding Einstein's twin paradox and time dilation as a black hole is approached was accomplished. The time vector magnitude associated with Einstein's gravity manifold was correlated to the "temporal slope value." The temporal slope value was used to identify a temporal wavelength, temporal frequency, and temporal density, see Assumption 4.1 "Temporal Density Summation."

The *temporal density* is the *rate that time cycles.* Earth was assumed to have a temporal density of 1 (*one*). In Einstein's twin paradox, the twin traveling close to the speed of light had a temporal density that was <1. Time cycled slower. When approaching a black hole, the temporal density is >1. Temporal density greater or less than Earth = 1 (one) is time dilation. It should not cause more or less clock time (t) to occur. Otherwise, as temporal density (temporal slope) increased or decreased the value of clock time (t) would increase or decrease. A higher density or lower density of the same clock time (t) value (magnitude) would result in different answers in the displacement equation or any other equation using clock time (t).

Therefore, clock time (t) times a temporal density greater than or less than Earth = 1 (*one*) is problematic. Our current technology does not typically address the concept of temporal frequency or time dilation.

According to general relativity, as a black hole is approached, the time vector of the gravity manifold becomes shorter. Einstein's interpretation of the time vector per general relativity is that time moves slower as a black hole is approached.

This paper assumes that light does not escape from a black hole because the speed of light is a constant that is only expressed as a positive value. When approaching a black hole, the temporal density (rate that time cycles) is higher. Therefore, light would have to travel backwards in time to escape a black hole. Light cannot exist at an earlier point than which it was created unless it travels backwards in time. Due to a black hole's temporal density, we would not share the same *current moment* with a black hole. The situation is the opposite of Einstein's twin paradox. Compared to the higher temporal density of a black hole, the Earth is like "the twin in a spaceship traveling close to the speed of light." Per this paper, time would cycle much slower on Earth than it does when a black hole is being approached.

4.4 Time Travel and Sierpinski Star Tetrahedron Particle

Question 4.1

Did the twin in a spaceship traveling close to the speed of light become ***dark matter*** and/or ***dark energy***?

Question 4.2

In the universe, has any mass ever been observed moving at a percentage or near the speed of light (at a relativistic speed)?

Question 4.1 and Question 4.2 are related to time dilation. Two clock time (t) clocks, in different locations, are needed to detect time dilation.

Time travel could be possible within the balanced system of systems, MD Fabric of Space if the concept of temporal density (time dilation) is being understood correctly. If "time" and "space" are continuous, then "time travel" would be the ability to relocate to an earlier or later point in "space." An earlier or later point in 4D Spatial Space would have a unique version of 4D Temporal Space and 4D EM Space (spatial, temporal, and EM coordinates) which defines its "current moment" or "Now." Instead of a "current moment" spatial portal, it would be a "past/future current moment" spatial portal. Einstein's twin paradox (a thought experiment) could be a time travel story.

Clock time (t) is a very accurate and linear method of *time keeping* per the Earth's orbital period, with its smallest unit of measure being 1 s. Clock time (t) is not a 4D Spatial Space, 4D Temporal Space, 4D EM Space, or MD Fabric of Space location that can be visited. Clock time (t) provides a temporal point of reference.

In developing a Basic Time Hypothesis, it became necessary to develop a basic theoretical structure of the "MD Fabric of Space." Figure 4.16 has the structure

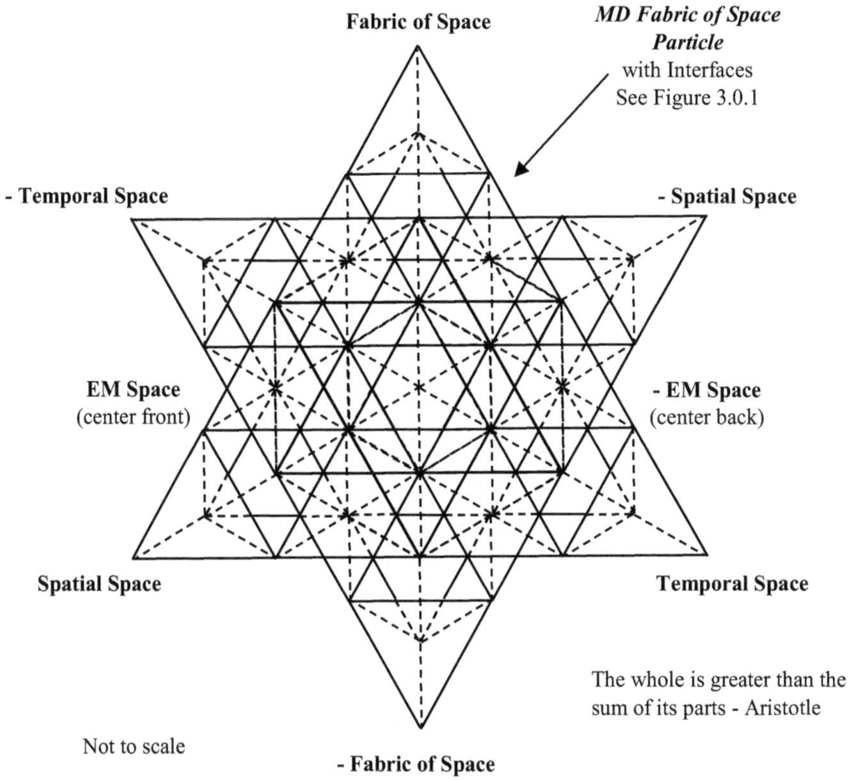

Fig. 4.16 Sierpinski Star Tetrahedron (crystal, fractal) particle

and orientation of 4D Spatial Space from one perspective. And the structure and orientation of 4D Temporal Space and 4D EM Space from another perspective.

Figure 4.16 is the star tetrahedron version of the Sierpinski Tetrahedron. It equates to the color hexagram, Jacob's Ladder, Circle of Fifths, color harmonic, color octave, and the Composite Color Wave.

Let Fig. 4.16 be a representative ***particle*** of the medium that defines the balance, system of systems, and MD Fabric of Space. The sub-spaces are positive and negative versions (polarity) of 4D Spatial Space, 4D Temporal Space, and 4D EM Space. Now, what are the *temporal mechanics* of this system of systems?

If the basic structural nature of Fig. 4.16 is crystalline, then the presence of mass (Spatial Space) within the structure would cause a piezoelectric effect. A crystal generates an electrical charge when a force is applied. A crystal vibrates when a charge is applied. Introducing mass within the structure of Fig. 4.16 could produce and maintain a charge and/or vibrations as the balanced system of systems of the MD Fabric of Space assumes a new balanced state. Time is a function of the curved MD Fabric of Space per the correlation of the time vector magnitude of Einstein's gravity manifold to a temporal wavelength.

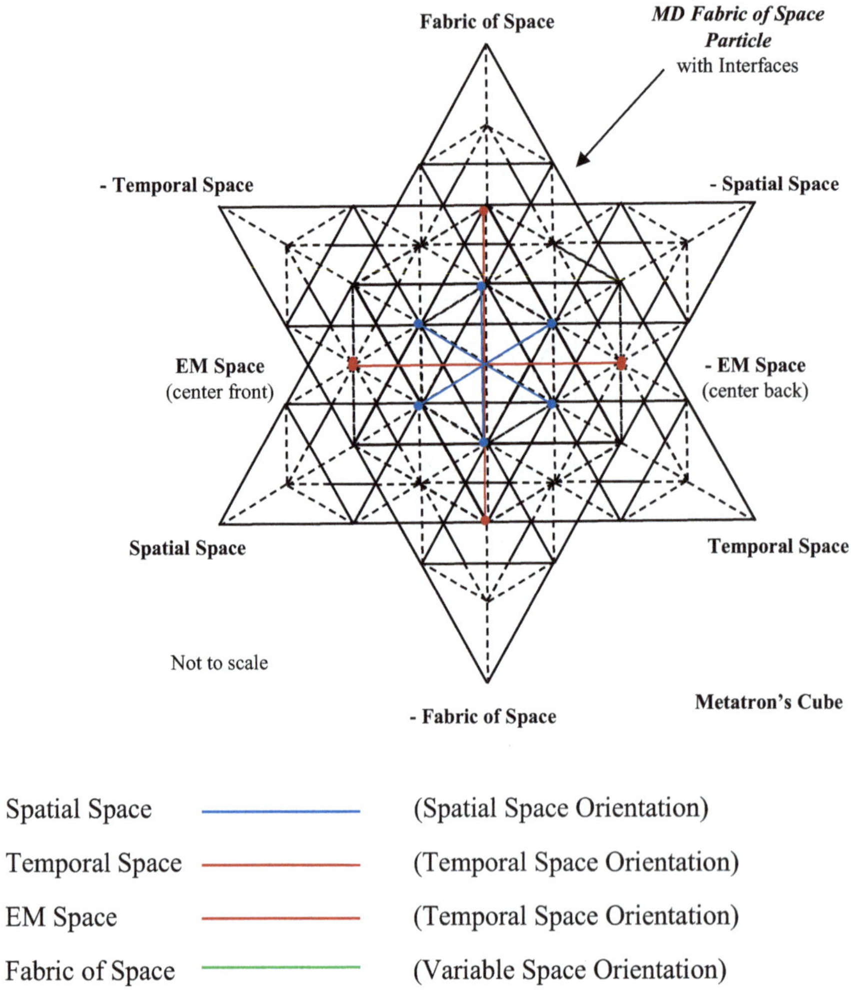

Fig. 4.17 Sierpinski Star Tetrahedron (crystal, fractal) particle coordinate systems

Temporal Space and EM Space Orientation:

- The view is down the x-axis at the center of Fig. 4.17 (Jacob's Ladder).
- The y-axis is from the front bottom corner to the top back corner.
- The z-axis runs from the left edge to the right edge of the cube.
- Consideration must be given to the characteristics and orientation of the mass (Spatial Space) creating the curved MD Fabric of Space. Flat MD Fabric of Space could have different characteristics and orientations.

Figure 4.17, the "*snowflake*," demonstrates the complexity of the balanced, system of systems of flat MD Fabric of Space at 1-tier below the surface. Adding concepts from general relativity, special relativity, quantum mechanics, Jacob's Ladder, basic music theory, etc. to the MD Fabric of Space makes Fig. 4.17 truly impressive.

4.5 Conclusion and Open Issues

Orbital mechanics is another method this paper used to help conceptualize "temporal mechanics." A spacecraft in a stable circular orbit around a mass is in freefall (microgravity). Its velocity is offsetting the downward pull (acceleration) of gravity.

Newton's First Law of Motion: An object will not change its motion unless a force acts on it.

For an orbiting spacecraft to slow down, it must accelerate. Acceleration pushes the spacecraft into a higher stable orbit, but its orbital period is longer. Acceleration pushes the spacecraft away from the gravitational pull of the mass it is orbiting. The orbital period (time vector, temporal wavelength) for the spacecraft becomes longer. Acceleration caused the spacecraft to relocate to a lesser curved area of the curved MD Fabric of Space of the mass it orbits. Its time vector became longer. Therefore, its temporal frequency (temporal density) becomes lower.

Note that the orbital period or unit value for clock time (t) has not been identified. On Earth, the orbital period of the planet is the basis for the International Standard (SI) for time being 1 s. The calculation of orbital velocity is different from the Velocity Formula Triangle. It is the gravitational constant that introduces the element of clock time (t) into the calculation for orbital velocity.

The calculation for orbital velocity is:

$$\textbf{V orbit} = \surd\left(\boldsymbol{GM}\right)/\boldsymbol{R} \;\; \text{[is measured in m/s]}$$

G = gravitational constant
M = mass of the body from the center
R = radius of the orbit from the center of the mass plus the distance above the mass.

At a high enough acceleration, a spacecraft will obtain its escape velocity and break free of the gravitational pull (curved MD Fabric of Space) of the mass it is orbiting. Its outwards velocity exceeds the downwards velocity of gravity.

Time is associated with the gravitational constant versus the radius or period of the orbit. As "R" increases "V orbit" decreases. As "R" decreases "V orbit" increases. Clock time (t) approximates the Earth = 1 (*one*) temporal frequency.

For an orbiting spacecraft to speed up, it must reduce its "R" value by falling deeper into the curved MD Fabric of Space caused by the mass it is orbiting. This is accomplished by slowing down (decelerating)—Einstein's twin paradox.

Calculating orbital velocity helps confirm the association between time and gravity. Just as it helps confirm the disassociation between clock time (t) and orbital period as it relates to "time." The weight of an orbiting mass is zero because gravity is zero. But the Earth is massive enough to curve the MD Fabric of Space and create its own gravity. This implies that gravity is separate from time, with both having a relationship to the balanced system of systems, of the curved MD Fabric of Space they are associated with.

General relativity relates to the very large. Quantum mechanics relates to the very small. Quantum mechanics is looking for gravity. If gravity is found at the quantum level, then there could also be a correlation to temporal frequency at the quantum level.

Clock time (t) is fixed and linear. Temporal frequency is subject to temporal density (rate that time cycles) due to changes in the balanced system of systems, of the curved MD Fabric of Space. Applying basic music theory concepts to 4D Temporal Space and temporal frequency could identify temporal overtones, undertones, color harmonics, and color octaves per Jacob's Ladder. The MD Fabric of Space includes the sub-spaces of 4D Spatial Space, 4D Temporal Space, 4D Electromagnetic (EM) Space, interfaces, and the influence of Primary Tone "F" as the fourth dimension of each sub-space.

In math, dilation means to compress or expand. In this paper, the concept of "time" is approached as "clock time (t)," "temporal frequency," and as "time dilation":

1. *Clock time (t), (SI) 1-s (particle)*, clock time (t) is typically expressed as an increasing positive linear value. In the displacement equation, $d = v \times t$, time is not subject to temporal density (time dilation). As the value of "t" increases towards infinity, displacement goes towards infinity. As the value of "t" decreases towards zero, displacement decreases towards zero. This is classical clock time (t) regarding the International Standard (SI) of 1 s. It is based on Earth's orbital period and should approximate the Earth = 1 (*one*) temporal frequency value regarding no time dilation. The circular constant 2π has a period of 1 s, clock time (t).
2. *Temporal frequency (spectrum)* has an Earth = 1 (*one*) reference for time vector magnitude, temporal wavelength, temporal frequency, temporal density, gravity per the associated temporal slope value. The temporal slope value equates to the time vector magnitude of Einstein's gravity manifold (general relativity). Temporal frequency identifies a spectrum that is associated with the balanced system of systems, curved MD Fabric of Space as defined by the temporal slope value (time vector magnitude). Correlating the temporal slope value with the time vector magnitude identifies a direct relationship between "temporal vibrations" and the curvature of the MD Fabric of Space. Variations in 4D Spatial Space, 4D Temporal Space, 4D EM Space, or the balanced system of systems, curved MD Fabric of Space could increase or decrease temporal density.

   Temporal density is the "rate that time cycles" as expressed by (temporal frequency_2)/(Earth = 1 frequency). It allows a different temporal frequency to be evaluated as an approximation of clock time (t). This is where "clock time (t)" × "temporal density" estimates the passage of clock time (t) within "temporal frequency_2". Temporal density with a value less than Earth = 1 (*one*) is where time cycles (vibrates) at a slower rate. A temporal density greater than Earth = 1 (*one*) is where time cycles (vibrates) at a faster rate. A clock time (t) year is still a clock time (t) year regardless of the rate that time cycles (vibrates).

The rate that time cycles identify the temporal vibrational rate of an area of Temporal Space within the MD Fabric of Space. Like the electromagnetic (EM) spectrum, there is a temporal spectrum. Earth = 1 (one) is the theoretical position along the linear temporal spectrum (x-axis) where Earth experiences no time dilation. It is the vibrational rate of the "space" we call home. Figure 4.12 is orthogonal to the linear temporal spectrum (x-axis). And Temporal Space has a different orientation than Spatial Space. See Figs. 4.13 and 4.17.

The circular constant 10(phi) has a period of [$2\pi/10$(phi)] or 1-long-second ($\approx$61/60). Pi is the ratio of a circle's circumference to its diameter. Jacob's Ladder is based on the golden ratio proportions phi: 1: PHI, where "1" is the Baseline. The Baseline equates to Earth = 1 (*one*). The color pattern is 10($\pm$phi), 10($\pm$1), 10($\pm$PHI). Color potential is orthogonal to Jacob's Ladder. The 10(x) is a family of circular constants where "x" is not equal to zero and "x" goes to $\pm\infty$. All 10(x) circular constants must use the 1-long-second time base. The circular constant 2π can be expressed as 10($\pi/5$). Jacob's Ladder identifies color harmonics, color octaves, and the associated overtones and undertones. Everything vibrates.

A correlation to Jacob's Ladder provides an interesting color harmonic and/or color octave insight into the basic nature of *temporal mechanics* and time dilation. Understanding "how things work" is the basis for science, mathematics, technology, etc. Time as a frequency does not work well with many of the equations used to describe how things work. Clock time (t) works with the equations used to describe how things work, but it does not work well with temporal mechanics or time dilation.

The circular constant 2π with a clock time (t) period of 1 s is a fundamental concept in mathematics. A circle drawn per a 10(x) circular constant also has a circumference to diameter ratio of π ($\pi = c/d$). Therefore, the circular constant 2π is based on the geometry of a circle and its clock time (t) period of 1-s. One second is based on the orbital period of the Earth. The circular constant 2π and its clock time (t) period of 1 s would have limited compatibility with the temporal mechanics theory of this paper unless clock time (t) is Earth = 1 (*one*). The temporal mechanics theory requires the flexibility of Jacob's Ladder and the MD Fabric of Space.

4.5.1 Open Issues

1. Assumption 4.1, Temporal Density Summation.
2. Figure 4.11, Slope of a Line, Curved vs Flat Space per Einstein's twin...
3. Hypothesis 4.1, Spatial Dilation.
4. Define a coordinate system for 4D Temporal Space based on its orientation.

5. Define a coordinate system for 4D Electromagnetic (EM) Space based on its orientation.
6. Define a coordinate system for MD Fabric of Space.
7. Define the coordinate system interface between 4D Spatial Space, 4D Temporal Space, 4D EM Space, and the MD Fabric of Space.
8. What is the MD Fabric of Space (the cosmic web)?
9. Is the MD Fabric of Space crystalline and like a piezoelectric material?
10. Determine the possible number of dimensions associated with the multi-dimensional (MD) Fabric of Space regarding 4D Spatial Space, 4D Temporal Space, 4D EM Space, and their interfaces.
11. Confirm the orientation and clocking of 4D Temporal Space and if the use of 10(x) circular constants [10(±phi), 10(±1), 10(±PHI)] and/or the use of the color equations of the Composite Color Wave causes that orientation and clocking.
12. Confirm the method for causing frequencies to combine in free space. Theory: unit circles (amplitudes) of the combining Primary Colors must be the same diameter (amplitude, wavelength) which will cause them to occupy the same space at the same time. Would apply to time by similarity.
13. Physical testing of concepts, as applicable.

"For objects and/or energies to affect each other they must interact."—unknown

"Similar equations, similar results"—unknown

"As above, so below; as within, so without; as the universe, so the soul…"—Hermes Trismegistus/Thoth

"Nothing rests; everything moves; everything vibrates"—The Kybalion or Kabalion (Hermes Trismegistus/Thoth)

Chapter 5
Original Lie Algebra Behind Real Structure, Models, and Interactive Computer System of the Chemical Elements, Compounds, and Compositions

Erik Trell

Abstract In earlier articles, I have completed an investigation of the Standard Model of the elementary particles by an interior Lie algebra realization where the infinitesimal generators are lattice elements of serial volume-preserving spheroidal transformations along their symmetry axes to exhaustive and exact replications of the observed evidence. I have then turned to the exterior algebra part as outlined in his original *Imaginärtheorie* and doctoral thesis *Over en Classe geometriske Transformationer* and when applied according to the Bohr *Aufbau* system found equally exhaustive and exact replication over the periodic table as here overviewed for especially the inorganic atoms and molecules and also the direct uses of their bits in modeling and online interactive computer programs and algorithms.

Keywords Algebraische Fläche · Alfabit · Betabit · Differential algebra · Differential equations · Infinitesimal transformations · Lie algebra · Linear algebra · Periodic System

5.1 Introduction

To get a fair counterbalance to the marginalizing misrepresentation of Marius Sophus Lie (1842–1899) in the double-dealing Wikipedia/Google broadcasting monopole, a congenial memoir of him in a comprehensive early 1900 Nordic encyclopedia[1] living up to its absolute commitment to objective and factual documentation informs that "he has forever tied his name to a string of new ideas and theories of the most far-reaching importance . . . already during his first

[1] Nordisk Familjebok, 1919

E. Trell (✉)
Department of Health, Medicine and Caring Sciences, University of Linköping, Linköping, Sweden

A. Adhikari et al. (eds.), *Quantitative Frontiers*, Industrial and Applied Mathematics, https://doi.org/10.1007/978-981-96-9435-8_5

eruptive period he enriched mathematics with a wealth of new material ... in 1869 he delivered the first draft of his *imaginärteori* [1] where the germ of many of his later discoveries was laid down ... in his thesis *Over a Classe geometriske Transformatiomer* 1871 [2, 3] he shed light on a hitherto unforeseen connection between the metric and projective properties of space ... and his disclosure of spherical geometry—one of the most beautiful discoveries in modern geometry according to Darboux ... and he advanced an entirely new mathematical subject, the theory of the transformation groups, a heroic feat that makes an epoch in the history of mathematics ... an entire system, revolutionizing a vast area of mathematics and which in origin, exposition, and completion in every detail is exclusively Lie's own achievement" ... well, and to that, originally and inceptively, linear algebra, differential equations, differential geometry, line congruences ... as *primum et ultimum* the instant structure and meaning of it all, that is, the direct fabric of Reality and "nature of Cartesian Geometry." [2, 3]

This must be anathema for Wikipedia's secluded principals, making it betray the publicist principles it flags, but for Lie was nothing strange per se. It was before the elementary particle era and the geometers of the time already regarded *der geraden Linie als Raum Element* [4]. And what Lie in a global vision saw and from which also his ensuing discoveries sprang were the conditions and mechanics of its perpetual running in successive cycles without obstruction or disintegration from the smallest to the largest scale, and not as a mere drawing but a real "*Ding*" [1]. Very briefly summarized he first to that end in *Ueber eine Darstellung des Imaginären in der Geometrie* [1] defines an *Imaginär-Punkt*, $Z = z + \text{pi}$, $X = x + \text{pi}$, where p stands for *das Gewicht* (later to get identity as the tangent), and *wenn* $p = 0$ *nenne ich den Punkt einen Null-Punkt*. Then he introduces $F(Z,X) = 0$: *eine Relation, zwischen Z und X*, and *jedes System (Z, X) welches der Gleichung genügt definirt einen Imaginär-Punkt* and *das Ensemble derselben constituirt die Imaginärcurve* whose *sinnliche* (here ≈ real) *Repräsentation ist eine gestreifte Fläche* (Fig. 5.1). Thereby, its infinitely tight amorphous points (in any coordinate system) get physical shape along its axes as *eine continuirliche Linie* of the infinitesimal distance between them into *eine algebraische Fläche* with here *den Nullstreifen* when $p = 0$. And when there is *eine lineäre Relation* (e.g., $BZ = X - A$) ... *ergibt sich das equivalente System; die Imaginär-Gerade,* whose *Streifen* ... over horizontal and vertical *Ebene* which *orthogonal* (Fig. 5.1a) or *um den Punkt (xyz)* ... *rotirend* (Fig. 5.1b) ... *ähnliche Figuren bestimmen.*

And so it goes on in increasing levels of interactions over and between *gestreifte Flächen*. Hopefully this very brief summary of some of its preliminaries together with Lie's pioneering projections may show that he created a rich both plane geometric and linear algebraic *System*, similar to a vector space (Fig. 5.1c), but more fundamental and dynamic. While a vector space just draws the blurring out resultants of its active limbs and thus no more than the tail of the dog, the original Lie algebra is the *Zitterbewegung*[2] walk of its separate *dx*, *dz*., and *dy* steps along or

[2] Term coined by Erwin Schrödinger in 1930.

a b c

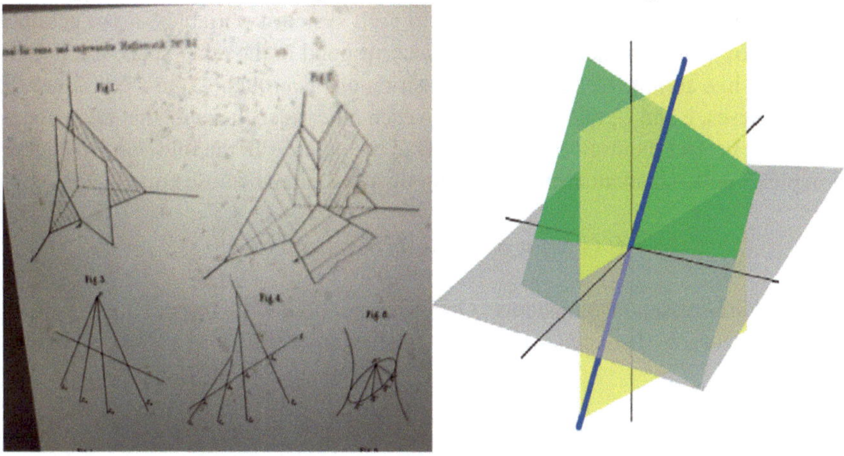

Fig. 5.1 (**a**, **b**) *Gestreifte Flächen* of ordered linear algebra and plane geometry *Imaginär-Curven* by straight lines of length equaling zero (distance between adjacent coordinate points). Note equivalent organization as in a present-day unordered linear vector space (**c**)

angulated to their axes to the detailed weaving of the world, *den man algebraisch aussprechen kann* [1].

Thus, *jeder plangeometrische Satz ist ein besonderer Fall eines stereometrischen Doppelsatzes in der Geometrie der Liniencongruenzen.* The gestreifte planes in Fig. 5.1a, b are projected through every point and the natural shape of the "line-element (dx, dy, dz)" [2, 3] is the respective side of bordering point neighborhoods duplicated in the transition from complex to real space. The further transferal of the algebra to the three (and higher)-dimensional space was presented by Lie in his Norwegian Ph.D. thesis *Over en Classe geometriske Transformationer* at Christiania (now Oslo) University in 1871 [2, 3]. In its radical simplicity, it was so innovative, however, that although it was given the highest marks in the dissertation, "not one ... had understood a single word of it." [5] Hence it soon went into oblivion in the faculty archives where I rediscovered it in the early 1970s and, with professor R. M. Santilli, translated it into English in 1988 [3].

Recapitulating in Lie's own words by quoting from the thesis in a somewhat repetitive and overlapping but largely consecutive order: "The Cartesian geometry ... renders ... of the geometry of space a representation of the algebra of three variable quantities." "Following Plückers trail I discuss the equation system.

$$[F_1\,(x\ y\ z\ X\ Y\ Z) = 0,\ F_2\,(x\ y\ z\ X\ Y\ Z) = 0]$$

which... defines a general reciprocity between two spaces. When in particular the two equations are linear in relation to each system's variables, a projection is obtained by which to each space's points correspond in the other space the lines of a Plücker line complex." "By a limit transition one is brought... to a consideration of two curves, c and C, that correspond to each other in such a way that the tangents of the one project themselves as the points of the other." "In the following, by curve-complex I understand an arbitrary system of space curves whose equations

$$f_1\,(x\ y\ z\ a\ b\ c) = 0,\ f_2\,(x\ y\ z\ a\ b\ c) = 0$$

contain three essential constants. On differentiation with respect to $x\ y\ z$ and elimination of $a\ b\ c$ between the two new and the initial equations a result is obtained of the form

$$f\ (x\ y\ z\ dx\ dy\ dz) = 0$$

If here $x\ y\ z$ are perceived as parameters, dx, dy, dz on the other hand as direction cosines, each point in the space is associated with a cone, namely, the collection of tangents to those complex-curves that go through the point in question." "To the curves c of the space r, thus unambiguously correspond the curves C of a definite curve-complex in R."

"It is now possible to show that the equations determine a general reciprocity between figures in the two spaces." "If it concerns a study of the space relative to a given line-complex... it may be particularly suitable to choose the straight lines of this complex as space-element"... "Descartes produces a quantity system $x\ y\ z$ ... he has among the unlimited manifolds of possible coordinate systems chosen a definite one... one realizes that the straight lines that meet a given condition may be used as the element of a geometry of the space, which gives a faithful representation of the algebra of three variables."

"To a given system or as one may also say to a given curve-complex corresponds a definite equation [$f = 0$]". "Any line-complex determines a complex of curves that are enveloped by the line-complex's lines and which have the property to be main tangent-curves on any surface that is generated by the system of these curves, each of which intersects the preceding."

"The equation system [$F_1(x\ y\ z\ X\ Y\ Z) = 0$, $F_2(x\ y\ z\ X\ Y\ Z) = 0$] determines a reciprocity between two spaces." "To the curves of the space r thus unambiguously correspond the curves C of a definite curve-complex in R, and likewise appear in r a complex of curves c which stand in the same relation to R's points—the two spaces are thus mapped into each other... in such a way that to r's points correspond in R the imaginary straight lines whose length equals zero." "Among the classes of transformations I obtain in this way... is a fundamental relation between the Plücker line geometry and a spatial geometry whose element is the sphere."

5.2 Methods

By that, Lie completed a "threefold infinity of linear transformations permutable between themselves" (and anticipating his forthcoming work on the continuous transformations). First there is the "unlimited manifold of possible co-ordinate systems" which are mutually exchangeable with their "elementary complex-cone" projections. Then there is the same infinity of linear transformations between "an arbitrary number of variables" in Plücker's *Neue Geometrie des Raumes* with the *geraden Linie als Raum-Element* [4], which is, however, confined to its flat curvature. And finally, Lie opened up the differential geometry infinity of "rounded curves" by his finding that "the Plücker line-geometry can be transformed into a sphere geometry."

This "particularly remarkable transformation," which was considered by Darboux to be one of the most beautiful discoveries in modern Geometry when it was first presented in *Comptes Rendues de l'Academie des Sciences* in 1870 [3], "takes place" between the "infinitely small... sphere's rectilinear generatrices" (= infinitesimal generators in today's, terminology) and those of the x y z "Cartesian Geometry quantity system." It is the "most general determination of space curves whose length equals zero and whose tangents thus possess the same property," and provides the Det = 1 Identity matrix [6–8] both of the principal "Nature of Cartesian Geometry" and the observed three-dimensional physical structure and events of the Euclidean Universe we live in. From both "philosophical reflections" and as immediate physical extension the infinitesimal as well as infinite straight line was thus understood as the elementary "quantity" of the things it could form, and its projections in various scales therefore as actual construction drawings. Unfortunately, while performed in the forerunner *Ueber eine Darstellung des Imaginären in der Geometrie* [1] (Fig. 5.1), they are lacking in the thesis so have to be inferred as faithfully as possible from the verbal prescriptions. Figure 5.2 aims to reconstruct the most basic of these, and their inward, surface, and outward parameter/main-tangent constitution and impact. The resulting hybrid of Nucleon identity (Fig. 5.2a, b) has been described in detail before [6–15] both in its Neutron-Proton duality and as gateway of the continuous geometrical transformations and other interactions of the elementary particles. Also the surface Muon "geodetic curve" appears in a static central coordination of its projection where the 90–180° inclined main-tangent lines (Fig. 5.2c) are linked to a single, duplicated, or helical orbit of length $(2\pi \times 2^{1/2})$ and resulting mass in inverse proportionality to the unit radius, that is, infinitesimal displacement of the Proton = $(2\pi \times 2^{½}) \times 938.27 = 105.59$ MeV in comparison with the measured $\mu^{\pm}$ mass of 105.66 MeV.

But when it comes to the outward-going 60–120° hexagonally inclined curve-complexes destined to the periodic table (Fig. 5.2d, e) they carry "running coordinates" and main-tangents with them as their physical members, which "arrange themselves pairwise as reciprocals" in the applied symmetry, in this case orthogonal versus spherical, to an entangled "curve congruence" structure. And here lies, albeit self-evident, the radical secret: the parametrization of the particular curve is not

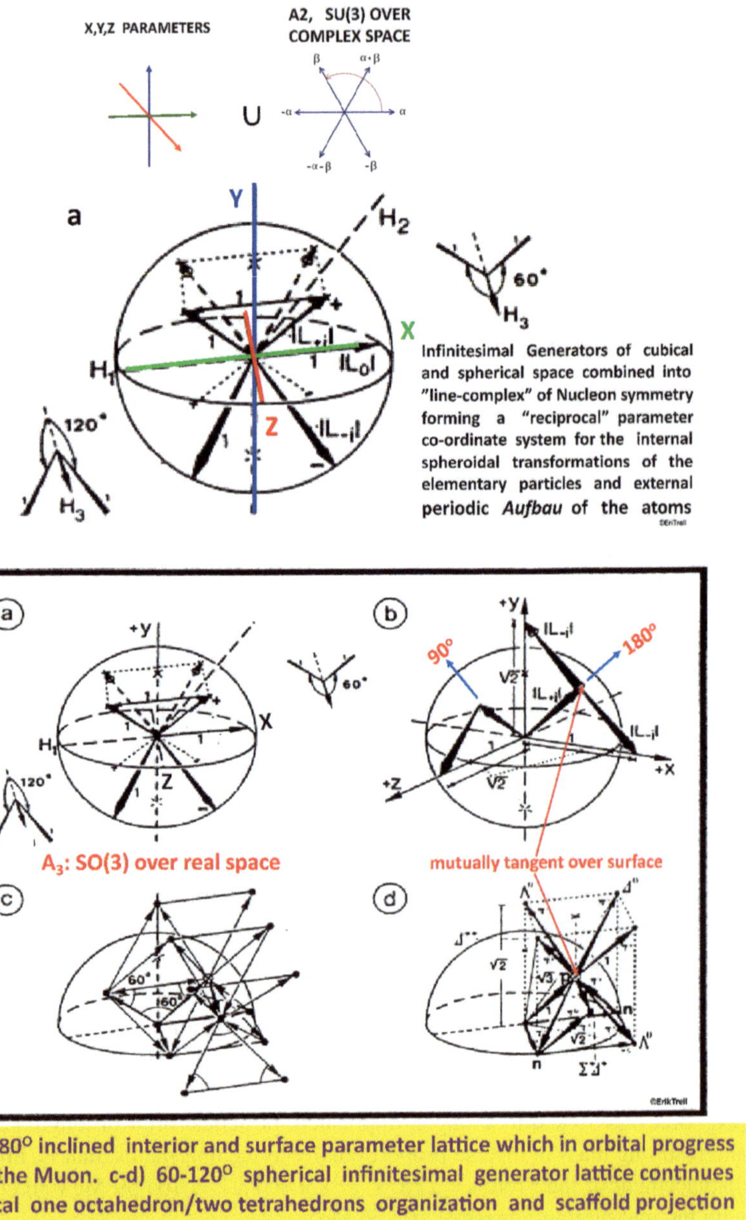

Fig. 5.2 (**a**, **b**) Cartesian geometry parameter/main-tangent line-complex of Nucleon identity for inwards projections of the elementary particle symmetries and events. (**c**–**e**) Continuing "curve net" for surface Muon and outward Periodic Table mapping

global and spanned in an external coordinate system, but part in the local realization, which is then unique and everything that exists as itself and necessarily space-filling its own field where the division is in the infinitesimal limit which also gives the universal dilution of each separate, not glove and hand but, rather, shoe and foot step in the own dual worldline's share of the totality.

And this is not described or exemplified in the thesis because implicit, but also and mainly "because... I don't want to deviate from the customary language of mathematics" becoming increasingly hard to formulate other than as "philosophical reflections" plausibly like the above. However, there are prescriptions in the groundbreaking work on the imaginaries [1] to keep within the *Streifen* which means that the line is drawn in separate steps of *dx*, *dz*, and *dy*, not necessarily in alternating order but to reach the goal of a continuous space-filling. This is shown in Fig. 5.3 and clearly the resulting complementary geodetic curves are infinitesimal, digital, and differential and the smallest and simplest possible; just straight lines of length asymptotically close to zero which in equally many steps "go through... the elementary complex-cones" along "elementary complex-directions... (*dx dy dz*) that belong to a complex-curve" and thus both digital and differential; and actually "to be considered as... the partial differential equation itself".

Now, as surveyed in Fig. 5.4, one may be released to a freer exposition in the ensuing deliberations, starting with the "two distinct sets" or strands of the reciprocal complex-cones here disclosed and how from their primordial line sequences in the early Universe they turn into full "material bodies." They proceed in isotropic distribution of their faint cosmic microwave radiation in which they cannot by themselves raise to volume and only very rarely encounter their own or complementary kind to various combination outcomes (Fig. 5.4a–e).

But there is due to be small fluctuations in the initial isotropy and they will escalate and accumulate, and eventually, over up to billions of years, reach the astrophysical stage, in the cataclysms of which all varieties of curves are forced against each other (Fig. 5.4f) to produce a multitude of compounds and fragments arguably even down to the combined *dx*/*dy*/*dz* quark root space counterparts.

5.3 Results

How to arrange the formed "material bodies" in the nucleogenetic ejection so that they not immediately jam in their continuing space-packing? As reported in detail before [9–15], Figs. 5.4f and 5.5a–c demonstrate in a series of hopefully self-explaining diagrams (enlargeable in the page) how this is solved by a biphasic Bohr *Aufbau* layout of the "two infinitesimally close-lying, each other intersecting... elementary complex cones whose apex points lie upon the... integral surface" for "tangent plane... differential equations."

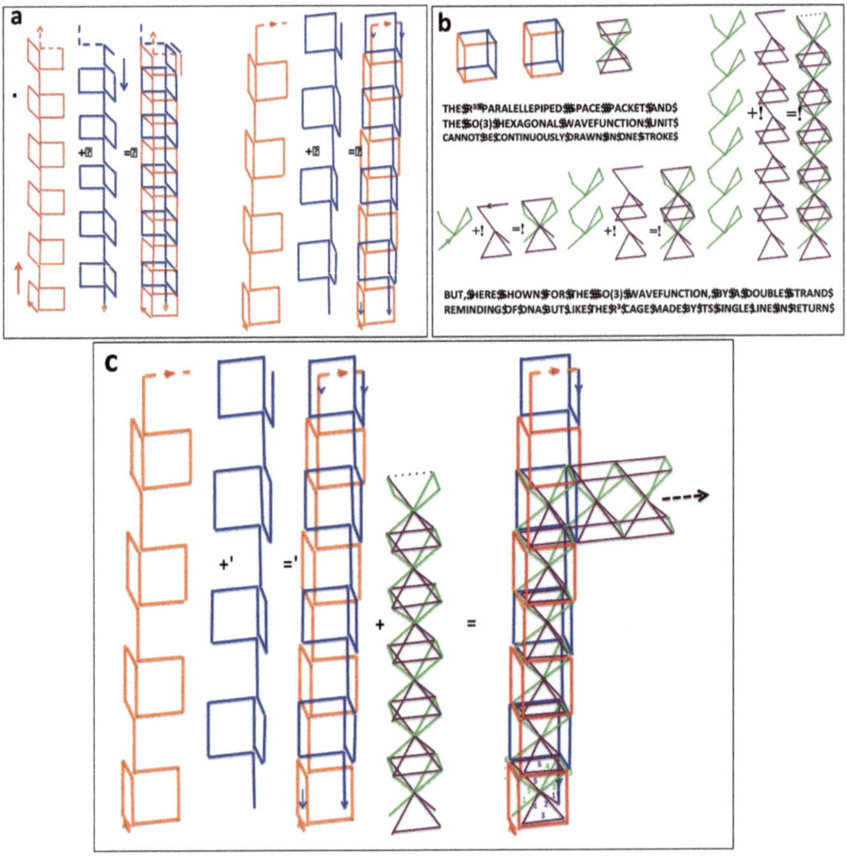

Fig. 5.3 The Figure displays how the original geometric Lie algebra "line element (dx, dy, dz)" works by direct steps of the shortest recurring sequence in divergent line congruence duplication to a) (mirror halves of) the 1,1,1 cubical 'space packet' of e.g. the Muon and $1{,}2^{1/2}{,}1$ of e.g. the periodic table, b) the same for the wave coil, and c) the combined 'wave-packets' (term coined, *Wellen-Paket*, by Erwin Schrödinger)

In translation, this gives a surface membrane of height $6 \times 2^{1/2}$ of the dual infinitesimal wave-packets and/or frame and wave bricks wafered against and wrapped in each other to two-storey modules which will thereby get a lower part with flat floor (Fig. 5.4f) and an upper part with flat roof (Fig. 5.5a), each of which containing 153 bricks holding 12 main-tangent lines so that the modular/linear inertia is $153 \times 12 = 1836 \approx$ the Proton/Electron mass ratio.

It can therefore be concluded that the main-tangent curve-complex in its various fractions from "singlet" and even smaller in the individual bricks to full particle in the module, embodies the Electron/Positron, and likewise the parameter counterparts the Neutron (Fig. 5.5b). And in Fig. 5.5c, it is indicated how in, e.g., collisions their walls may be cracked to, respectively, zigzag Photon (photoelectric effect

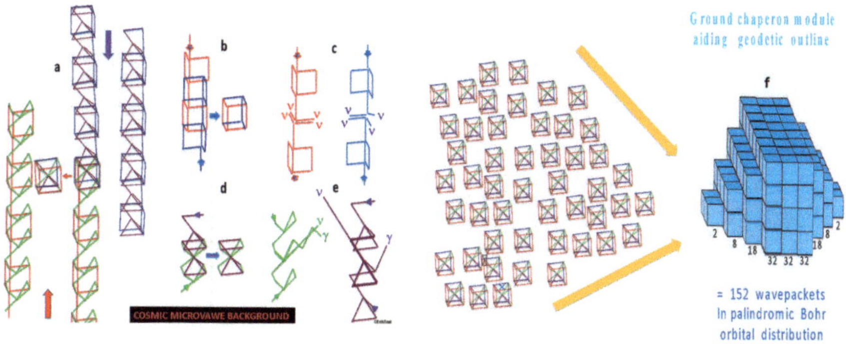

Fig. 5.4 So, in primordial Universe the separated mates of (a) mutually tangential complex-cone hybrids besides (b-e) the varieties in Figure 3 and others would just harmonically float around incomplete in their isotropic space? However, small interferences are inevitable with occasional annihilations or superposition which generate a faint cosmic background radiation and (f) at places over up to billions of years may escalate to stellar formations and processes in the colossal compression of which innumerous line congruence moieties are squeezed together and come out as escaping Photons and Neutrinos when bursting in ipsilateral confrontations, or as the described space, wave and wave-packet embodiments in contralateral superpositions; here shown for the wave-packets. f) They then need a distinct space-filling packing order to avoid instant jamming - which is offered by the Bohr *Aufbau* system forming a layerable module in a lower half with flat bottom shown in the figure, and a reciprocal line congruence upper half with flat roof, and together shaping the crystal seeds, all of equal size, of the periodic table

if reversed) and straight Neutrino endless line sequences with consequential zero mass.

More in detail, Fig. 5.6 describes the modular *Aufbau* and tessellation of Lie's "fundamental relation between the Plücker line geometry and a spatial geometry whose element is the sphere." This is realized by the integration of both the reciprocal straight and spherical infinitesimal generators—which can also run separately—within the same differential volume element, whose plane-filling modular stashing and hierarchical further proceedings are illustrated; more direct and complete than the earlier Grassmann geometric algebra [17], which is solely along the Plücker lines [4], and the somewhat later (1878) Clifford geometric algebra [18] which implements duplicity and curvature by the quaternions while Lie manages this and *Darstellung des Imaginären in die Geometrie* [1], too, direct.

The realization continues with the second (Fig. 5.7a) and third (Fig. 5.7b) period atoms. Up till Iron the nucleosynthesis is mainly fusion under the huge pressure of the stellar processes, and therefore maximally centered. In the first, Lithium to Boron states this brings in a marked single-excess Neutron isotope preponderance. This suddenly ceases from Carbon where mid-symmetry is reached, and wholly vanishes in Fluorine, which diverges from the neighbors, however, while Carbon, Nitrogen, and Oxygen can mingle and together with Hydrogen from first period form an interlocking "organic" jigsaw set which may fill space by serrated tessellation. Otherwise, as opening up in period three by the sole $_{11}$Sodium^{23} isotope

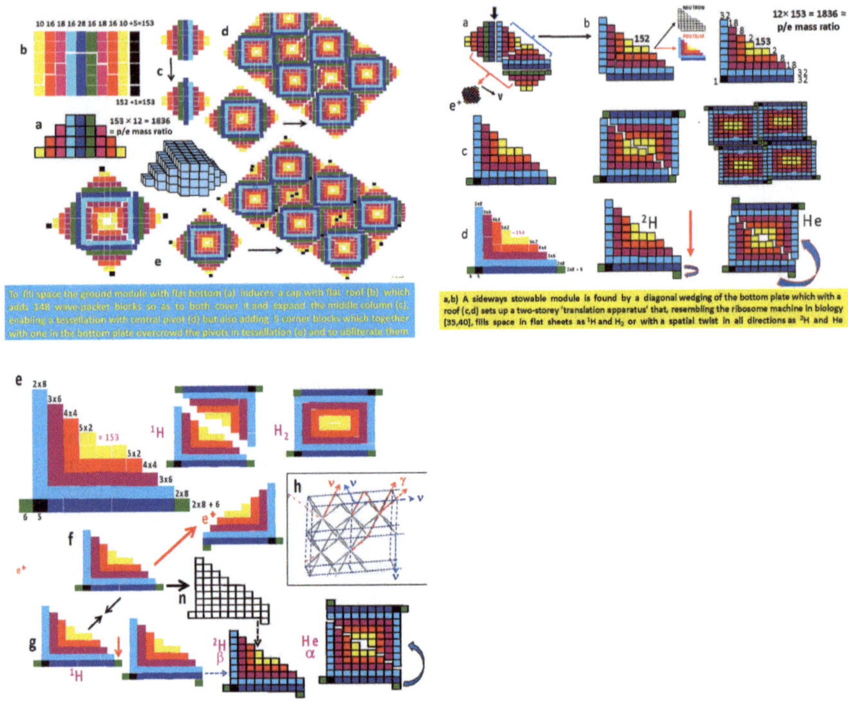

Fig. 5.5 (**a**–**c**) Panorama of Bohr *Aufbau* of the atomic tessellation layers

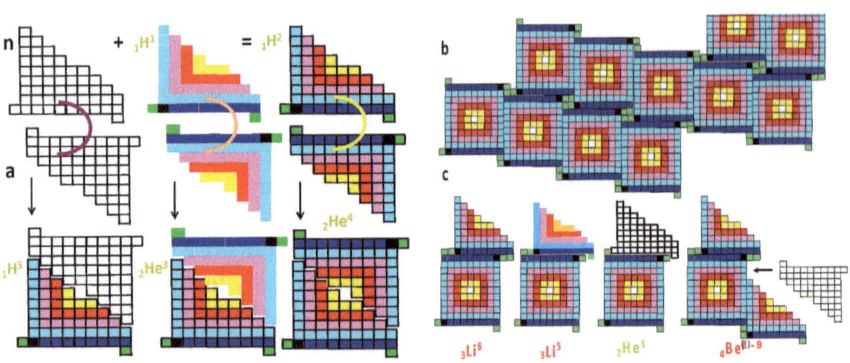

Fig. 5.6 (**a**, upper row) Combination of Neutron and Electron/Positron in the Bohr *Aufbau* modules generates $_1H^2$ Deuterium (β particle). (Columns) Superposition of Deuterium with Neutron gives $_1H^3$ Tritium, with Electron $_2He^3$ and with Deuterium $_2He^4$ Helium. (**b**) Space-filling tessellation of Helium = α particle over the plane. (**c**) Further basic isotopes

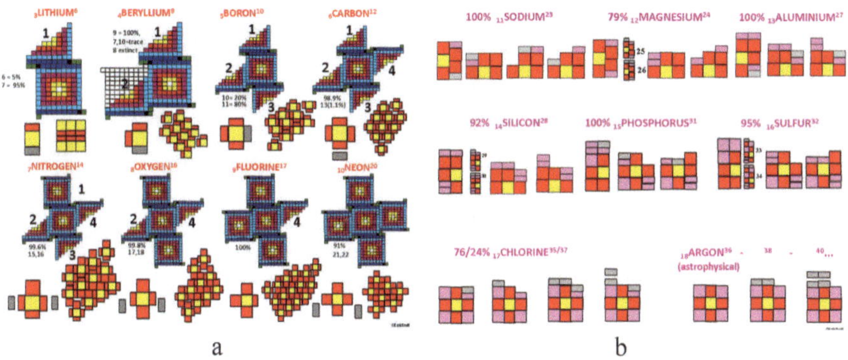

Fig. 5.7 (**a**) Space-filling module and simplified "Duplo" [19] representations of second period atoms. Numbers in bold are oxidation states. In Lithium to Boron single-excess neutron (gray bricks) isotopes dominate (% given). (**b**) In Duplo representation alone: Third period states. Note rectangular form approximation enabling space-filling aggregations. Colors are those of the noble gases in successive periods

(Fig. 5.6b), the forwards, "inorganic" layout is square and, except in $_{20}Ca^{40}$, needs increasing Neutron excess for maintaining flat side contours. It is hoped that these lacking ready terminology vague descriptions will be clarified in the following figures and text.

The distinction between organic and inorganic is thus constitutional and of course Hydrogen, Carbon, Nitrogen, and Oxygen in their native or straight isomeric forms can assume taking part in inorganic structures when fitting. And vice versa, inorganic atoms may rearrange to fill up gaps in organic compounds, e.g., Phosphorous into ATP and the phosphate deoxyribose backbone of DNA, and Sulfur into the amino acids, cysteine and methionine [20, 21]. So far, observed events occur in the realization as in reality, and in the continuation this is substantiated by a combined tabular/graphical overview of the periodic table.

This should be most interesting and useful per se and verifies the completeness and exactitude of the system. Other real-world features that may be noted but not further commented in the forthcoming are the alkaline, donor character of the first member of every period and the halogen towards the last, and the possibility to shift the marginal modules in the intermediate to find suitable molecular combinations in a range of oxidation states.

Next out is the fourth period from Potassium to Zinc, passing Iron (Fig. 5.8a), after which the nucleosynthesis mainly proceeds by marginal accretion (Fig. 5.8b) through slow or in larger atoms rapid Neutron capture. A special case is Scandium whose sole isotope can only be generated by rapid neutron capture, probably because this is the only way of getting sufficiently many to fill the rectangular contour. Note that also $_{20}$Calcium^{40} is the largest atom not needing a neutron to fill its straight contour.

Then come the rest of the fourth period elements to Krypton (Fig. 5.9a), and the first part of the fifth period (Fig. 5.9b).

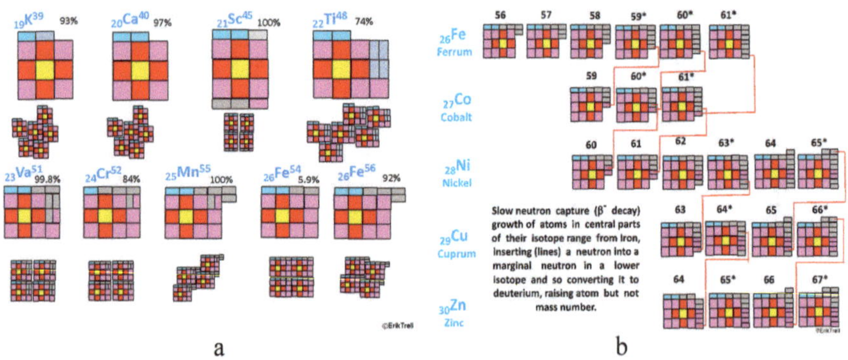

Fig. 5.8 (**a**) Fourth period atoms from potassium to iron. (**b**) Continuation of fourth period from iron to zinc

Fifth period is fulfilled in Fig. 5.10a and sixth period begins in Fig. 5.10b. Note start of Lanthanides with Cerium.

Sixth period continues in Fig. 5.11.

And concludes in Fig. 5.12.

In the seventh period, most states are synthetic and extremely short-lived. But their outlines can all be step by step reproduced exactly as they appear in observations and experiments, next in Uranium to Hassium (Fig. 5.13).

Now approaching the end of the cavalcade it must be emphasized that it contains a lot of information not further commented upon in the text but marked out in the figures; the Lanthanides, the Actinides, the covering of subshells, etc. in the same order as in reality and bringing so strong evidence that it is indeed the true "Nature of Cartesian Geometry" which is literally surfacing that it is beyond any denial. A larger jubilee portrait of this finale is accordingly granted (Fig. 5.14)

In its successive completion, Oganesson is therefore both integral and icon of the periodic table; canvas as well as palette of all the patterns within its perfectly rectangular frame; all the shells in penultimate *Aufbau* with $1 + 4 + 9 + 18 + 18 + 9 + 4$ saturated alpha bricks and $2 \times 29 = 58$ excess neutron contour fillers in right K L M N O P Q half of its Bohr shell positions hierarchically distributed over the plane, and all tiled by just two brick sorts; a quadratic "Alphabit"© (Helium) and a rectangular "Betabit"© (Deuterium) in alternating, fermionic and bosonic order.

This opens for important discoveries and uses, and half the fun of getting there, first from the infinitesimal to the atomic size, is also guided by Lie in his "general geometric interpretation of partial differential equations of the first order between x y z is obtained by . . . the general surface which at all its points has a three-point contact with . . . a given curve complex," i.e., "that one always gets an integral surface by adjoining to a surface . . . characteristic curves each of which intersects the nearest preceding one." Translating this to the initial infinitesimal modules (Figs. 5.4, 5.5, and 5.6) and onwards of either the combined wave-packets or separate space cages one can see that the complementary roof and bottom halves come

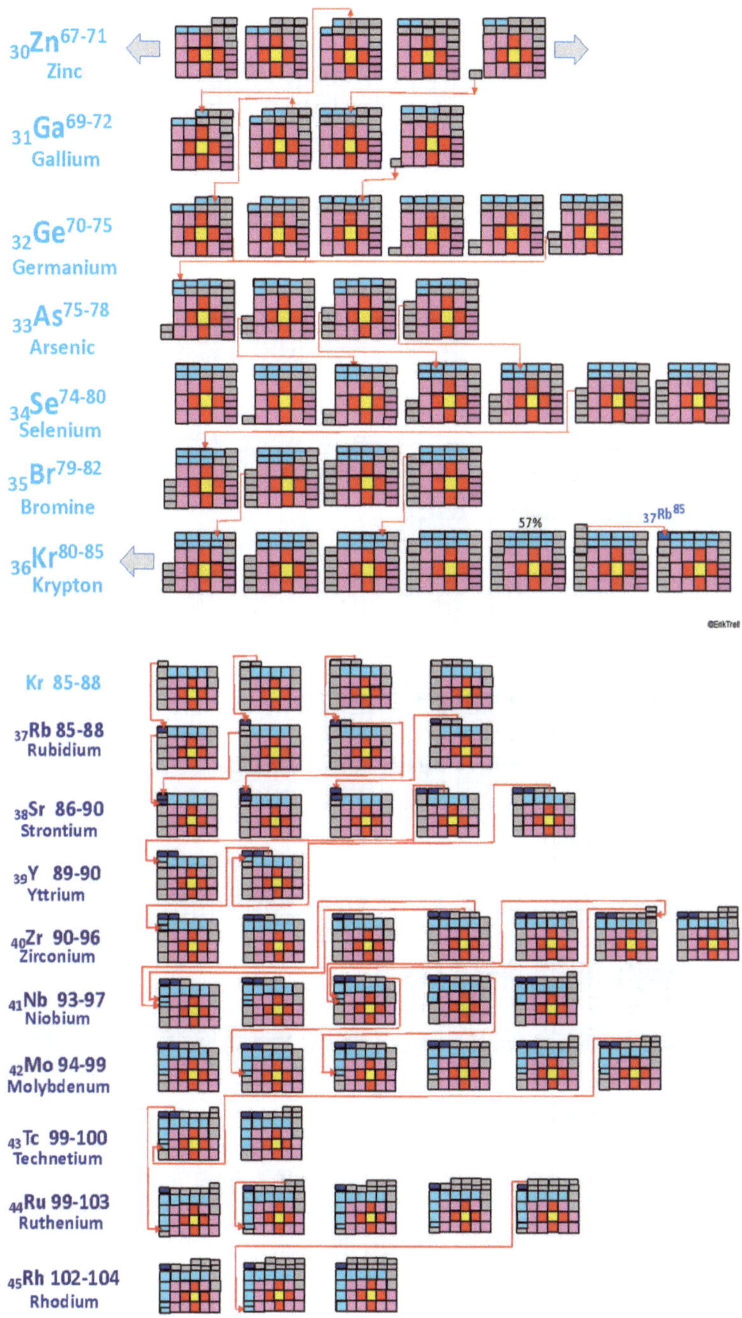

Fig. 5.9 (**a**) Remaining fourth period atoms and their central isotopes. Like in all other atoms they are plane-filling by aggregation which does not mean they are stable. Slow neutron capture beta channels are shown like in other figures. Note complete alpha filling in Krypton. (**b**) Initial states in the fifth period, including Technetium which has no stable isotopes

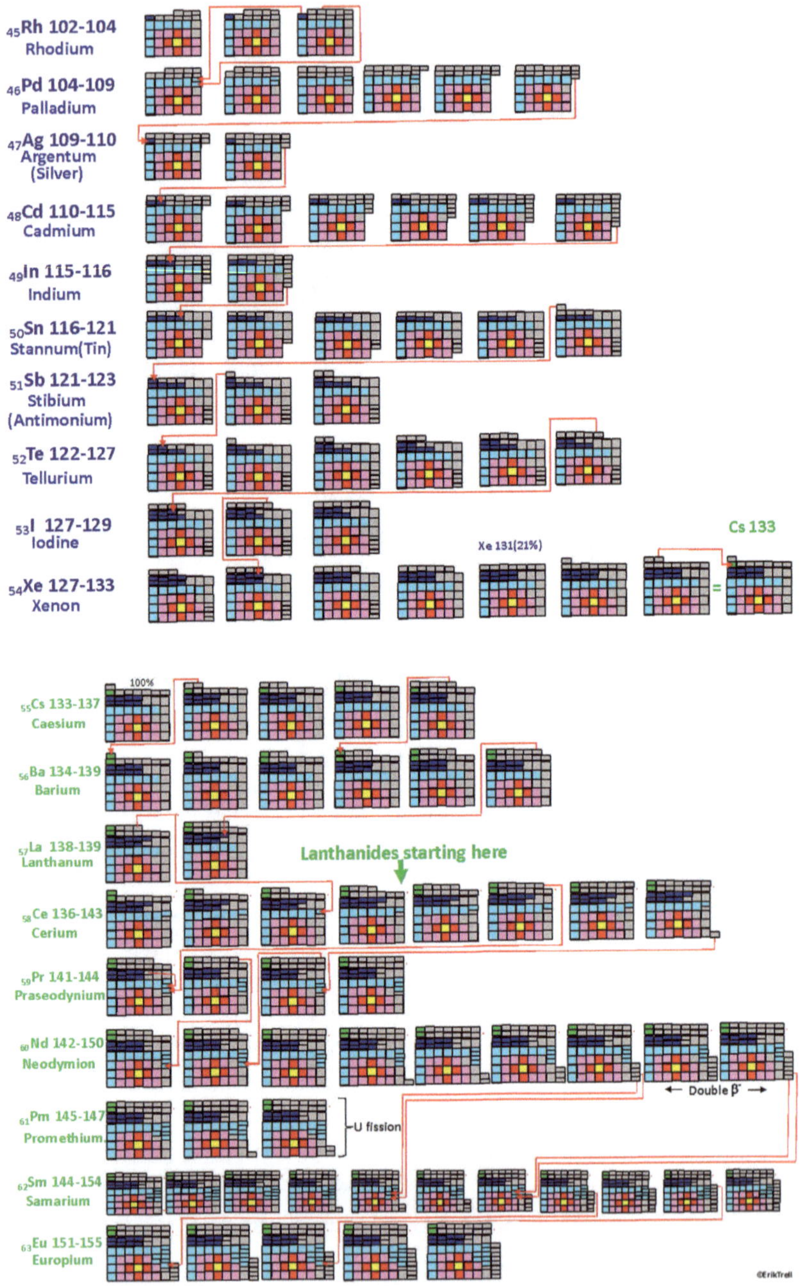

Fig. 5.10 (**a**) Fifth period concludes its layer with full alpha blocks in Xenon. (**b**) start of sixth period

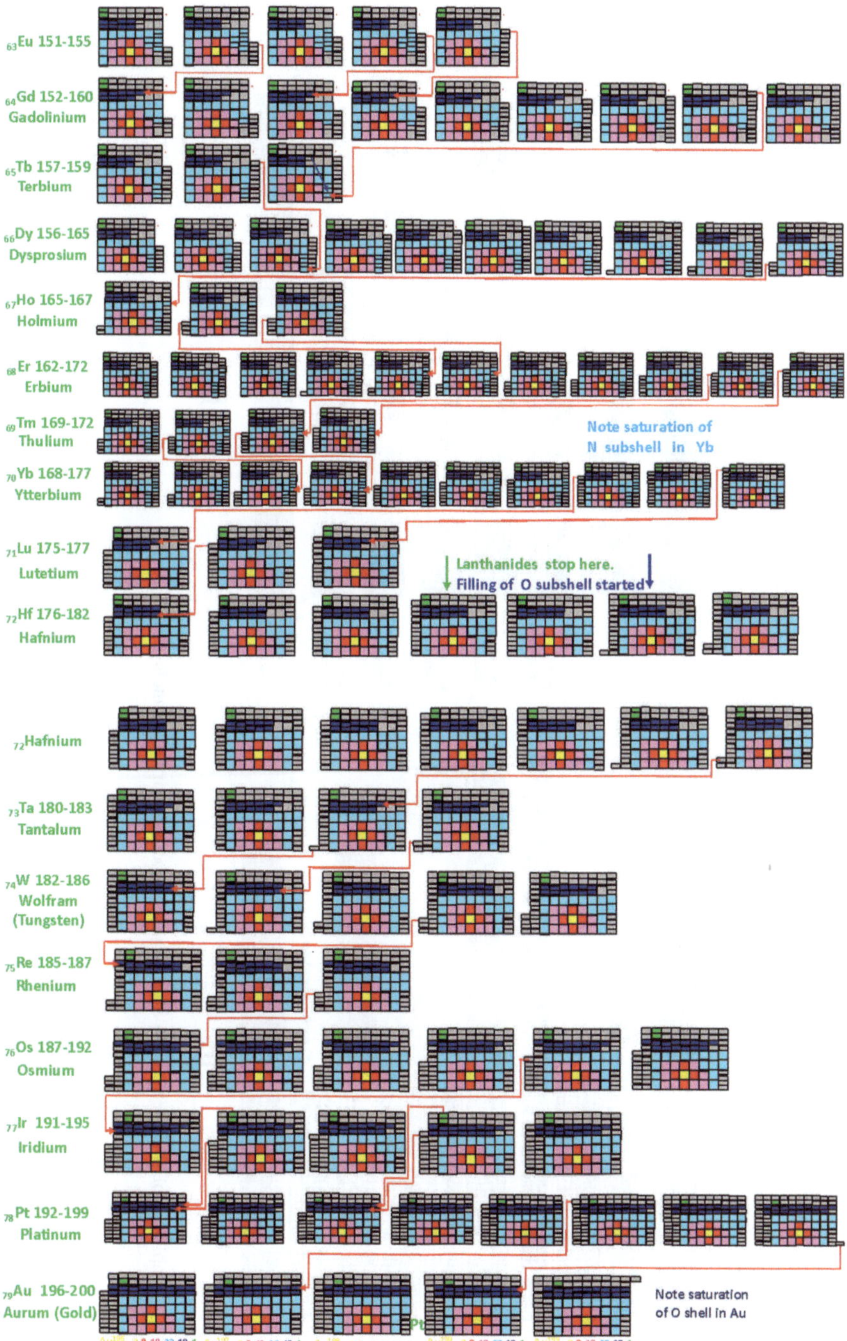

Fig. 5.11 Continuation of sixth period till gold

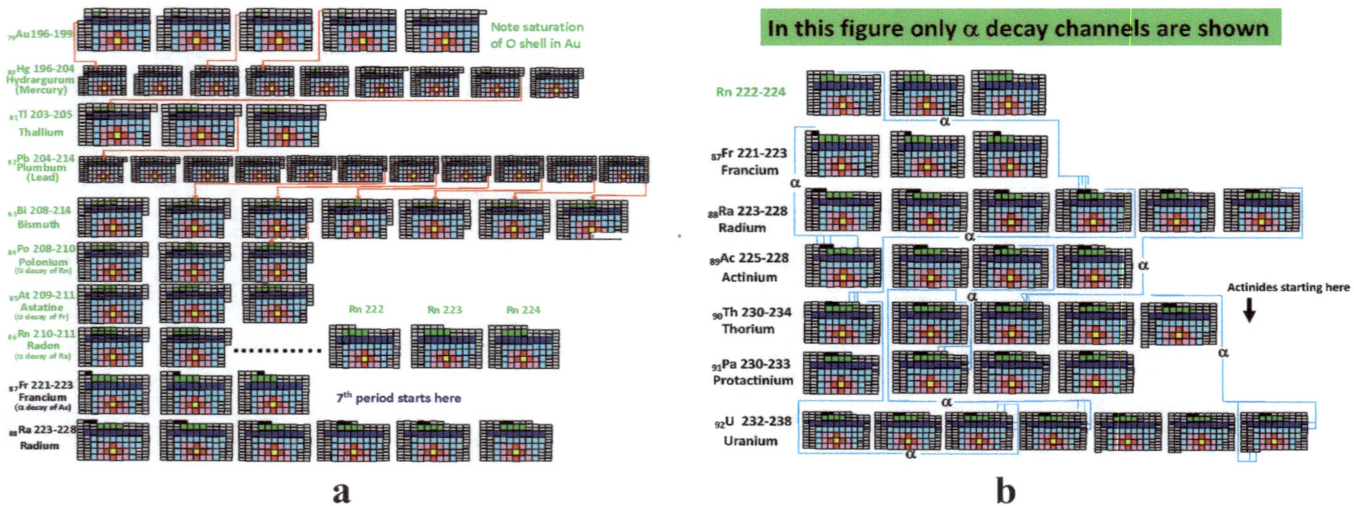

Fig. 5.12 (**a**) Completion of sixth period. (**b**) Start of seventh period. Alpha channels are shown

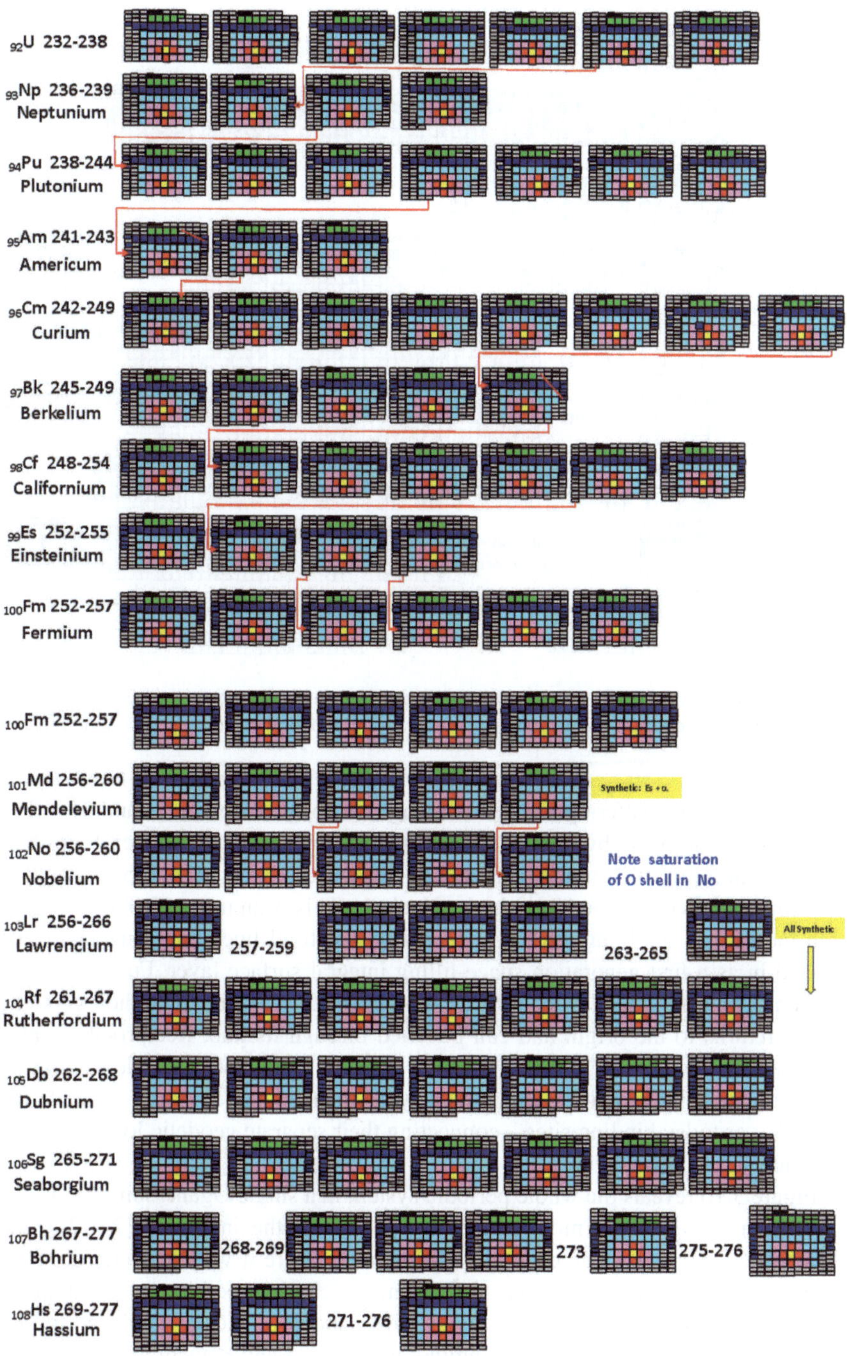

Fig. 5.13 Uranium to Hassium

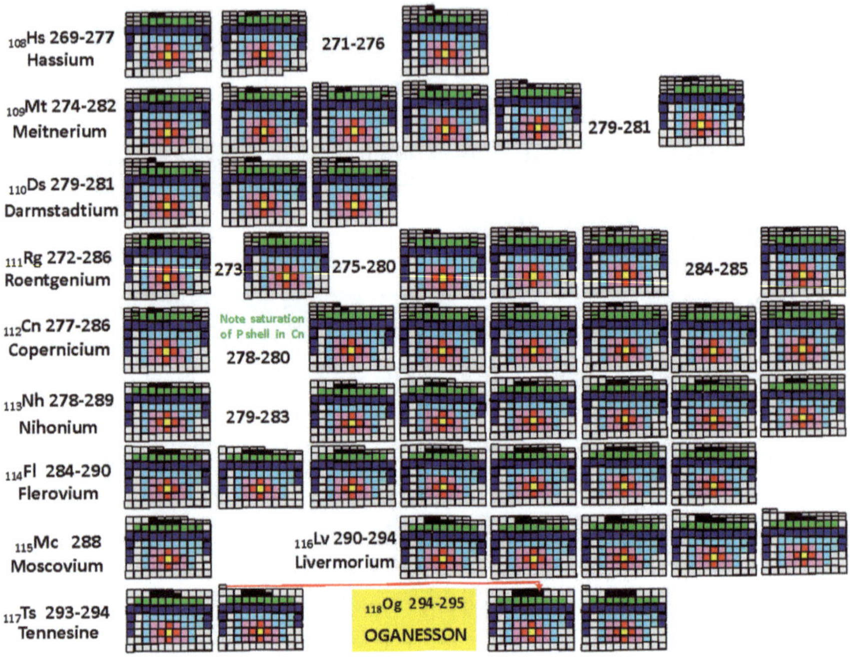

Fig. 5.14 Finale of period 7, concluding the stepwise meandering about the periodic table to exactly reach and replicate Oganesson—and the first step of period eight

naturally from the wafering of the mirroring line congruence curves and how they expand over the plane by continuation of their reciprocal line-complex loops like the knots of a gigantic oriental carpet. The atom width is between 1–5 $\times$ 10^{-10} m in comparison with 1.7 $\times$ 10^{-15} in the Nucleon. This is an up to 10,000 $\times$ 10,000 area increase of the brane of height 6 $\times$ $2^{1/2}$ which all line complexes enter and expand in as a first generation space-filling integral surface layer. Further growth happens by stacking the layers upon each other (Fig. 5.15a) in which their geodetic circuit returns to the origin and can be lifted through its hub. Reciprocity is also the key when as shown in Fig. 5.15b two or more atoms meet and depending on their compatibility may form compounds by at "intersecting . . . polar . . . osculating planes"—or today binding sites—connecting their separate geodetic loops into one common space-filling figure.

Figure 5.16 reveals the whole periodic system in a single Oganesson crystal.

When coming to the molecular compounds both the inorganic [20, 21] and organic [21, 22] have been exemplified separately. Here it will be focused on the remarkable aggregations by and of Carbon, which by repetition of itself in all directions can link into fullerenes, buckyballs, nanotubes, etc. as displayed in Fig. 5.17a while Fig. 5.17b illustrates longitudinal iteration and branching practiced by amino acids to build the proteins. Integration in compound structures like vessel walls is demonstrated by cholesterol in Fig. 5.17c and functional effects of shape by

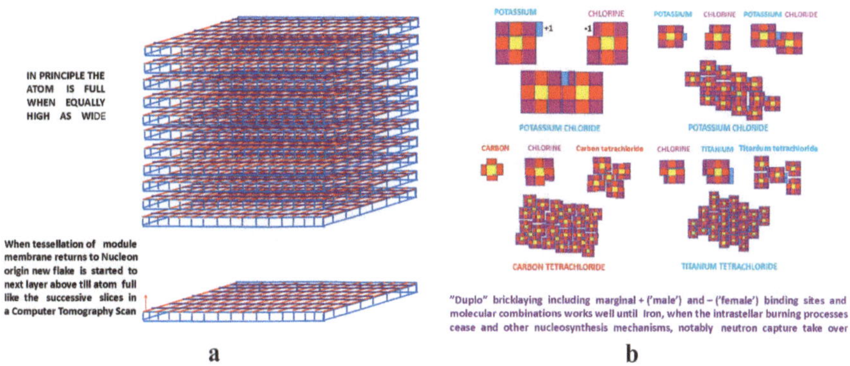

Fig. 5.15 (**a**) Formation and propagation of atom crystal by stacking its horizontal expansion integral surface. (**b**) Examples of cogging in separate atom cross-section geodetic circuits at reciprocal osculation sites to one space-filling figure

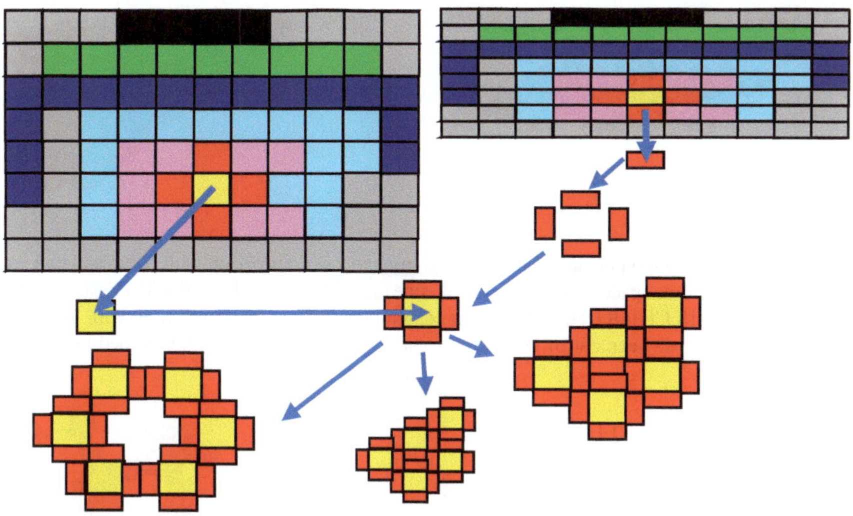

Fig. 5.16 Oganesson: Icon and integral for reference of whole known periodic table; canvas and palette of its realization; catalog of its patterns and supply of its building pieces, construction models and templates, and keyboard, bits and algorithms of its online interactive computer system

DNA in Fig. 5.17d. These are only a few examples of the organic chemistry findings, notably on the amino acids and DNA [20–22], and which add to the overwhelming evidence of their direct structural reality.

But as discussed at length in [22], existing means and models of describing them (Fig. 5.18) don't match the real data any longer and will remain insufficient and misleading. So the conclusion is that the best, "natural shape" [22] templates and

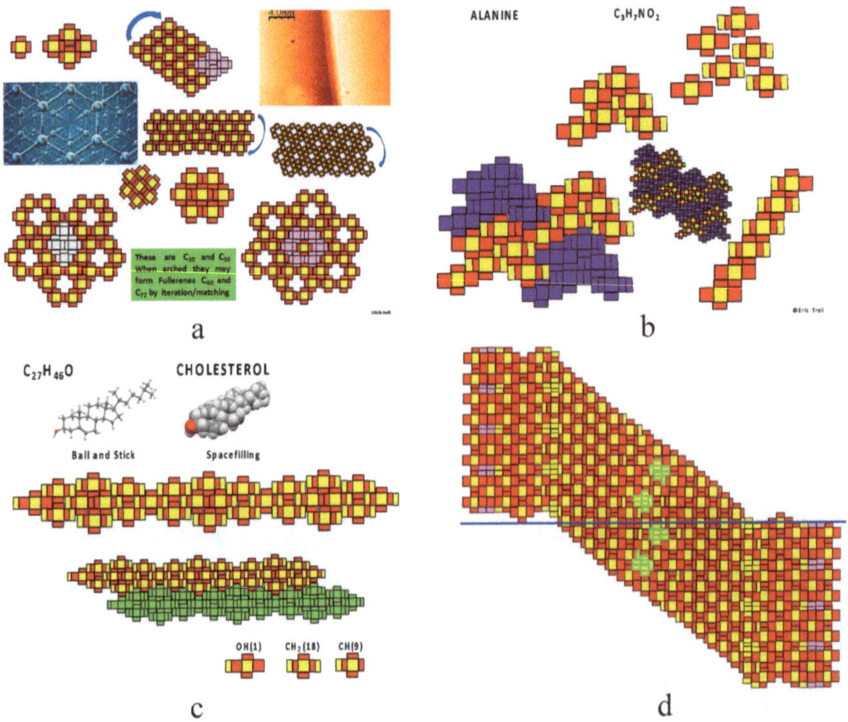

Fig. 5.17 (**a**) aggregation of carbon to observed structures in comparison with photo and ball and stick model. (**b**) Some space-filling forms of alanine including bendable fiber with constant width. (**c**) Assembly of 18 CH_2, 9 CH, and 1 OH to longitudinally layerable cholesterol with hooks for tissue integration. (**d**) Segment of DNA whose backbone ends zip-lock when joined under tube formation

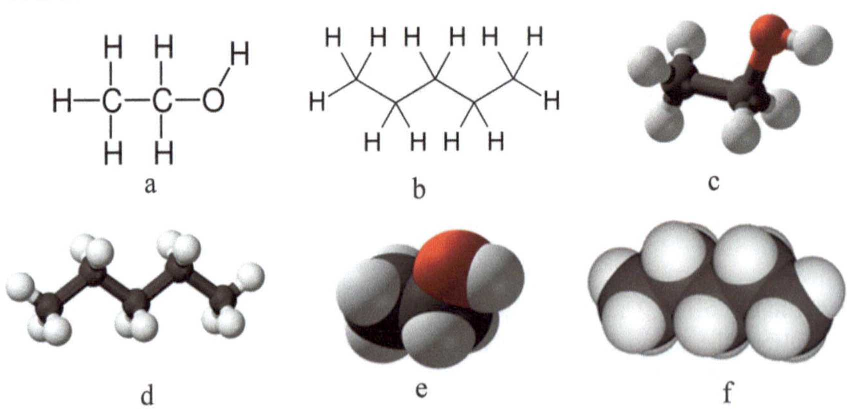

Fig. 5.18 Examples of models/templates of ethanol and pentane. (**a, b**) Skeletal model putting the atoms in a bond network whose lines lack physical reality. (**c, d**) Ball and Stick model gives a three-dimensional framework but in principle nothing new. The atoms are equal-sized balls attached to the ends of the skeleton sticks so that they are separated in space rather than united, with large areas of void between them. (**e, f**) Space-filling, or Calotte model, gives the atoms a spherical shape of their electron cloud with different size determined by their van der Waals radius, and kind indicated by arbitrary color. When the atoms are brought together in molecules they are flattened inwards but continue to bulge out on the surface like adjacent calottes which impedes tight contact and compounds between them

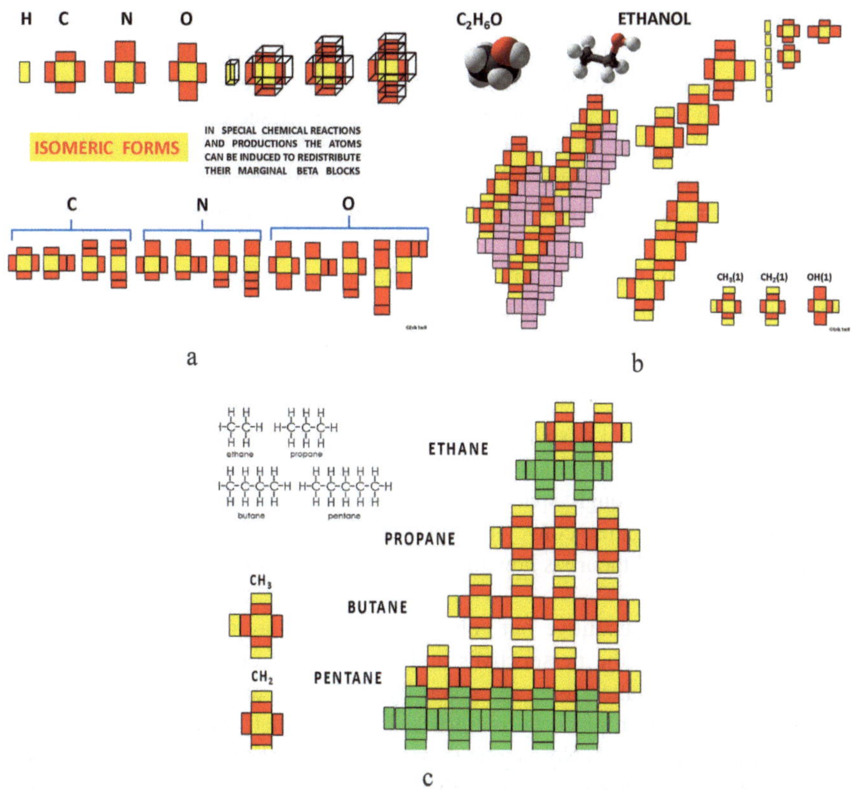

Fig. 5.19 (**a**) "Natural structure" templates of hydrogen, carbon, nitrogen, and oxygen. (**b**, **c**) Same in ethanol and pentane

bits of them are themselves to be directly applied in structural modeling and online interactive computing where the algorithms are their actual transactions. This is illustrated in Fig. 5.19 in some notable organic cases but applies also in the inorganic domain.

5.4 Discussion

It is actually quite fantastic that the discovery of the real forms of the atoms and their two species has not been made long before. Probably, their great simplicity has kept them under the radar of current string and other ultrasophisticated research and theory build, far above the ordinary Cartesian event horizon. On the other hand their essential digital and crystal nature is so in line with modern computer technology and games like Minecraft [13–15] that they virtually lie in time. The "Game of Lie" was due to come from the ground with its simplicity of terms and operations [15]. Two complementary bits; the Betabit equivalent with Deuterium, and the Alphabit, equivalent with Helium, and their groupings as decomposable pixels and templates direct searchable on the keyboard and moved to click-in tessellation as alluded to in Fig. 5.16 form the software of an interactive computer program already under implementation, which allows detailed research and generation over a universal range.

It is interesting, therefore, to compare the present article with corresponding thoughts in a modern approach to the same subject. Loop Quantum Gravity [23] also aims to "build the geometry of spacetime from scratch." "Two aspects of a single, stretchy thing . . . measuring a few billion billion billionths of a meter . . . resembles a frothy quantum foam." "warping of the very fabric of spacetime." "Also space comes in discrete chunks, so that there is a smallest possible area and smallest possible volume"... "tiny indivisible bits." "Recipe for transporting direction-indicating vectors through spacetime . . . encodes all the information about the physical geometry . . . area-conveying links connect little chunks of space." "Thus, the spin networks tell theorists how to put space together, one little patch at a time." [23]

These somewhat repetitive but coherent quotations are congenial to Lie's but yet a bit diffident to the perspectives and potentials of classical and differential geometry, where smallest bits are given by $dx\ dy\ dz$. and "direction-indicating vectors" are already in place by reciprocal local parametrization—but lost in the lack of translation of Lie's Norwegian thesis for over a transitional century. But now they may still confer a vital recipe to the mainstream even including "the strings" who "move in a spacetime whose shape has been chosen from the beginning, as if they were actors on a previously constructed stage." However, "A truly fundamental theory of gravity, everyone agrees, would build the stage itself." [23] That could well be by putting the strings in a hierarchically expanding *Aufbau* order [16].

And the *Aufbau* principle plays an irreplaceable instrumental role also in "the quantum mechanical explanation of the periodic system" [24] without which it doesn't reach longer than Helium [12–15]. However, there is no contradiction there; Quantum Mechanics is the ruler and caretaker of the "frothy quantum foam" of "tiny indivisible bits" [23] within and without and around and in the currents before and debris after they appear in the *Aufbau* ordered atom accumulations to work by these as described not only here but in the universal evidence of a direct real structure of our world; from the oldest far East and Veda sources, the Greek and Arab cultures, the remarkably parallel primordial African cube-hexagon linguistic cosmography [12–15, 25], Rowland's single operator quantum mechanics rewrite system [26], the platonic solids, oriental tiling, cubism, geometric art [25], Minecraft . . . And to the present objective results; simple, exact, exhaustive, reproducible, and of a statistic significance beyond calculation.

References

1. Lie, M.S.: Ueber eine Darstellung des Imaginären in der Geometrie. Journal für die reine und angewandte Mathematik. **70**, 346–353 (1869)
2. Lie, M.S.: Over en Classe geometriske Transformationer. Ph.D. thesis, Kristiania (now Oslo) University (1871)
3. Trell, E., Santilli, R.M.: Marius Sophus Lie's doctoral thesis over en Classe Geometriske Transformationer. Algebras Groups Geom. **15**, 395–445 and hadronicpress.com/lie.pdf (internet open accessible) (1998)
4. Pluecker, J.: Neue Geometrie des Raumes gegründet auf die Betrachtung der geraden Linie als Raum Element. Leibzig Druck und Verlag von B G Teubner (1868)
5. Stubhaug, A.: The Mathematician Sophus Lie. It was the Audacity of My Thinking. Springer Verlag, Berlin, Heidelberg (2003)
6. Trell, E.: Representation of particle masses in hadronic SU(3) transformation diagram. Acta Phys. Austriaca. **55**, 97–110 (1983)
7. Trell, E.: On rotational symmetry and real geometrical representations of the elementary particles with special reference to the N and Δ series. Phys. Essays. **4**, 272–283 (1991)
8. Trell, E.: Real forms of the elementary particles with a report of the Σ resonances. Phys. Essays. **5**, 362–373 (1992)
9. Trell, E.: Elementary particle spectroscopy in regular solid rewrite. AIP Conf. Proc. **1051**, 127–141 (2008)
10. Trell, E.: Digital outline of the elementary particles via a root space diagram approach. J. Comput. Methods Sci. Eng. **13**, 245–270 (2013)
11. Trell, E.: Lie, Santilli, and nanotechnology. AIP Conf. Proc. **1637**, 1100–1109 (2014)
12. Trell, E., Akpojotor, G., Edeagu, S., Animalu, A.: Structural wave-packet tessellation of the periodic table and atomic constitution in real $R^3 \times SO(3)$ configuration space. J. Phys. Conf. Ser. **1251**(012047), 1–15 (2019)
13. Trell, E.: A space-frame periodic table representation system testing relativity in nucleosynthesis of the elements. J. Phys. Conf. Ser. **1557**, 012006 (2020)
14. Trell, E.: From photon to Oganesson. Lie algebra realization of the standard model extending over the periodic table. J. Phys. Conf. Ser. **2081**, 012034 (2021)
15. Trell, E.: A bottom-up 'Game of Lie' cellular automaton evolution in SO(3) root space both nucleating, crystallizing and space-filling the complete atomic realm. J. Phys. Conf. Ser. **2197**, 012025 (2022)
16. Bohr, N.: Der Bau der Atome und die physikalischen und chemischen Eigenschaften der Elemente. Z. Phys. **9**, 1–67 (1922)
17. Grassmann, H.: Die Ausdehnungslehre von 1844 oder die liniale Ausdehnungslehre. Leipzig Otto Wigand (1844 and 1878)
18. Clifford, W.: Applications of Grassmann's extensive algebra. Am. J. Math. **1**, 350–358 (1878)
19. Portegeis Zwart, S.: Computational astrophysics for the future. Science. **36**, 79–80 (2018)
20. Trell, E.: Self-assembled nanostructures of atoms and molecules with examples of various organic compounds including the proteinogenic amino acids and DNA. Proc. IMBIC. **11**, 1–9 (2022)
21. Trell, E.: Lie differential geometry Aufbau of the atoms and molecules. Space Time Fundam. Interact. **3–4**, 299–313 (2023)
22. Trell, E.: Back to the future of natural structure. Proc. IMBIC. **12**, 1–8 (2024)
23. Cho, A.: Constructing spacetime-no strings attached. Science. **298**, 1166–1167 (2002)
24. Scerri, E.R.: How good is the quantum mechanical explanation of the periodic system? J. Chem. Educ. **75**, 1384–1385 (1998)
25. Trell, E., Edeagu, S., Animalu, A.: Geometric Lie algebra in matter, arts and mathematics with incubation of the periodic system of the elements. AIP Conf. Proc. **1798**, 020162 (2017)
26. Rowlands, P.: Zero to Infinity. The Foundations of Physics. World Scientific, New Jersey, London, Singapore, Beijing, Shanghai, Hong Kong, Taipei, Chennai (2007)

Chapter 6
Secret Sharing Schemes for General Access Structures with Leakage Resilience

Sabyasachi Dutta

Abstract A secret sharing scheme generates shares of a secret for a set of participants such that any authorized set can reconstruct the secret, and shares of an *unauthorized* subset of participants do not leak any information about the secret. A recent line of research, motivated by leakage of shares in real-life implementation of secret sharing, is security against a powerful adversary that can receive leakage information about *all shares*. Srinivasan-Vasudevan in Crypto'19 and Kumar-Meka-Sahai in FOCS'19 proposed two models of leakage, static and adaptive leakage for secret sharing schemes, respectively, and proposed compilers that convert traditional secret sharing schemes into leakage-resilient ones. In this chapter we revisit these leakage functions for general access structures.

Keywords Secret sharing · Leakage · Adaptive · Extractors · Bounded collusion protocol · Communication complexity · Number on forehead · Access structure

6.1 Introduction

Secret sharing scheme was independently proposed by Blakely [6] and Shamir [20] and since then has found diverse applications in cryptography, for example, threshold cryptography [9, 19], multiparty computation [4, 13], and many more. In a secret sharing scheme there is a set of n parties $\mathcal{P} = \{1, 2, \ldots, n\}$ that receive the shares of a secret $s \in \mathcal{S}$ from a trusted *dealer* with the property that *qualified* subsets of parties can recover the secret, and *nonqualified* (forbidden) subsets learn nothing about the secret. The collection of qualified sets is denoted by $\mathcal{A}$ and is called an *access structure* on the set of parties $\mathcal{P}$. The collection of forbidden sets is denoted by $\mathcal{Z}$ where we assume that $\mathcal{Z} = 2^{\mathcal{P}} \setminus \mathcal{A}$. For example, in a (t, n)-threshold access structure, any subset of parties with size at least t constitutes a qualified set

---

S. Dutta (✉)
Department of Computer Science and Engineering, SRM University, Amaravati, AP, India

A. Adhikari et al. (eds.), *Quantitative Frontiers*, Industrial and Applied Mathematics, https://doi.org/10.1007/978-981-96-9435-8_6

and the rest, i.e., subsets of size at most $t-1$ form the collection of forbidden sets. However, in case of a general access structure there may not exist such nice separation between $\mathcal{A}$ and $\mathcal{Z}$ other than $\mathcal{Z} = 2^{\mathcal{P}} \setminus \mathcal{A}$.

A secret sharing scheme, in its basic form, provides two types of guarantees: perfect (information theoretic) secrecy of the secret for *passive* adversaries who capture shares corresponding to forbidden sets and perfect recovery of the secret by the qualified sets of shares.

A *passive* adversary is one who follows the prescribed algorithms but is interested in acquiring more information than what it is allowed to according to the access structure. The follow-up works extended this model to consider *active* or *malicious* adversaries who would not follow the protocol and present modified shares (or even no shares) in the reconstruction phase [7, 18, 22]. In both of the above described models, an adversary is allowed to get hold of a forbidden set of shares but not have any information (not even partial) about the rest of the shares.

Recently, motivated by real-life applications of secret sharing schemes, [1, 17, 21] considered an adversary that has access to the shares of a forbidden set and leakages from the shares of other parties.

In order to understand how devastating and powerful these leakage attacks can be, we consider the following simple example: Consider a k-out-of-n Shamir secret sharing scheme over $\mathbb{F}_{2^m}$. Thus, from k many shares it is possible to reconstruct the secret, but from $k-1$ many shares no information about the secret is revealed. If the share of each party is λ-bits long, then even if $(k-1)$ shares, i.e., $\lambda(k-1)$ bits are given to the adversary, it is not possible to guess the secret better than uniform. However, if an adversary leaks only the least significant bits from each of k many shares, then it can compute the least significant bit of the secret by taking a linear combination of these leaked bits. The coefficients of the linear combination are Lagrange's coefficients which are public information and hence known to the adversary. So, only k bits of leakage gives more power to the adversary to attack a scheme.

In fact, Guruswami and Wootters [15] showed the surprising result that, for Shamir threshold scheme [20] over a field $\mathbb{F}_{2^m}$ of characteristic 2, a single-bit leakage from each share would recover the secret.

Therefore, studying secret sharing schemes that can prevent such kind of leakage attacks is important. Goyal and Kumar [14] gave a formal definition of *leakage-resilient secret sharing* (LR-SS) (as a building block of *non-malleable secret sharing*) which was further developed in [1, 5, 17]. The notion of *leakage-resilient secret sharing scheme* (LR-SSS) was first formally introduced by Goyal and Kumar [14], toward building a *non-malleable secret sharing* (which relaxes the protection goal against active tampering by requiring that tampering with the shares either does not change the secret or completely "destroys" the secret). They also noted that LR-SSS may be of independent interest, and indeed LR-SSS has generated a significant amount of interest and follow-up works [1, 5, 12, 17]. The notion of secrecy in all these works is *statistical indistinguishability* of two adversarially chosen secrets.

In an LR-SSS, the leakages from the shares are modeled as adversary querying the parties with leakage functions (adversarially chosen) to be evaluated on their shares. The leakage functions can be applied individually to single shares [1, 5, 21] or groups of shares [17]. In the former model, the leakage functions $\{\mathsf{Leak}_i\}_{1 \le i \le n}$ are chosen independent of each other giving rise to *independent leakages*. However, each Leak_i may or may not depend on a fixed subset of adversarially chosen (forbidden) subset of shares. If Leak_is do not depend on other shares, then the leakage is called *local*. Otherwise, it is called *joint* or *strong-local leakage*. On the other hand, in the second model of leakage defined in [17], the Leak functions queries on a group of shares in multiple rounds to obtain leakage known as joint and adaptive leakage. In the following we describe the leakage models.

6.1.1 Models of Leakage

The leakage models used in the existing literature assume leakages are from stored shares, and computations (e.g., share generation and secret reconstruction) are leak-free. The leakage models are adaptations of the "`memory attacks`" model by Akavia et al. [2]. In this model, everything that is stored in the memory is subject to leakage, and the only restriction is that the output of leakage function is shrinking.

Local (Independent) Leakage The *local* leakage model [5] assumes that leakage functions acting on shares are specified, independent from each other, by the adversary before the secret is chosen, and each function outputs a bounded number of bits. This notion was strengthened by Aggarwal et al. [1] and Srinivasan and Vasudevan [21] where the adversary gets hold of a forbidden set of shares in the clear and local leakages from the rest of the shares.

Strong-Local (Nonadaptive) Leakage In this model of leakage [21], the adversary has access to shares of a forbidden set, and depending on them it chooses the leakage functions for the rest of the shares. However, the leakage functions are independent of each other. Thus, at the end of the leakage protocol the adversary has a set of forbidden shares in the clear and leakages from the rest. This class of leakage functions is called a *semi-local* leakage family (*see* Definition 2).

Joint and Adaptive Leakage Kumar-Meka-Sahai [17] define an adaptive adversary that obtains the leakage through running a multi-round protocol, at the end of which μ bits will be obtained. The leaked bit(s) in each round depends on p shares (out of a total of n) of the secret. These leakage protocols are called *bounded collusion protocols (BCPs)* and are denoted by a (p, n, μ)-BCP (*see* Sect. 6.2.3).

In this chapter we consider the strong-local leakage model of Srinivasan and Vasudevan (referred to as *SV leakage model*) [21] and joint-adaptive leakage model of Kumar, Meka, and Sahai (referred to as *KMS leakage model*) [17]. In both these models leakage functions depend on a set of shares, but the leakage functions are chosen independently in the former model, and in the latter case they are adaptively chosen. Moreover, in the first model leakage functions depend on a fixed subset of shares, whereas in the second the subsets are chosen adaptively by the adversary.

Both papers propose compilers and constructions that convert a threshold SSS to an LR-threshold SSS in the corresponding leakage models. The authors [21] remarked that their approach can further be extended to produce a compiler for constructing strong-local leakage-resilient SSS realizing a t-monotone general access structure in which every qualified set is of size at least t. On the other hand, Kumar et al. [17] gave a concrete compiler to obtain an adaptive leakage-resilient SSS for a t-monotone general access structure.

However, in both the papers privacy requirement against leaking adversary holds for at most $t - 1$ captured or colluded shares. This essentially makes the schemes behave in a ramp-like manner where there is a gap between "privacy requirement" and "reconstructability requirement" against leaking adversary. We note that when there is no leakage, they satisfy the conditions of perfect schemes.

In this chapter we show that these compilers although achieve the highest protection (in terms of the total number of leaked bits for the given leakage model) for threshold schemes, when extended to general access structure can provide protection for much smaller amount of leakage than what the corresponding leakage models can allow (see Note 2, Example 1 and Example 2 in Sect. 6.1.3 for more clarification) and is permissible in a general secret sharing scheme. Our goal is to close this gap and provide compilers for general access structures that can leak maximal amount of permissible bits in each of these two leakage models.

6.1.2 Background

Srinivasan and Vasudevan [21] In this work, the authors provided two compilers for obtaining leakage-resilient secret sharing schemes—-one in local leakage model and another in strong-local leakage model. More precisely, they provided two compilers that can transform any (t, n)-threshold secret sharing scheme into leakage-resilient ones with respect to corresponding leakage models.

In the local leakage model, the adversary gets hold of $t-1$ shares in the clear and (independent) leakages from the rest of the shares. Their proposed scheme resists any adversary to infer (statistical secrecy) about the underlying secret message.

We are interested in the strong-local (nonadaptive) leakage model and so we describe it more elaborately. This model is proposed by the authors in which the adversary $\mathcal{A}$ is allowed to choose two secrets m_0, m_1 of the same length from the space of secret messages. It then sends the messages to the dealer/challenger $\mathcal{D}$. The challenger $\mathcal{D}$ randomly selects one of the secrets, say, m_b, where $b \in \{0, 1\}$, and runs share generation algorithm on m_b. $\mathcal{A}$ is then provided with $t - 2$ shares (chosen by $\mathcal{A}$). With the help of these shares, $\mathcal{A}$ then selects leakage functions to be applied on the rest of the $n - t + 2$ shares and sends them to $\mathcal{D}$. In the next step, $\mathcal{D}$ applies these leakage functions on the rest of the shares and sends $\mathcal{A}$ leakages. The authors showed that after this leakage protocol, when the adversary is given one more share in the clear (i.e., $\mathcal{A}$ now possesses $t - 1$ many shares in the clear and leakages from the rest), the adversary has negligible advantage over the random output while guessing the secret that is chosen by $\mathcal{D}$. We summarize the idea of the compiler (for more details we refer to Theorem 3.8 in [21]).

Their compiler takes any (t, n)-threshold secret sharing scheme and generates the shares $(sh[1], \ldots, sh[n])$ for a secret m. Then an extractor is used to generate random pads that will be used to mask the shares. Suppose Ext is an average case strong seeded extractor. Let $w_1, \ldots, w_n$ be uniformly sampled "sources" and s be a uniform seed from appropriate domains. Each share $\mathsf{share}[i]$ is then masked by XOR-ing the corresponding extractor output and a random pad r_i as $sh[i] \oplus \mathsf{Ext}(w_i, s) \oplus r_i$. Lastly, s and all r_i's are shared using the same threshold scheme to all the parties. The full share of party i is $\mathsf{share}_i = (w_i, \mathsf{share}[i] \oplus Ext(w_i, s) \oplus r_i, s_i, (r_{(1,i)}, \ldots, r_{(n,i)}))$.

The authors remarked that this construction of theirs can be directly generalized—if the adversary can capture at most $t - 1$ many shares and leakages from the rest of the shares (in the strong-local leakage model), then it is possible to find a compiler that admits a leakage-resilient secret sharing scheme for any t-monotone access structure.

Our goal is to expose a maximal set of shares to the adversary including leakages from the rest such that the system is not trivially compromised.

Kumar-Meka-Sahai [17] The authors considered (t, n) threshold SSS and defined leakage in terms of the transcript of a multi-round communication protocol where in each round a message is generated as a (leakage) function of the share set of $t-1$ parties. The $(t-1, n, \mu)$-BCP protocol class allows a total of μ bits to be leaked by using multiple messages.

More specifically, $(t - 1, n, \mu)$-BCP are protocols with this structure where in each round the shares of at most $t - 1$ many parties are used, and a total of μ bits is leaked (*see* Sect. 6.2.3). The leakage function can be any function belonging to the class $(t - 1, n, \mu)$-BCP. An LR-SSS in this model provides statistical secrecy of secret for an adversary who has access to the leakage in this model, and the assumption on the access structure is that it should be t-monotone.

Kumar et al. [17] (i) construct a leakage-resilient (t, t) threshold scheme tolerating against $t - 1$ bounded collusion [notation: $(t - 1, t, t)$-LRSSS], (ii) use $(t - 1, t, t)$-LRSSS and perfect-hash family to construct a $(t - 1, t, n)$-LRSSS, and finally (iii) construct an $\mathcal{A}[t]$-LRSSS from $(\mathsf{ShareGen}_{\mathcal{A}[t]}, \mathsf{Rec}_{\mathcal{A}[t]})$ and a $(t - 1, t, n)$-LRSSS, where $\mathcal{A}[t]$ is a t-monotone general access structure.

The construction of LR-SSS for t-monotone general access structures is as follows: For a secret bit m, two bits L, R are randomly chosen such that $m = L \oplus R$. Now L is secret shared among the parties using a share generation algorithm realizing the access structure and R is secret shared using a $(t - 1, t, n)$-LRSSS. Thus, R is shared in such a way that it gives leakage resilience with respect to the class $(t - 1, n, \mu)$-BCP. They stated the following result which provides stronger security guarantee against an adaptive adversary.

Result: [(Lemma 7 **from [17])]** Any secret sharing scheme on n parties that is ϵ-leakage resilient with respect to $(t - 1, n, \mu + 1)$-BCP is also ϵ-leakage resilient against an adaptive adversary who completely leaks any $t-1$ shares after executing a leakage protocol $\mathsf{Leak} \in (t-1, n, \mu)$-BCP.

We observe that once again the restriction put on the general access structure is that it has to be t-monotone and the security is against leakages from the class $(t-1, n, \mu)$-BCP.

6.1.3 Our Work

In the works of [17, 21] the assumption is that if the adversary can capture/control up to $t-1$ parties, then their compilers/constructions can admit an LR-SSS for t-monotone general access structures. However, in a general access structure there can be forbidden sets of size greater than $t-1$. Thus the above assumptions severely limit the adversary from gaining information to the maximum possible extent which it can utilize.

In this chapter we consider the following question: what would be strongest security model against a leaking adversary for general access structures?

We note that in general access structures there may not be any explicit relationship (in terms of the sizes) between a qualified set and a forbidden set. This makes defining the security condition tricky. Intuitively, for any type of leakage model the maximum security we can expect is when a leaking adversary sees the shares corresponding to a maximal forbidden set in the clear and leakages from the rest. Leakages can vary according to different leakage models. Without further ado we define the secrecy of secret messages as follows.
(**Statistical Secrecy w.r.t Leaking Adversary**) For every (maximal) forbidden set $F \in Z_M$, for all leakage functions $f \in \mathcal{L}$ and for any two distinct secrets $m_0 \neq m_1$ in S, it must hold that

$$\{\{\mathsf{ShareGen}(m_0)_i\}_{i \in F}, f\{\mathsf{ShareGen}(m_0)\}\}$$

$$\approx_\epsilon$$

$$\{\{\mathsf{ShareGen}(m_1)_i\}_{i \in F}, f\{\mathsf{ShareGen}(m_1))\}\}.$$

Note 3 In the above definition Z_M denotes the collection of all maximal forbidden sets (see Sect. 6.2). For example, in a (t, n)-threshold access structure any subset of size $t-1$ is a maximal forbidden set. However, for general access structures they can be of different sizes. Note that for (t, n)-threshold access structures the above formulation allows an adversary to see a set of $t-1$ shares in the clear and the leakages from the rest of $n-t+1$ shares. This is the maximum an adversary can have in its view before proceeding to break the security of the system.

Note 4 In the previous works [17, 21] the above definition of statistical secrecy with respect to a leaking adversary holds only when F is of size at most $t-1$ even in the case of general access structures.

- **Initial Setup & Input:** Secret message m & a t-monotone access structure $(\mathcal{A}_0, Z_M)$. Also suppose that $(\mathsf{ShareGen}_{\mathcal{A}_0}, \mathsf{Rec}_{\mathcal{A}_0})$ denote a perfect secret sharing scheme realizing $(\mathcal{A}_0, Z_M)$.

- **Share Generation (LRShareGen):**
  1. Run $\mathsf{ShareGen}_{\mathcal{A}_0}(m) \longrightarrow (\mathsf{share}[1], \ldots, \mathsf{share}[n])$.
  2. Choose a uniform seed $s \in_R \{0,1\}^d$.
  3. For each $i = 1, 2, \ldots, n$ do:
     $w_i \in_R \{0,1\}^\eta$
     (masking string) $r_i \in_R \{0,1\}^\rho$
     Compute: $\mathsf{share}[i] \oplus Ext(w_i, s) \oplus r_i$
     Run $\mathsf{ShareGen}_{\mathcal{A}_0}(r_i) \longrightarrow (r_{(i,1)}, \ldots, r_{(i,n)})$
  4. Run $\mathsf{ShareGen}_{\mathcal{A}_0}(s) \longrightarrow (s_1, \ldots, s_n)$
  5. Output: $\mathsf{share}_i = (w_i, \mathsf{share}[i] \oplus Ext(w_i, s) \oplus r_i, s_i, (r_{(1,i)}, \ldots, r_{(n,i)}))$

- **Secret Reconstruction (LRRec):**
  1. Input : a qualified set of shares $\{\mathsf{share}_i\}_{i \in X}$ where $X \in \mathcal{A}_0$
  2. for every $i \in X$ do:
     Parse $\mathsf{share}_i = (w_i, \mathsf{share}[i] \oplus Ext(w_i, s) \oplus r_i, s_i, (r_{(1,i)}, \ldots, r_{(n,i)}))$
  3. Run $\mathsf{Rec}_{\mathcal{A}_0}(s_i)_{i \in X} \longrightarrow s$ (Recall: $|X| \geq t$)
  4. for every $i \in X$ do:
     Run $\mathsf{Rec}_{\mathcal{A}_0}(r_{(i,j)})_{j \in X} \longrightarrow r_i$
     Retrieve $\mathsf{share}[i] = (\mathsf{share}_i)_2 \oplus Ext(w_i, s) \oplus r_i$
  5. Output $m \longleftarrow \mathsf{Rec}_{\mathcal{A}_0}(\mathsf{share}[i])_{i \in X}$.

Fig. 6.1 Leakage-resilient secret sharing for general access structures with respect to extended semi-local leakage family described in Definition 10

In Sect. 6.3, we define the strongest nontrivial class of leakage functions for t-monotone general access structures (*see* Definition 10) using the strong-local leakage model. This can be seen as the general access structure counterpart of Srinivasan-Vasudevan's leakage class for threshold case. We also give a compiler (*see* Fig. 6.1) that can convert a secret sharing for a general access structure to one that is leakage resilient against this class (*see* Definition 10).

Extending Srinivasan et al.'s approach to define the admissible share set of an adversary in threshold schemes would suggest that the leakage functions could be determined after seeing $|F| - 1$ shares, where F is a maximal forbidden set F, and $|F|$ denotes its cardinality. This however does not work. Let $F' \subset F$ and $F' = F \setminus \{x\}$. We have $|F'| = |F| - 1$ which satisfies the requirement of being an admissible set. However the adversary can trivially win the distinguishability game if there exist a party u and a qualified set A such that $A \subset (F' \cup \{u\})$. The example below illustrates this point.

Example 1 Consider the monotone access structure on seven parties with the collection of minimal qualified sets $\mathcal{A}_0 = \{123, 134, 234567\}$. Thus the access structure is 3-monotone, and $\{23456\}$ is a maximal forbidden set. Suppose an adversary is allowed to leak from share_1 depending on fixed (but adversarially chosen) four shares from the set of five shares $\{\mathsf{share}_2, \mathsf{share}_3, \ldots, \mathsf{share}_6\}$. We see that if adversary specifies the leakage function L_1 as $\text{LSB}(\mathsf{Rec}(\mathsf{share}_1, \{\mathsf{share}_2, \mathsf{share}_3\}))$, then the secrecy of the scheme is compromised. This is a trivial/pathological attack which is possible on the system.

We want to ensure that even if the adversary gets $\{\mathsf{share}_2, \mathsf{share}_3, \ldots, \mathsf{share}_7\}$ in the clear and leakage from share_1, it should not be able to break the secrecy. At the same time, we must ensure that the leakage from share_1 can only depend on share_2 and share_4 (keeping in mind the qualified sets $\{123\}$ and $\{134\}$) so that the leaking adversary does not trivially attack the system. We note that extending the scheme of Srinivasan-Vasudevan (we present the extension in Appendix I) does not give us leakage resilience with respect to stronger security model as described above. This is due to the fact that the leakage resilience of the extended scheme depends on a (t, n)-threshold secret sharing, and thus if a forbidden set of size greater than t is exposed to the adversary, then adversary can attack the system.

In Sect. 6.4 we extend the notion of bounded-collusion (leakage) protocol to the case of general access structure and allow each protocol message (i.e., leakage) to be constructed by using the shares of a forbidden set, with the limitation that the total leaked bit in the transcript is bounded by μ. Note that it is in contrast to the model of Kumar-Meka-Sahai [17] where the adversary could specify only $t - 1$ shares to compute leakage bits.

For $\mathcal{F} \subseteq \mathcal{Z}$, we define $(\mathcal{F}, n, \mu)$-BCP as a leakage protocol, where in each round of the protocol a subset of shares of a forbidden set $F \in \mathcal{F}$ is used (*see* Sect. 6.4). Our definition includes all leakage functions (to the maximum extent) that do not lead to trivial success of the adversary in indistinguishability game.

We give a construction of leakage-resilient secret sharing scheme for general access structure where the leakage resilience is against $(\mathcal{F}, n, \mu)$-BCP protocols. In fact, in our construction we consider the family of forbidden sets $\mathcal{F}$ to be the family of maximal forbidden sets Z_M (*see* Sect. 6.2). This enables the adversary to leak more information bits as leakage at the end of leakage protocol.

Example 2 (Secret Sharing on Compartmental Access Structure) One of the widely used access structures is compartmental access structure in which parties belong to different organizations and they need to collaborate in order to recover a secret—e.g., if majority of the parties from each of the organization agree to pull their shares, then only the secret can be recovered.

Consider an example of such an access structure on a set of $n_1 + n_2$ parties in which the first n_1 parties belong to a compartment $\mathcal{C}_1$ and the rest n_2 belong to another compartment $\mathcal{C}_2$. Any minimal qualified set has to contain at least t_1 parties from $\mathcal{C}_1$ **and** at least t_2 parties from $\mathcal{C}_2$. Without loss of generality, let us assume $t_1 < t_2$. We note that the access structure is a $(t_1 + t_2)$-monotone access structure.

Srinivasan–Vasudevan [21] restrict the adversary to obtain $t_1 + t_2 - 1$ shares in the clear and leakages from the rest of the shares depending on any fixed $t_1 + t_2 - 2$ shares out of the obtained $t_1 + t_2 - 1$ shares. If we assume that the shares are α bits long and at most μ bits of leakages are obtained from each of the uncaptured shares, then at the end of leakage protocol, the adversary (concretely) obtains at most $\alpha(t_1 + t_2 - 1) + (n_1 + n_2 - t_1 - t_2 + 1)\mu$ bits of leakage information.

In our model, adversary can obtain any list of maximal forbidden set of shares in the clear and μ bits of leakages from each of the rest of the shares depending on fixed $(t_1 + t_2 - 2)$ shares from the list. Thus at the end of the leakage protocol the adversary can obtain (concretely) up to $\max.\{(t_1+n_2-1)\alpha+(n_1-t_1+1)\mu, (t_2+n_1-1)\alpha+(n_2-t_2+1)\mu\}$ bits of leaked information. Since μ is less than the share size α, we see that in our model adversary can exploit more information in its favor.

Kumar–Meka–Sahai [17] considered an (joint) adaptive leakage model in which the adversary can get at most $\alpha(t_1 - 1) + \mu$ bits of leakage information. Note that in this case μ is the total number of leaked bits obtained by the adversary after executing a $(t_1 - 1, n_1 + n_2, \mu)$-BCP.

On the other hand, in our extended model of adaptive leakage of [17] an adversary can receive up to $max.\{(t_1 + n_2 - 1)\alpha, (t_2 + n_1 - 1)\alpha\} + \mu$ bits of information as leakage.

In short, we observe that our model preserves secrecy for stronger adversaries.

Example 3 (Secret Sharing on Hierarchical Access Structure) Another important class of access structure is a hierarchical access structure in which parties belong to different levels of the same organization and they need to collaborate in order to recover a secret.

Consider an access structure on a set of $n_1 + n_2$ parties in which the first n_1 parties belong to a level $\mathcal{L}_1$ and the rest n_2 belong to another level $\mathcal{L}_2$. Any minimal qualified set has to contain at least t_1 parties from $\mathcal{L}_1$ **or** at least t_2 parties from $\mathcal{L}_1 \cup \mathcal{L}_2$. Without loss of generality, let us assume $t_1 < t_2$, and this makes the access structure a t_1-monotone access structure.

Srinivasan–Vasudevan [21] enable the adversary to obtain leakage information (at the end of leakage protocol) bounded by $(t_1 - 1)\alpha + (n_1 + n_2 - t_1 + 1)\mu$ bits, whereas in our model the adversary is able to capture $(t_2-1)\alpha+(n_1+n_2-t_2+1)\mu$ bits of leakage.

On the other hand, in the (joint) adaptive leakage model of Kumar-Meka-Sahai [17], the adversary can get at most $\alpha(t_1-1)+\mu$ bits of leakage information, whereas in our extension it can have $\alpha(t_2 - 1) + \mu$ bits.

More Related Works

The study of leakage resilience in the context of secret sharing was studied by Dziembowski-Pietrzak [11]. They called it intrusion resilience where the leakage was adaptive and the adversary leaked from each share by an iterative process.

However, their reconstruction procedure was interactive and required many rounds which depended on the leakage tolerated.

Benhamouda et al. [5] studied "local" leakage resilience of Shamir secret sharing scheme (in fact, any linear secret sharing scheme) and showed that (a) if Shamir's secret sharing is implemented over field of characteristic 2, then it is *not* leakage resilient, and (b) however, it becomes leakage resilient if implemented over a field with characteristic large prime. The scheme is leakage resilient if the threshold is $n - o(\log n)$, where n denotes the number of parties. They left open the problem for the case when threshold value is $\frac{n}{2}$.

Aggarwal et al. [1] considered nonadaptive individual *local* leakage model and gave a compiler that turns any secret sharing scheme into a leakage-resilient one. Their work is subsumed by the independent work of Srinivasan and Vasudevan [21].

6.2 Preliminaries

We follow standard notations and symbols throughout. Let $\mathcal{P} = \{1, 2, 3, \ldots, n\}$ denote a set of parties. Let $2^{\mathcal{P}}$ denote the set of all subsets of $\mathcal{P}$. Let $\mathcal{A} \subseteq 2^{\mathcal{P}}$ and $\mathcal{Z} \subseteq 2^{\mathcal{P}}$, where $\mathcal{A} \cap \mathcal{Z} = \emptyset$, respectively, denote the set of all qualified sets and the set of all forbidden sets. The pair $(\mathcal{A}, \mathcal{Z})$ constitutes an access structure on $\mathcal{P}$. An access structure is called **monotone** if every superset of a qualified set is a qualified set. We only consider monotone access structures.

An *access structure* is completely defined by the collection of so-called **minimal** qualified sets. We will denote the collection of minimal qualified sets by $\mathcal{A}_0$, which is defined by $\mathcal{A}_0 = \{X \subseteq \mathcal{P} : B \in \mathcal{Z} \; \forall \; B \subsetneq X\}$. An access structure is called **t-monotone** if every minimal qualified set is of size at least t.

Furthermore, we consider another important collection of all **maximal** forbidden sets $Z_M = \{F \in \mathcal{Z} : \forall i \notin F \; \exists X \in \mathcal{A}_0 \; such \; that \; X \subseteq F \cup \{i\}\}$. That is, if any party joins a maximal forbidden set, it becomes a qualified (may not be minimal) set.

In a (k, n)-threshold access structure, any subset of size k is a minimal qualified set, and any $k-1$ sized subset is a maximal forbidden set.

The definition of a secret sharing scheme realizing a monotone access structure $(\mathcal{A}_0, Z_M)$ is as follows. Here $\mathcal{A}_0$ means the collection of minimal authorized sets, and Z_M denotes the collection of all maximal forbidden sets.

Definition 1 (Secret Sharing Scheme [3]) A secret sharing scheme $\mathcal{S}$ for an access structure $(\mathcal{A}_0, Z_M)$ consists of a pair of algorithms (ShareGen, Rec). ShareGen is a probabilistic algorithm that gets as input a secret m (from a domain of secrets S) and a number of parties n and generates n shares $(\mathsf{share}_1, \ldots, \mathsf{share}_n) \longleftarrow \mathsf{ShareGen}(m)$. Rec is a deterministic algorithm that gets as input the shares of a subset B of parties and outputs a string. The requirements for defining a secret sharing scheme are as follows:

1. (Correctness) For every secret $m \in S$ and every qualified set $B \in \mathcal{A}$, it must hold that $Pr[\mathsf{Rec}(\{\mathsf{share}_i\}_{i \in B}, B) = m] = 1$.

2. (Perfect Secrecy) For every forbidden set $F \in Z_M$ and for any two distinct secrets $m_0 \neq m_1$ in S, it must hold the following two distributions are identical:

$$\{\mathsf{ShareGen}(m_0)_i\}_{i \in F} \textit{ and } \{\mathsf{ShareGen}(m_1)_i\}_{i \in F}.$$

Note 5 The definition above can be relaxed by introducing negligible error probabilities in *correctness* and *secrecy* conditions. We mention that the existing works on leakage-resilient secret sharing accommodate these relaxations in their protocols and so does our work.

For completeness we define the semi-local leakage family of functions which models the joint but nonadaptive leakages from the honest shares.

Definition 2 (Semi-Local Leakage Family [21]) Let $\mathcal{S}_1 \times \cdots \times \mathcal{S}_n$ be the domain of shares for some secret sharing scheme and $1 \leq t \leq t' \leq n$ and μ be natural numbers. The parameters t, t', μ defining a semi-local function family are adaptive threshold, corruption threshold, and amount of leakage, respectively. The family $\mathcal{H}_{t,t',\mu}$ consists of functions $h_{T,T',\bar{\tau}}$ (where $\bar{\tau} = (\tau_1, \ldots, \tau_n)$) such that:

- $T \subseteq T' \subseteq \{1, \ldots, n\}$
- $|T| = t$ and $|T'| = t'$
- The function $h_{T,T',\bar{\tau}}$ when given input $(\mathsf{share}_1, \ldots, \mathsf{share}_n)$ outputs:
  - share_i for all $i \in T'$
  - Leakages $\tau_j((\mathsf{share}_{i_1}, \ldots, \mathsf{share}_{i_t}), \mathsf{share}_j)$ for all $j \in \{1, \ldots, n\} \setminus T'$. That is, the leakages on the honest shares are adaptively chosen by the Adv depending on the shares of T.

For a (k, n)-threshold scheme, $t' = k - 1$ and $t = k - 2$ are an optimal choice of parameters [21].

Definition 3 (Leakage-Resilient Secret Sharing Scheme) A secret sharing scheme $\mathcal{S}$ for an access structure $(\mathcal{A}_0, Z_M)$ consists of a pair of algorithms (ShareGen, Rec). ShareGen is a probabilistic algorithm that gets as input a secret m (from a domain of secrets S) and a number of parties n and generates n shares $(\mathsf{share}_1, \ldots, \mathsf{share}_n) \longleftarrow \mathsf{ShareGen}(m)$. Rec is a deterministic algorithm that gets as input the shares of a subset B of parties and outputs a string. The secret sharing scheme (ShareGen, Rec) is said to be ϵ-leakage-resilient against a leakage family $\mathcal{L}$ if the following conditions are met:

1. (Correctness) For every secret $m \in S$ and every qualified set $B \in \mathcal{A}$, it must hold that $Pr[\mathsf{Rec}(\{\mathsf{share}_i\}_{i \in B}, B) = m] = 1$.
2. (Perfect Secrecy w.r.t Non-leaking Adversary) For every forbidden set $F \in Z_M$ and for any two distinct secrets $m_0 \neq m_1$ in S, it must hold the following two distributions:

$$\{\mathsf{ShareGen}(m_0)_i\}_{i \in F} \textit{ and } \{\mathsf{ShareGen}(m_1)_i\}_{i \in F} are identical.$$

3. (Statistical Secrecy w.r.t Leaking Adversary) For every forbidden set $F \in Z_M$, for all $f \in \mathcal{H}$, and for any two distinct secrets $m_0 \neq m_1$ in S, it holds that

$$\{\{\mathsf{ShareGen}(m_0)_i\}_{i \in F}, f(\mathsf{ShareGen}(m_0)_i)\}$$
$$\approx_\epsilon \{\{\mathsf{ShareGen}(m_1)_i\}_{i \in F}, f(\mathsf{ShareGen}(m_1)_i)\}.$$

6.2.1 *Cumulative Array for an Access Structure*

Let $Z_M = \{F_1, F_2, \ldots, F_\rho\}$, *i.e.*, there are ρ many maximal forbidden sets written in some order. Let us now recall the idea of cumulative array (see [16]) for $\mathcal{A}_0$. The **cumulative array** (CA) is an $n \times \rho$ Boolean matrix such that $CA(i, j) = 1$ if and only if $i \notin F_j$, where n is the number of parties.

Example 4 Let us consider the following access structure on a set of $\mathcal{P} = \{1, 2, 3, 4, 5, 6\}$ six parties with $\mathcal{A}_0 = \{134, 135, 136, 145, 146, 156, 234, 235, 236, 245, 246, 256\}$ and $Z_M = \{123, 124, 125, 126, 3456\}$.

Then the corresponding cumulative array is given by

Parties	$F_1 = \{123\}$	$F_2 = \{124\}$	$F_3 = \{125\}$	$F_4 = \{126\}$	$F_5 = \{3456\}$
1	0	0	0	0	1
2	0	0	0	0	1
3	0	1	1	1	0
4	1	0	1	1	0
5	1	1	0	1	0
6	1	1	1	0	0

Note 6 We note the following:

- If F is any forbidden/unauthorized set of parties, then it is contained in at least one maximal forbidden set F_t, say, $1 \leq t \leq \rho$. When the Boolean matrix CA is restricted to the rows indexed by the parties in F, then the tth column in the restricted matrix does not contain 1 in any position.
- On the other hand, if X is any (minimal) qualified set, then it is contained in none of the F_is. In this case, when the Boolean matrix CA is restricted to the rows indexed by the parties in X, then every column in the restricted matrix contains at least one 1.

A cumulative array for a general access structure can be used to construct a secret sharing for the access structure. Suppose we want to share a secret m to realize the access structure described in Example 4. Toward this, we first share the secret using a (5, 5)-threshold secret sharing, e.g., Shamir's scheme to obtain $(\mathsf{share}[1], \ldots, \mathsf{share}[5])$. This number 5 corresponds to the size of Z_M. Now we output the share for party i, $1 \leq i \leq 6$, as follows: $\mathsf{share}_i = \{\mathsf{share}[j] : CA(i, j) = 1\}$, i.e., we look at the ith row of the array, find those j for which $CA(i, j) = 1$, and assign those $\mathsf{share}[j]$s to party i. One can verify that this "assignment" gives a secret sharing scheme realizing the given access structure.

More generally, we have the following theorem.

Theorem 1 ([16]) *For any monotone access structure* $(\mathcal{A}_0, Z_M)$ *there exists a secret sharing scheme realizing the access structure and share size of the scheme is bounded by* $|Z_M|.\log\{max(|m|, |Z_M|)$, *where* $|m|$ *denotes the length of the secret.*

6.2.2 Extractors

We state some standard definitions and results which will be required in the following sections.

Definition 4 Let us suppose X is a random variable with support S. The min-entropy of X is defined as

$$H_\infty = \min\{\log \frac{1}{Pr[X = s]} : s \in S\}.$$

An (n, k)-source is a distribution on $\{0, 1\}^n$ with min-entropy k.

Definition 5 Suppose X and Y be two distributions on a set S. The statistical distance between the distributions is defined as

$$SD(X, Y) = \frac{1}{2} \sum_{s \in S} |Pr[X = s] - Pr[Y = s]|.$$

X and Y are said to be ϵ-close if $SD(X, Y) \leq \epsilon$. A standard notation to capture that these distributions are ϵ-close is $X \approx_\epsilon Y$.

Result: If $X \approx_{\epsilon_1} Y$ and $Y \approx_{\epsilon_2} Z$, then $X \approx_{\epsilon_1+\epsilon_2} Z$ (Triangle Inequality).

The definition of (average) conditional min-entropy [10] is defined as follows.

Definition 6 ([10]) The average conditional min-entropy of X given W is defined as

$$\tilde{H}_\infty(X|W) = \log(E_{w \longleftarrow W}[\max_x Pr[X = x|W = w]]).$$

Result: ([10]) If a random variable W can take at most l values, then $\tilde{H}_\infty(X|W) \geq H_\infty(X) - \log_2 l$.

Definition 7 (Strong Seeded Extractor) A function $\mathsf{Ext} : \{0, 1\}^n \times \{0, 1\}^d \longrightarrow \{0, 1\}^m$ is called a strong seeded extractor for sources with min-entropy k and error

ϵ if for any (n, k)-source X and an independently and uniformly chosen random string U_d, we must have

$$\mathsf{Ext}(X, U_d)||U_d \approx_\epsilon U_m||U_d,$$

where U_m and U_d are independent.

The connection between an average case seeded extractor and a strong seeded extractor was proved in [10]. We mention the result here.
Result: [10] For any $\delta > 0$, if Ext is a (k, ϵ)-strong seeded extractor, then it is also a $(k + \log(\frac{1}{\delta}), \epsilon + \delta)$-average case strong extractor.

6.2.3 Bounded Collusion Protocols and NoF Communication Complexity

Kumar et al. [17] modeled adaptive leakage using bounded collusion protocols and constructed leakage-resilient secret sharing whose security relies on communication complexity in the "number-on-forehead" model of communication. We recall some definitions and results from [17].
(p, n, μ)-**Bounded Collusion Protocols** [17]: Consider n parties, and let p be an integer $p < n$, called the collusion bound. The leakage to the adversary is through messages that are constructed and broadcasted (and hence seen by the adversary) by a subset of at most p parties in each round. The message is constructed using the shares of the parties and all previous communications. The process continues until a maximum of μ bits are leaked to the adversary. The collection of all such possible (leakage) protocols Leak constitutes the family (p, n, μ)-BCP. The adversary must succeed in indistinguishability game using the leakage of a protocol from this family. More formally:

- Let $\mathsf{share}_1, \ldots, \mathsf{share}_n$ be the n shares, and let τ denote the transcript of the leakage protocol Leak. At the start of the protocol, τ is set $\emptyset$, and at the end of each round the message of that round is appended to τ, until the length of τ is up to μ.
- In each round, the Next function takes as input the current transcript τ and determines a subset S of size p of parties and the description of a leakage function f that takes $\{\mathsf{share}_i\}_{i \in S}$ as input. At the end of the round the transcript is updated as $\tau \longleftarrow \tau || f\{\mathsf{share}_i\}_{i \in S}$.
- The above process is repeated until the Next function outputs $\perp$. Output the final transcript, which is at most μ bits long, as leakage.

Definition 8 ([17]) A t-out-of-n threshold secret sharing scheme is called a (p, t, n, μ, ϵ)-LRSS if the scheme is ϵ-leakage resilient (*see* Definition 3) against (p, n, μ)-BCP. If the parameters μ and ϵ are clear from the context, then we simply denote it by (p, t, n)-LRSS.

Note 7 For an n-out-of-n secret sharing scheme and $p = n-1$, the communication protocol corresponds to the "number on forehead" NoF model introduced in [8]. This is the best we can hope for because collusion of n parties will reveal the secret. In other words, leakage resilience is achieved against the collusion of maximal forbidden sets.

We now describe the communication complexity of a function in NoF model introduced by Chandra et al. [8].

Definition 9 ([17]) Let us suppose there are n many parties, and an element of a set S is written on the forehead of each party. Each party can see the number on the forehead of all other parties and has no idea of the number written on its own forehead. Suppose these parties wish to compute any arbitrary n party Boolean-valued function $f : S^n \longrightarrow \{0, 1\}$. These parties are allowed to communicate among themselves using a black-board. At the beginning, the black-board is empty, and each party is only allowed to append information to it (without any erasing). Their goal is to compute f while minimizing the number of bits that needs to be written on the black-board. The *NoF communication complexity* refers to the minimum number of bits of communication required in order to gain ϵ advantage in computing f using the protocol described above.

The communication complexity in the NoF model is denoted by $CC_n^{NoF}(f)$.

Result: [Lemma 1 of [17]] For any $\rho \geq 2$, any leakage bound $\mu \geq 0$, and any $\epsilon > 0$, if there is an efficient ρ-party function $f : (\{0, 1\}^b)^\rho \longrightarrow \{0, 1\}$ with $Comm.Complexity_\rho^{NOF}(f) \geq \mu$, then there is an efficient ρ-out-of-ρ secret sharing scheme which is ϵ leakage resilient with respect to $(\rho-1, \rho, \rho)$-BCP. For single-bit secrets, the resulting scheme has share size $1+b$ bits. In particular, if the function f is the *generalized inner product* function, then $b = O(2^\rho(\mu + \log(\frac{1}{\epsilon})))$. We present such a construction given in [17] in Appendix II.

6.3 Leakage-Resilient Secret Sharing for General Access Structures (Joint but Nonadaptive Leakage)

This section follows (nonadaptive) joint leakage model of [21] for (t, n)-threshold access structures. In this model, called strong (semi-local) leakage, the adversary captures $t-1$ shares, chooses and fixes $t-2$ of them, and uses them to select individual leakage functions to leak from honest shares. The authors presented a compiler for converting a (t, n)-threshold scheme to an LR (t, n)-threshold scheme. They also *suggested* that their construction can be modified for general access structures that are t-monotone, resulting in a leakage-resilient scheme against the semi-local leakage family $\mathcal{H}_{(t-2,t-1,\mu)}$ (*see* Definition 2). In the Appendix, we give an explicit construction based on this suggested direction (see Appendix I).

The class of leakage functions $\mathcal{H}_{(t-2,t-1,\mu)}$, however, excludes many leakage functions that are admissible for a t-monotone general access structure. We note that a general access structure is always t-monotone when t is chosen to be the size of the smallest qualified set. To clarify, consider Example 1 where the access structure is 3-monotone, but there is a forbidden set of size 5. The strong leakage model proposed by Srinivasan and Vasudevan [21] restricts the adversary to capture at most two shares and leakages from honest shares can depend on at most one of them. So, it excludes the scenario when the adversary is able to capture the *five* shares corresponding to the forbidden set and define leakages from the honest shares which can depend on one of these captured shares.

We note there is no direct relationship between the size of the smallest qualified set and the size of the largest unqualified set, and so to define all admissible share sets for defining leakage functions, one needs to consider the actual forbidden sets.

In Definition 10 we give a natural extension of the semi-local leakage function family in [21] for general access structures and in Fig. 6.1 give a compiler that takes a secret sharing scheme for a t-monotone general access structure and convert it into a leakage-resilient scheme with respect to the *extended family of semi-local leakage functions*.

We fix the following notations and definitions. Let $(\mathcal{A}_0, Z_M)$ denote a t-monotone access structure on a set $\mathcal{P}$ of n many parties, $(\mathsf{ShareGen}_{\mathcal{A}_0}, \mathsf{Rec}_{\mathcal{A}_0})$ denote a secret sharing scheme realizing the access structure, ρ be the bit-length of each share under the above scheme (here we assume that the length of the largest share is ρ bits and the smaller shares are padded with 0's to make their size exactly ρ bits), and η and d be integers such that there exists an $(\eta - \mu, \epsilon)$-average case strong seeded extractor $Ext : \{0,1\}^\eta \times \{0,1\}^d \longrightarrow \{0,1\}^\rho$.

The two main results of this section are stated in Theorems 2 and 3. The details are given afterward.

Theorem 2 *Let $n, t, \mu \in \mathbb{N}$ and $(\mathcal{A}_0, Z_M)$ denote a t-monotone access structure on a set $\mathcal{P}$ of n many parties where $t \leq n$ and a secret domain $\mathcal{S}$. Suppose for some η, d and $0 \leq \epsilon < 1$ the following exist:*

- *A perfect secret sharing scheme $(\mathsf{ShareGen}_{\mathcal{A}_0}, \mathsf{Rec}_{\mathcal{A}_0})$ realizing the access structure with share size ρ for secrets in $\mathcal{S}$*
- *A perfect (t, n)-threshold secret sharing scheme $(\mathsf{ShareGen}_{(t,n)}, \mathsf{Rec}_{(t,n)})$*
- *A $(\eta - \mu, \epsilon)$-average case strong seeded extractor $\mathsf{Ext} : \{0,1\}^\eta \times \{0,1\}^d \longrightarrow \{0,1\}^\rho$*

Then the construction in Appendix I (see the Appendix) gives rise to a secret sharing scheme realizing the access structure $(\mathcal{A}_0, Z_M)$ and is $2n\epsilon$-leakage resilient against the semi-local leakage family $\mathcal{H}_{(t-2,t-1,\mu)}$. The share size of the scheme is $\eta + \rho + d + n\rho$.

Note 8 Appendix I describes a secret sharing scheme realizing any t-monotone access structure. The *correctness* and *secrecy* (when there is no leakage) properties follow directly from the same properties enjoyed by the underlying $(\mathsf{ShareGen}_{\mathcal{A}_0}, \mathsf{Rec}_{\mathcal{A}_0})$ scheme. Leakage resilience with respect to the semi-local leakage function family $\mathcal{H}_{(t-2,t-1,\mu)}$ (*see* Definition 2) holds, and the proof is similar to that of Theorem 3.8 from [21].

We note that if there is a forbidden set F of size $\geq t$ and the shares are given in the clear to an adversary, then he/she is able to retrieve all $\mathsf{share}[i]$ in F, all r_j's, and s. At this point (even if the leakages from the rest of the honest shares depend only on a size $(t-2)$ subset of shares in F), the leakage resilience of the scheme fails.

Theorem 3 *Given the same conditions described in Theorem 2 and using the same notations, the construction in Fig. 6.1 admits a secret sharing scheme realizing the access structure* $(\mathcal{A}_0, Z_M)$ *and is* $2n\epsilon$*-leakage resilient against the extended semi-local leakage family* $\mathcal{H}_{(\mathcal{A}_0, Z_M, \mu)}$ *(see Definition 10).*

Our aim is to extend the *semi-local leakage function family* approach of Srinivasan and Vasudevan [21] to general access structures and consider the largest class of functions that can be available to the adversary before the indistinguishability game can be trivially won. Toward this, let us first define the following.

Definition 10 (Semi-Local Leakage Family for General Access Structure) Let $(\mathcal{A}_0, Z_M)$ be a t-monotone access structure, and there exists a secret sharing scheme S such that the domain of shares is $S_1 \times \cdots \times S_n$, where n denotes the number of parties. Suppose μ denotes the number of bits that is leaked from each share. An extended semi-local leakage function family $\mathcal{H}_{(\mathcal{A}_0, Z_M, \mu)}$ consists of functions $\{h_{T,T',\tilde{\tau}} : T' \in Z_M, T \subsetneq T', |T| \leq t-2\}$.

For every $i \notin T'$, the function τ_i takes input $((\mathsf{share}_\alpha)_{\alpha \in T}, \mathsf{share}_i)$ and leaks μ bits. Overall, every $h_{T,T',\tilde{\tau}} \in \mathcal{H}_{(\mathcal{A}_0, Z_M, \mu)}$ when given input $(\mathsf{share}_1, \ldots, \mathsf{share}_n)$ outputs $(\mathsf{share}_\beta)_{\beta \in T'}$ and $\tau_i(((\mathsf{share}_\alpha)_{\alpha \in T}, \mathsf{share}_i))$ for every $i \notin T'$.

Note 9 The above definition enables us to expose a maximal forbidden set of shares to the adversary that will be used for choosing the leakage function. The definition removes the trivial cases that access to a set of shares by the adversary will immediately break the leakage resilience. In this definition there is no requirement on the number of shares that are given to the adversary to be at most $t-1$.

Proof of Theorem 3 *Correctness and Secrecy.* The secret sharing protocol described in Fig. 6.1 satisfies *correctness* and *secrecy* when there is no leakage. They follow directly from the underlying secret sharing protocol $(\mathsf{ShareGen}_{\mathcal{A}_0}, \mathsf{Rec}_{\mathcal{A}_0})$. *Leakage Resilience.* The proof of leakage resilience against the extended family of semi-local leakage family is similar to the proof of Theorem 3.8 of [21].

We recall that for any $h_{T,T',\tilde{\tau}}$ in the family $\mathcal{H}_{(\mathcal{A}_0, Z_M, \mu)}$ when given the input $(\mathsf{share}_1, \ldots, \mathsf{share}_n)$ it outputs share_i for all $i \in T'$, and for every other $j \notin T'$, it outputs $\tau_j((\mathsf{share}_\alpha)_{\alpha \in T}, \mathsf{share}_j)$. We mention here that $|T| \leq t-2$, and thus $T \cup \{j\} \notin \mathcal{A}$. We need to show that for any two distinct secrets m_1, m_2, $SD(h_{T,T',\tilde{\tau}}(m_1), h_{T,T',\tilde{\tau}}(m_2)) \leq 2n\epsilon$ for some prefixed negligible ϵ.

The proof strategy is to construct a sequence of hybrid sharing procedures $\mathsf{LRShareGen}_0, \ldots, \mathsf{LRShareGen}_n$, where the $\mathsf{LRShareGen}_0(m)$ denotes the actual share generation as in Fig. 6.1 and the last $\mathsf{LRShareGen}_n(m)$ contains no information about m. For each $i \notin T'$, in the ith hybrid we will replace the part $\mathsf{share}_i \oplus \mathsf{Ext}(w_i, s) \oplus r_i$ (occurring in the ith share) by a random string random. We then show that the distributions of the outputs of $h_{T,T',\tilde{\tau}}$ between the consecutive hybrids are within a statistical distance ϵ—thanks to the property of underlying extractor. In this way, we replace all the honest shares $\mathsf{share}_i \oplus \mathsf{Ext}(w_i, s) \oplus r_i$ by random strings random for all $i \notin T'$. Once this process is over, the leakage distributions are statistically close, and now the entire secrecy property is reduced to the secrecy property of underlying $(\mathsf{ShareGen}_{\mathcal{A}_0}, \mathsf{Rec}_{\mathcal{A}_0})$.

Consider a total ordering "$\prec$" on the set of honest shares, i.e., $\{1, \ldots, n\} \setminus T'$. For any secret m, two consecutive hybrid sharing procedures $\mathsf{LRShareGen}_i$ and $\mathsf{LRShareGen}_{i'}$ (where i' is the successor of i with respect to $\prec$) differ in exactly one position—the position of ith share. In $\mathsf{LRShareGen}_i$, $\mathsf{share}[i] \oplus \mathsf{Ext}(w_i, s) \oplus r_i$ is present, whereas in $\mathsf{LRShareGen}_{i'}$ it is replaced by a random string random.

Suppose the hybrid Hyb_i^m denotes the distribution of $h_{T,T',\tau}$ when it is applied to the list of "shares" in $\mathsf{LRShareGen}_i$ with underlying secret message m. Our aim is to show that Hyb_i^m and $\mathsf{Hyb}_{i'}^m$ are ϵ-close in the statistical distance.

Let us suppose that the statistical distance between Hyb_i^m and $\mathsf{Hyb}_{i'}^m$ is greater than ϵ. We will use this to break the property of underlying Ext. More precisely, we are going to propose a reduction which can specify a leakage function l such that the distributions of the form $(s, l(w), \mathsf{Ext}(w, s))$ and $(s, l(w), \mathsf{random})$ are statistically far.

In order to create such a view, the reduction first sets the shares share_j for all $j \in T$ to be randomly chosen strings. That is, it chooses $w_j, sh'_j, s_j, r_{\bullet j}$ uniformly at random from the appropriate domains and set share_j to be $(w_j, sh'_j, s_j, r_{\bullet j})$. For brevity, we have used the notation sh'_j to denote the second entry of the shares, i.e., the part corresponding to $\mathsf{share}_j \oplus \mathsf{Ext}(w_j, s) \oplus r_j$ and also $r_{\bullet j}$ instead of the vector $(r_{(1,j)}, r_{(2,j)}, \ldots, r_{(n,j)})$.

For i', the reduction picks $sh'_{i'}, s_{i'}$ and the $r_{\bullet i'}$ at random and defines a leakage function $l : \{0,1\}^\eta \longrightarrow \{0,1\}^\mu$ which on input w outputs $\tau_{i'}((\mathsf{share}_\alpha)_{\alpha \in T}, (w, sh'_{i'}, s_{i'}, r_{\bullet i'}))$. Note that this leakage function is in compatibility with the leakage model we are considering.

The extractor challenge that reduction receives is of the form $(s, l(w), z)$, where either $z = \mathsf{Ext}(w, s)$ or $z = \mathsf{random}$.

The reduction now sets up the feed as follows: It runs $\mathsf{ShareGen}_{\mathcal{A}_0}(m)$ to get $(\mathsf{share}[1], \ldots, \mathsf{share}[n])$.

For each $j \in T$ it sets $r_j = sh'_j \oplus \mathsf{share}[j] \oplus \mathsf{Ext}(w_j, s)$ and for i' sets $r_{i'} = sh'_{i'} \oplus \mathsf{share}[i'] \oplus z$. Now if $j \notin T' \cup i'$ and $j \prec i'$, the reduction sets w_j, sh'_j, r_j to be random and for all other j picks w_j, r_j at random and sets $sh'_j = \mathsf{share}[j] \oplus \mathsf{Ext}(w_j, s) \oplus r_j$. Lastly, it picks all the randomness $r_{jj'}$'s and s_j's that are not yet set in such a manner so that they form a valid random sharing of the r_j's and s. The reduction now simulates the output of $h_{T,T',\tau}$ in the following manner:

- For every $j \in T'$, it outputs share_j.
- For i', it outputs $l(w)$ (as the output of $\tau_{i'}$).
- For every other $j \notin T' \cup \{i'\}$, outputs $\tau_j((\mathsf{share}_\alpha)_{\alpha \in T}, (w_j, sh'_j, s_j, r_{\bullet j}))$.

We note that if $z = \mathsf{Ext}(w, s)$, then the above list of outputs corresponds to the distribution Hyb_i^m, and if $z = \mathsf{random}$, then it corresponds to the distribution of $\mathsf{Hyb}_{i'}^m$. Since the function l outputs μ bits of leakage, thus for a randomly chosen $w \in_R \{0, 1\}^\eta$ we must have $\tilde{H}_\infty(w|l(w)) \geq \eta - \mu = k$ (by the result following Definition 6).

Therefore, if the statistical distance between $\mathsf{Hyb}_{i'}^m$ and Hyb_i^m is greater than ϵ, then the above reduction contradicts the fact that the underlying extractor is a (k, ϵ)-strong average-case seeded extractor. This contradiction shows that our initial assumption was wrong, and thus, we must have $\mathsf{Hyb}_{i'}^m \approx_\epsilon \mathsf{Hyb}_i^m$. So now a repeated application of the above argument we infer that $\mathsf{Hyb}_0^m \approx_{n\epsilon} \mathsf{Hyb}_n^m$. And now by triangle inequality we have $\mathsf{Hyb}_0^m \approx_{2n\epsilon} \mathsf{Hyb}_0^{m'}$ for any two distinct secrets m, m'.■

6.4 LR Secret Sharing for General Access Structures: Joint and Adaptive Leakage

In this section we consider the model of joint and adaptive leakage on the shares as described in Kumar et al. [17]. The leakage from shares happens in multiple rounds following a (p, n, μ)-BCP (Sect. 6.2.3), where in each round the adversary can choose at most p parties (out of n). The authors [17] provided a construction of LR-SS scheme for $(p + 1)$-monotone general access structure against (p, n, μ)-BCP. However, in a general access structure the size of a forbidden set can be higher than p, and so the collusion bound does not allow the adversary to exploit the maximum possible resources it can have. We extend Kumar et al.'s definition of adaptive collusion for general access structures by allowing the adversary to select a (maximal) forbidden set of parties in each round of the protocol and use their joint computation and the protocol transcript to that point for constructing the next message (leaked information bits).

$(\mathcal{F}, n, \mu)$-Bounded Collusion Protocol Suppose $(\mathcal{A}_0, Z_M)$ is a monotone access structure on n parties, where $\mathcal{A}_0$ denotes the collection of minimal authorized sets and Z_M denotes the collection of maximal forbidden sets. The definition of the family $(\mathcal{F}, n, \mu)$-BCP is the same as (p, n, μ)-BCP except the fact that in every round of the leakage protocols among n parties the leakage arbitrarily depends on (maximal) forbidden sets of shares, *i.e.*, $\mathcal{F} = Z_M$.

Our main construction is based on the existence of an (n, n)-LRSS against $(n - 1, n, \mu)$-BCP. The construction of such a scheme is given in Section 3.2 of [17]. For the sake of completeness we describe their scheme in Appendix.

6.4.1 LRSS for General Access Structures: (Joint and Adaptive Leakage)

We use a new approach to constructing a leakage-resilient compiler for the leakage function class $(\mathcal{F}, n, \mu)$-BCP, using cumulative arrays. In Appendix I, we outline a construction of a secret sharing scheme for a general access structure. The construction uses a (ρ, ρ) threshold secret sharing, where ρ is the number of maximal forbidden sets. In the following construction the (ρ, ρ) threshold scheme is a $(\rho - 1, \rho, \rho)$-LRSS. We will use the $(\rho - 1, \rho, \rho)$-LRSS from [17] for this. In Fig. 6.2 we describe our scheme. Our construction uses a leakage-resilient $(\rho - 1, \rho, \rho)$ secret sharing scheme at the base level in which unavailability of just one share makes it leakage resilient. Using some basic combinatorics the shares of the $(\rho - 1, \rho, \rho)$-LRSS are distributed among the actual parties so that every maximal forbidden set misses at least one share of this underlying $(\rho - 1, \rho, \rho)$-LRSS. Now the leakage resilience of our protocol is directly linked to that of the underlying $(\rho - 1, \rho, \rho)$-LRSS.

Theorem 4 *Let $(\mathcal{A}_0, Z_M)$ be a monotone access structure on a set of n parties, where $\mathcal{A}_0$ and Z_M denote the collections of minimal qualified sets and maximal forbidden sets, respectively. Also, assume $|Z_M| = \rho$, and let $(\mathsf{LRShareGen}_{(\rho-1,\rho,\rho)}, \mathsf{LRRec}_{(\rho-1,\rho,\rho)})$ be a $(\rho - 1, \rho, \rho)$-LRSS, which can be obtained as described in Appendix III.*

Then the construction in Fig. 6.2 gives rise to a secret sharing scheme (LRShareGen, LRRec) realizing the given access structure and is leakage resilient against any (F, n, μ)-BCP for $F \in Z_M$. Moreover, if share size of underlying $(\rho - 1, \rho, \rho))$-LRSS is c, then share size of the new scheme is bounded by $c.\rho$.

- **Initial Setup & Input** : secret message m & access structure $(\mathcal{A}_0, Z_M)$ where $Z_M = \{F_1, \ldots, F_\rho\}$. Let $(\rho-1, \rho, \rho)$-LRSS be the secret sharing scheme described in Section 4.1.
- **Share Generation (LRShareGen)**:
  1. Run $\mathsf{LRShareGen}_{(\rho-1,\rho,\rho)}(m) \longrightarrow (\mathsf{share}[1], \ldots, \mathsf{share}[\rho])$.
  2. Compute Cumulative Array $CA[i, j]$: $1 \le i \le n$ and $1 \le j \le \rho$ for the given access structure.
  3. Output: $\mathsf{share}_i = \{\mathsf{share}[j] : CA[i, j] = 1\}$ for all party i.
- **Secret Reconstruction (LRRec)**:
  1. Input : a qualified set of shares $\{\mathsf{share}_i\}_{i \in X}$ where $X \in \mathcal{A}_0$
  2. Output $m \longleftarrow \mathsf{LRRec}_{(\rho-1,\rho,\rho)}(\mathsf{share}[i])_{i \in X}$.

Fig. 6.2 Joint and adaptive leakage-resilient secret sharing for general access structures

Proof of Theorem 4 *Correctness and Secrecy.* When a minimal (and hence any) qualified subset of parties comes together to retrieve the secret, collectively they are in possession of all the ρ shares $\mathsf{share}[1], \ldots, \mathsf{share}[\rho]$ of the underlying (ρ, ρ)-threshold secret sharing scheme. Hence the correctness of our scheme follows.

When there is no leakage and a maximal (and hence any) forbidden set of shares is obtained by an adversary, then we observe that it misses at least one share of the underlying (ρ, ρ)-scheme. Perfect privacy of our scheme now follows directly.
Leakage Resilience. Suppose the secret sharing protocol described in Fig. 6.2 is not leakage resilient. Toward this, let us suppose there exists an adversary $\mathcal{B}$ who can specify a $\mathsf{Next} \in (F, n, \mu)$-BCP to distinguish between shares of m_1 and m_2 ($m_1 \neq m_2$). We use $\mathcal{B}$ to construct an adversary Adv which can violate the leakage resilience of the underlying $(\rho - 1, \rho, \rho)$-LRSS.

The reduction works as follows:

- Randomly select m_b, where $b \in \{1, 2\}$.
- Run $\mathsf{LRShareGen}_{(\rho-1,\rho,\rho)}(m_b) \longrightarrow (\mathsf{share}[1], \ldots, \mathsf{share}[\rho])$, and fix these shares.
- On input transcript τ, $\mathcal{B}$ executes $\mathsf{Next}(\tau)$ and specifies a leakage function Leak
- If $\mathsf{Next}(\tau)$ outputs $\perp$, then Adv outputs $\perp$; otherwise, if $\mathsf{Next}(\tau)$ outputs a forbidden set F (can be maximal forbidden), then construct a leakage function g which works as follows—for every $i \in F$ find $j \notin F_i$, and execute $\mathsf{Leak}\{\mathsf{share}[j]\}_{j \notin F_i}$ and output the value. At the end of this step output F, g.

We observe that if $\mathcal{B}$ can distinguish the shares between m_1 and m_2 with advantage ϵ, then by the above reduction Adv can distinguish between the shares of m_1 and m_2 with the same advantage. This contradicts the leakage resilience of the underlying $(\rho - 1, \rho, \rho)$-LRSS. ■

Appendix

I. Explicit construction of Leakage-Resilient Secret Sharing for t-monotone general access structures, following the approach of [21]:

- **Initial Setup & Input :** secret message m & a t-monotone access structure $(\mathcal{A}_0, Z_M)$. Also suppose that $(\mathsf{ShareGen}_{\mathcal{A}_0}, \mathsf{Rec}_{\mathcal{A}_0})$ denote a perfect secret sharing scheme realizing $(\mathcal{A}_0, Z_M)$ and there exists a perfect (t, n)-threshold secret sharing scheme $(\mathsf{ShareGen}_{(t,n)}, \mathsf{Rec}_{(t,n)})$.
- **Share Generation** (LRShareGen):

  1. Run $\mathsf{ShareGen}_{\mathcal{A}_0}(m) \longrightarrow (\mathsf{share}[1], \ldots, \mathsf{share}[n])$.
  2. Choose a uniform seed $s \in_R \{0, 1\}^d$.

(continued)

3. For each $i = 1, 2, \ldots, n$ do:
   $w_i \in_R \{0, 1\}^\eta$
   (masking string) $r_i \in_R \{0, 1\}^\rho$
   Compute: $\mathsf{share}[i] \oplus Ext(w_i, s) \oplus r_i$
   Run $\mathsf{ShareGen}_{(t,n)}(r_i) \longrightarrow (r_{(i,1)}, \ldots, r_{(i,n)})$
4. Run $\mathsf{ShareGen}_{(t,n)}(s) \longrightarrow (s_1, \ldots, s_n)$
5. Output: $\mathsf{share}_i = (w_i, \mathsf{share}[i] \oplus Ext(w_i, s) \oplus r_i, s_i, (r_{(1,i)}, \ldots, r_{(n,i)}))$

- **Secret Reconstruction** (LRRec):
  1. Input : a qualified set of shares $\{\mathsf{share}_i\}_{i \in X}$ where $X \in \mathcal{A}_0$
  2. for every $i \in X$ do:
     Parse $\mathsf{share}_i = (w_i, \mathsf{share}[i] \oplus Ext(w_i, s) \oplus r_i, s_i, (r_{(1,i)}, \ldots, r_{(n,i)}))$
  3. Run $\mathsf{Rec}_{(t,n)}(s_i)_{i \in X} \longrightarrow s$ (Recall: $|X| \geq t$)
  4. for every $i \in X$ do:
     Run $\mathsf{Rec}_{(t,n)}(r_{(i,j)})_{j \in X} \longrightarrow r_i$
     Retrieve $\mathsf{share}[i] = (\mathsf{share}_i)_2 \oplus Ext(w_i, s) \oplus r_i$
  5. Output $m \longleftarrow \mathsf{Rec}_{\mathcal{A}_0}(\mathsf{share}[i])_{i \in X}$.

II. $(\rho - 1, \rho, \rho)$**-leakage-resilient secret sharing scheme from [17]**
The scheme consists of two phases—share generation and reconstruction. It assumes the existence and black-box access to an efficient n party function $f : (\{0, 1\}^b)^n \longrightarrow \{0, 1\}$ with communication complexity of f to be greater or equal to μ in the NOF model.

1. **Share generation:** On input a secret bit m
   - sample uniformly and independently $r_i \in \{0, 1\}^b$ for all $i = 1, \ldots, \rho$
   - compute the bit $r \longleftarrow f(r_1, \ldots, r_\rho)$
   - compute s such that $s \oplus r = m$
   - sample uniformly and independently $s_1, \ldots, s_\rho \in \{0, 1\}$ and find s_ρ such that $s_1 \oplus \cdots \oplus s_\rho = s$
   - Output $\mathsf{share}_i = (r_i, s_i)$ for all $i = 1, \ldots, \rho$
2. **Reconstruction:** On input ρ shares $\mathsf{share}_1, \ldots, \mathsf{share}_\rho$
   - parse share_i as (r_i, s_i) for all $i = 1, \ldots, \rho$
   - compute $r \longleftarrow f(r_1, \ldots, r_\rho)$
   - compute $s_1 \oplus \cdots \oplus s_\rho = s$
   - Output $m \longleftarrow s \oplus r$

References

1. Aggarwal, D., Damgård, I., Nielsen, J.B., Obremski, M., Purwanto, E., Ribeiro, J., Simkin, M.: Stronger leakage-resilient and non-malleable secret sharing schemes for general access structures. In: Advances in Cryptology—CRYPTO 2019, pp. 510–539. Springer International Publishing, Cham (2019)
2. Akavia, A., Goldwasser, S., Vaikuntanathan, V.: Simultaneous hardcore bits and cryptography against memory attacks. In: Reingold, O. (ed.) Theory of Cryptography, pp. 474–495. Springer, Berlin (2009)
3. Beimel, A.: Secret-sharing schemes: A survey. In: Coding and Cryptology, pp. 11–46. Springer, Berlin (2011)
4. Ben-Or, M., Goldwasser, S., Wigderson, A.: Completeness theorems for non-cryptographic fault-tolerant distributed computation. STOC '88, pp. 1–10. ACM, New York (1988)
5. Benhamouda, F., Degwekar, A., Ishai, Y., Rabin, T.: On the local leakage resilience of linear secret sharing schemes. In: Advances in Cryptology—CRYPTO 2018, pp. 531–561. Springer International Publishing, Cham (2018)
6. Blakley, G.R.: Safeguarding cryptographic keys. In: AFIPS 1979, pp. 313–317 (1997)
7. Carpentieri, M.: A perfect threshold secret sharing scheme to identify cheaters. Des. Codes Cryptography **5**(3), 183–187 (1995)
8. Chandra, A.K., Furst, M.L., Lipton, R.J.: Multi-party protocols. In: STOC'83, 25–27 April, 1983, Boston, Massachusetts, USA, pp. 94–99. ACM, New York (1983)
9. Desmedt, Y., Frankel, Y.: Threshold cryptosystems. In: Advances in Cryptology—CRYPTO' 89 Proceedings, pp. 307–315. Springer, New York (1990)
10. Dodis, Y., Ostrovsky, R., Reyzin, L., Smith, A.: Fuzzy extractors: how to generate strong keys from biometrics and other noisy data. SIAM J. Comput. **38**(1), 97–139 (2008)
11. Dziembowski, S., Pietrzak, K.: Intrusion-resilient secret sharing. In: 48th Annual IEEE Symposium on Foundations of Computer Science (FOCS'07), pp. 227–237 (2007)
12. Faonio, A., Venturi, D.: Non-malleable secret sharing in the computational setting: adaptive tampering, noisy-leakage resilience, and improved rate. In: Advances in Cryptology—CRYPTO 2019, pp. 448–479. Springer International Publishing, Cham (2019)
13. Goldreich, O., Micali, S., Wigderson, A.: How to play any mental game. In: Proceedings of the Nineteenth Annual ACM Symposium on Theory of Computing (STOC '87), pp. 218–229. Association for Computing Machinery, New York (1987)
14. Goyal, V., Kumar, A.: Non-malleable secret sharing. In: Proceedings of the 50th Annual ACM SIGACT Symposium on Theory of Computing, STOC 2018, pp. 685–698. ACM, New York (2018)
15. Guruswami, V., Wootters, M.: Repairing Reed-Solomon codes. IEEE Trans. Inf. Theory **63**(9), 5684–5698 (2017)
16. Ito, M., Saito, A., Nishizeki, T.: Multiple assignment scheme for sharing secret. J. Cryptology **6**(1), 15–20 (1993)
17. Kumar, A., Meka, R., Sahai, A.: Leakage-resilient secret sharing against colluding parties. In: 2019 IEEE 60th Annual Symposium on Foundations of Computer Science (FOCS), pp. 636–660 (2019)
18. Rabin, T.: Robust sharing of secrets when the dealer is honest or cheating. J. ACM **41**(6), 1089–1109 (1994)
19. De Santis, A., Desmedt, Y., Frankel, Y., Yung, M.: How to share a function securely. In: STOC '94 (1994)
20. Shamir, A.: How to share a secret. Commun. ACM **22**(11), 612–613 (1979)
21. Srinivasan, A., Vasudevan, P.N.: Leakage resilient secret sharing and applications. In: Advances in Cryptology—CRYPTO 2019, pp. 480–509. Springer International Publishing, Cham (2019)
22. Tompa, M., Woll, H.: How to share a secret with cheaters. J. Cryptology **1**(2), 133–138 (1988)

Chapter 7
Approximation of Functions in Generalized Hölder Space by Means of Picard Singular Integral

Abhay Pratap Singh and Uaday Singh

Abstract In this chapter, we study the approximation properties of the Picard singular integral for the functions belonging to generalized Hölder spaces. We measure the rate of approximation in a generalized norm and obtained the estimates in terms of modulus of continuity.

Keywords Modulus of continuity · Generalized Hölder space · Rate of approximation · Picard singular integral

7.1 Introduction

For $1 \leq p < \infty$, let $\mathcal{L}^p$ be the set of all real-valued functions for which $|f|^p$ is Lebesgue integrable on $\mathbb{R}$ and $\mathcal{L}^\infty$ be the set of all uniformly continuous and bounded real-valued functions on $\mathbb{R}$. The norm of $f \in \mathcal{L}^p (1 \leq p \leq \infty)$ is defined as

$$\|f\|_p = \begin{cases} (\int_{\mathbb{R}} |f(x)|^p \, dx)^{1/p}, & 1 \leq p < \infty; \\ \sup\limits_{x \in \mathbb{R}} |f(x)|, & p = \infty. \end{cases}$$

The singular integral of Picard associated with a function $f \in \mathcal{L}^p$ is defined by

$$\mathcal{P}_r(f, x) = \frac{1}{2r} \int_{\mathbb{R}} f(x+t) e^{\frac{-|t|}{r}} \, dt,$$

for $x \in \mathbb{R}$, $0 < r \leq 1$, and $r \to 0$.

A. P. Singh (✉) · U. Singh
Department of Mathematics, IIT Roorkee, Roorkee, India
e-mail: apratapsingh@ma.iitr.ac.in; uadayfma@iitr.ac.in

A. Adhikari et al. (eds.), *Quantitative Frontiers*, Industrial and Applied Mathematics, https://doi.org/10.1007/978-981-96-9435-8_7

It can be observed that for each $0 < r \leq 1$, the Picard singular integral is a well-defined positive linear operator on $\mathcal{L}^p$ space.

For $f \in \mathcal{L}^p, 1 \leq p \leq \infty$, its s-th modulus of continuity is defined by

$$\omega_s(f, t) = \sup_{0 \leq |k| \leq t} \|\Delta_k^s f(x)\|_p, s \in \mathbb{Z}^+,$$

where $\Delta_k^s f(x) = \sum_{m=0}^{s} (-1)^{s-m} \binom{s}{m} f(x + mk)$.

Let $\mathcal{M}$ be the set of modulus of continuity type functions ω, where ω is continuous and nondecreasing and satisfies $\omega(0) = 0$ and $\frac{\omega(t)}{t}$ is nonincreasing for all positive values of t [8, pp. 93–97].

For a given $1 \leq p \leq \infty$ and $\omega \in \mathcal{M}$, generalized Hölder space $\mathcal{H}(p, \omega)$ is the class of functions $f \in \mathcal{L}^p$ for which

$$\|f\|_\omega = \sup_{h>0} \frac{\|\Delta_h f\|_p}{\omega(h)} < \infty. \tag{7.1}$$

The generalized Hölder norm on $\mathcal{H}(p, \omega)$ is defined by

$$\|f\|_{\mathcal{H}(p,\omega)} = \|f\|_p + \sup_{h>0} \frac{\|\Delta_h f\|_p}{\omega(h)}. \tag{7.2}$$

We say that $f \in \overline{\mathcal{H}}(p, \omega) \subset \mathcal{H}(p, \omega)$, if

$$\lim_{h \to 0} \frac{\|\Delta_h f\|_p}{\omega(h)} = 0. \tag{7.3}$$

$$\mathcal{H}(p, \omega, m) = \{f \in \mathcal{H}(p, \omega) \mid \mathcal{P}_r^{(n)}(f, x) \in \mathcal{H}(p, \omega), 2 \leq n \leq m\}.$$

The norm in $\overline{\mathcal{H}}(p, \omega)$ and $\mathcal{H}(p, \omega, m)$ is the same as in $\mathcal{H}(p, \omega)$.

Remark 7.1.1 If $\omega(t) = t^\alpha$, $0 < \alpha \leq 1$, then

$$\mathcal{H}(p, \omega) = \{f \in \mathcal{L}^p, 1 \leq p \leq \infty, \|\Delta_p f\|_p = O(t^\alpha)\},$$

with the norm

$$\|f\|_{\mathcal{H}(p,\omega)} = \|f\|_p + \sup_{t>0} \frac{\|\Delta_p f\|_p}{t^\alpha}.$$

Remark 7.1.2 If $\omega, \mu \in \mathcal{M}$ such that $\frac{\omega(t)}{\mu(t)}$ is nondecreasing, then

$$\mathcal{H}(p, \omega) \subset \mathcal{H}(p, \mu) \subset \mathcal{L}^p, 1 \leq p \leq \infty.$$

Throughout the chapter, we shall write

$$\Lambda_x(t) = f(x+t) - 2f(x) + f(x-t)$$

and

$$E(x,r) = \mathcal{P}_r(f,x) - f(x).$$

7.2 Known Results

The rate of approximation for functions by using singular integrals has been studied by many authors such as R. N. Mohapatra and R. S. Rodriguez [1], B. Firlejy [3], L. Rempulska and Z. Walczak [6], and S. G. Gal [4, 5]. By using uniform norm the following results are obtained by R. N Mohapatra et al. [1] and S. G. Gal [4].

Theorem 7.2.1 ([4]) *If $f \in C_{2\pi}$, the space of all continuous 2π periodic functions on R. then*

$$\|E(x,r)\|_C = O(\omega(f,r)).$$

where $\|.\|_C$ denotes the uniform norm.

Further this estimate is improved by S. G. Gal [5] in form of higher order modulus of continuity. The approximation properties for $f \in \mathcal{L}^p$ by using Picard singular integral are given in [6].

Theorem 7.2.2 ([6]) *Let $f \in \mathcal{L}^p, 1 \leq p \leq \infty$. Then*

$$\|E(x,r)\|_p \leq \frac{5}{2}\omega_2(f,r),$$

for all $r > 0$.

The study has further been extended to exponential weighted spaces. In this direction some quantitative type theorems have been studied in [7].

It can be noted that if a function belongs to Hölder continuous class, then a better approximation order can be found under Hölder norm [3]. Mainly our work is to extend this study for $\mathcal{L}^p$ spaces and obtain the rate of approximation.

7.3 Lemmas

In this section, we discuss certain useful properties in the form of lemmas.

Lemma 7.3.1 *Let $\omega, \mu \in \mathcal{M}$. Then*

$$||\Lambda_{x+h}(t) - \Lambda_x(t)||_p = O\left[\mu(h)\left(\frac{\omega(t)}{\mu(t)}\right)\right].$$

Proof Applying the generalized Minkowski's inequality, we have

$$\begin{aligned}||\Lambda_{x+h}(t) - \Lambda_x(t)||_p &\le ||f(x+h-t) - f(x+h)||_p + ||f(x+t) - f(x)||_p \\ &+ ||f(x-t) - f(x)||_p + ||f(x+h-t) - f(x+h)||_p \\ &= O(\omega(t)).\end{aligned}$$

Also

$$\begin{aligned}||\Lambda_{x+h}(t) - \Lambda_x(t)||_p &= ||f(x+h+t) - 2f(x+h) + f(x+h-t) \\ &- f(x+t) + 2f(x) - f(x-t)||_p \\ &\le ||f(x+h+t) - f(x+t)||_p + 2||f(x+h) - f(x)||_p \\ &+ ||f(x+h-t) - f(x-t)||_p \\ &= O(\omega(h)).\end{aligned}$$

If μ is positive and nondecreasing, then for $t \le h$, we obtain

$$||\Lambda_{x+h}(t) - \Lambda_x(t)||_p = O(\omega(t)) = O\left[\mu(t)\left(\frac{\omega(t)}{\mu(t)}\right)\right] = O\left[\mu(h)\left(\frac{\omega(t)}{\mu(t)}\right)\right].$$

Since $\frac{\omega(t)}{\mu(t)}$ is positive and nondecreasing, if $t \ge h$, then $\frac{\omega(t)}{\mu(t)} \ge \frac{\omega(h)}{\mu(h)}$. Therefore

$$||\Lambda_{x+h}(t) - \Lambda_x(t)||_p = O(\omega(h)) = O\left[\mu(h)\left(\frac{\omega(t)}{\mu(t)}\right)\right].$$

Lemma 7.3.2 *Let $f \in \mathcal{H}(p, \omega)$, $1 \le p \le \infty$. Then for every $r \in (0, 1]$:*

1. $||\mathcal{P}_r(f, x)||_p \le ||f||_p$.
2. $||\mathcal{P}_r(f, x)||_\omega \le ||f||_\omega$.
3. $||\mathcal{P}_r(f, x)||_{\mathcal{H}(p,\omega)} \le ||f||_{\mathcal{H}(p,\omega)}$.

Proof For $1 \le p < \infty$, we have

$$\begin{aligned}||\mathcal{P}_r(f, x)||_p &= \left|\left|\frac{1}{2r}\int_{\mathbb{R}} f(x+t)e^{\frac{-|t|}{r}}dt\right|\right|_p \\ &= \left(\int_{\mathbb{R}}\left|\frac{1}{2r}\int_{\mathbb{R}} f(x+t)e^{\frac{-|t|}{r}}dt\right|^p dx\right)^{\frac{1}{p}} \\ &\le \frac{1}{2r}\left(\int_{\mathbb{R}} e^{\frac{-|t|}{r}}dt\right)\left(\int_R |f(x+t)|^p dt\right)^{\frac{1}{p}}\end{aligned}$$

$$\le \left(\int_{\mathbb{R}} |f(u)|^p du\right)^{\frac{1}{p}}$$
$$\le ||f||_p.$$

For $p = \infty$, we have

$$\begin{aligned}
||\mathcal{P}_r(f,x)||_\infty &= \sup_{x\in\mathbb{R}} \left|\frac{1}{2r}\int_{\mathbb{R}} f(x+t)e^{\frac{-|t|}{r}}dt\right| \\
&\le \frac{1}{2r}\left(\int_{\mathbb{R}} e^{\frac{-|t|}{r}}dt\right)\left(\sup_{u\in\mathbb{R}} |f(u)|\right) \\
&\le ||f||_\infty.
\end{aligned}$$

2. $$\begin{aligned}
||\mathcal{P}_r(f,x)||_\omega &= \sup_{t>0} \frac{||\Delta_h \mathcal{P}_r(f,x)||_p}{\omega(t)} \\
&= \sup_{t>0} \frac{||\mathcal{P}_r(\Delta_h f,x)||_p}{\omega(t)} \\
&\le \sup_{t>0} \frac{||\Delta_h f||_p}{\omega(t)} \\
&\le ||f||_\omega.
\end{aligned}$$

3. $$\begin{aligned}
||\mathcal{P}_r(f,x)||_{\mathcal{H}(p,\omega)} &= ||\mathcal{P}_r(f,x)||_p + ||\mathcal{P}_r(f,x)||_\omega \\
&\le ||f||_p + ||f||_\omega \\
&\le ||f||_{\mathcal{H}(p,\omega)}.
\end{aligned}$$

The proof is ended.

Lemma 7.3.3 *Let $f \in \mathcal{H}(p,\omega,m)$. Then*

$$||\mathcal{P}_r^{(2)}(f,x)||_{\mathcal{H}(p,\omega,m)} \le \frac{1}{r^2}K||f||_{\mathcal{H}(p,\omega,m)}, \tag{7.4}$$

where K is some constant and $r \in (0,1]$.

Proof Following [2, p. 129], we have

$$\begin{aligned}
||\mathcal{P}_r^{(2)}(f,x)||_{\mathcal{H}(p,\omega,m)} &= ||\mathcal{P}_r^{(2)}(f,x)||_{\mathcal{H}(p,\omega)} \\
&= ||\mathcal{P}_r^{(2)}(f,x)||_p + ||\mathcal{P}_r^{(2)}(f,x)||_\omega \\
&\le r^2\left(||\chi^{(2)}||_{\mathcal{L}^1}||f||_p + ||\chi^{(2)}||_{\mathcal{L}^1}||f||_\omega\right),
\end{aligned}$$

where $\chi(t) = e^{\frac{-|t|}{r}}$. Let $K = ||\chi^{(2)}||_{\mathcal{L}^1}$, so we get

$$
\begin{aligned}
||\mathcal{P}_r^{(2)}(f,x)||_{\mathcal{H}(p,\omega,m)} &\leq \frac{1}{r^2}K\left(||f||_p + ||f||_\omega\right)\\
&\leq \frac{1}{r^2}K||f||_{\mathcal{H}(p,\omega,m)}.
\end{aligned}
$$

Hence the proof is completed.

7.4 Direct Approximation Theorem

In this part, we prove the following direct approximation theorem for the functions $f \in \mathcal{H}(p,\omega)$, $1 \leq p \leq \infty$, and obtain the rate of approximation.

Theorem 7.4.1 *Let ω and μ be the modulus of continuity type functions such that $\frac{\omega(t)}{\mu(t)}$ is nondecreasing and the function*

$$\gamma(t) = e^{-\frac{\epsilon t}{r^2}}\frac{\omega(t)}{\mu(t)}$$

be nonincreasing for some $0 < \epsilon < r$. Then, for $f \in \mathcal{H}(p,\omega)$, $1 \leq p \leq \infty$, we have

$$\|E(x,r)\|_{\mathcal{H}(p,\mu)} \leq \frac{5}{2}\omega_2(f,r) + \frac{1}{2}\frac{\omega(r)}{\mu(r)}\left[1 + \frac{\epsilon}{e(r-\epsilon)}\right], \tag{7.5}$$

for every $0 < r \leq 1$.

Proof We have

$$
\begin{aligned}
E(x,r) &= \mathcal{P}_r(f,x) - f(x)\\
&= \frac{1}{2r}\int_{\mathbb{R}} f(x+t)e^{\frac{-|t|}{r}}dt - \frac{1}{r}\int_0^\infty e^{\frac{-|t|}{r}} f(x)dt\\
&= \frac{1}{2r}\left(\int_{-\infty}^0 f(x+t)e^{\frac{-|t|}{r}}dt + \int_0^\infty f(x+t)e^{\frac{-|t|}{r}}dt\right)\\
&\quad + \frac{1}{r}\int_0^\infty e^{\frac{-|t|}{r}} f(x)dt\\
&= \frac{1}{2r}\int_0^\infty (f(x+t) - 2f(x) + f(x-t))e^{\frac{-|t|}{r}}dt\\
&= \frac{1}{2r}\int_0^\infty \Lambda_x(t)e^{\frac{-|t|}{r}}dt.
\end{aligned}
$$

From Theorem 7.2.2, we get

$$\|E(x,r)\|_p \leq \frac{5}{2}\omega_2(f,r). \tag{7.6}$$

Now

$$\begin{aligned} E(x+h,r)-E(x,r) &= \frac{1}{2r}\int_0^\infty \Lambda_{x+h}(t)e^{\frac{-|t|}{r}}\,dt - \frac{1}{2r}\int_0^\infty \Lambda_x(t)e^{\frac{-|t|}{r}}\,dt \\ &= \frac{1}{2r}\int_0^\infty (\Lambda_{x+h}(t)-\Lambda_x(t))e^{\frac{-|t|}{r}}\,dt. \end{aligned}$$

Therefore, we get

$$\begin{aligned} \|E(x+h,r)-E(x,r)\|_p &= \left\| \frac{1}{2r}\int_0^\infty (\Lambda_{x+h}(t)-\Lambda_x(t))e^{\frac{-|t|}{r}}\,dt \right\|_p \\ &\leq \frac{1}{2r}\int_0^\infty ||\Lambda_{x+h}(t)-\Lambda_x(t)||_p e^{\frac{-|t|}{r}}\,dt \\ &\leq \frac{1}{2r}\left(\int_0^r ||\Lambda_{x+h}(t)-\Lambda_x(t)||_p e^{\frac{-|t|}{r}}\,dt \right. \\ &\quad \left. + \int_r^\infty ||\Lambda_{x+h}(t)-\Lambda_x(t)||_p e^{\frac{-|t|}{r}}\,dt \right) \\ &\leq I_1 + I_2, \end{aligned} \tag{7.7}$$

where

$$I_1 = \frac{1}{2r}\int_0^r ||\Lambda_{x+h}(t)-\Lambda_x(t)||_p e^{\frac{-|t|}{r}}\,dt,$$

and

$$I_2 = \frac{1}{2r}\int_r^\infty ||\Lambda_{x+h}(t)-\Lambda_x(t)||_p e^{\frac{-|t|}{r}}\,dt.$$

Now, by using Lemma 7.3.1, we have

$$\begin{aligned} I_1 &= \frac{1}{2r}\int_0^r ||\Lambda_{x+h}(t)-\Lambda_x(t)||_p \, e^{\frac{-|t|}{r}}\,dt \\ &\leq \frac{1}{2r}\int_0^r \mu(h)\frac{\omega(t)}{\mu(t)}e^{\frac{-|t|}{r}}\,dt \end{aligned}$$

$$\leq \frac{\mu(h)}{2r}\frac{\omega(r)}{\mu(r)}\int_0^r e^{\frac{-|t|}{r}}dt$$

$$\leq \frac{\mu(h)}{2}\frac{\omega(r)}{\mu(r)}\left(1-\frac{1}{e}\right). \tag{7.8}$$

Similarly, we get

$$I_2 = \frac{1}{2r}\int_r^\infty ||\Lambda_{x+h}(t) - \Lambda_x(t)||_p e^{\frac{-|t|}{r}}dt$$

$$\leq \frac{1}{2r}\int_r^\infty \mu(h)\frac{\omega(t)}{\mu(t)}e^{\frac{-|t|}{r}}dt$$

$$\leq \frac{\mu(h)}{2r}\int_r^\infty \gamma(t)e^{\left(\frac{\epsilon}{r^2}-\frac{1}{r}\right)t}dt$$

$$\leq \frac{\mu(h)\gamma(r)}{2r}\int_r^\infty e^{\left(\frac{\epsilon}{r^2}-\frac{1}{r}\right)t}dt$$

$$\leq \frac{\mu(h)\gamma(r)}{2}\frac{r}{r-\epsilon}e^{\frac{\epsilon-r}{r}}, 0<\epsilon<r. \tag{7.9}$$

From (7.8) and (7.9), we get

$$I_1 + I_2 \leq \mu(h)\left[\frac{1}{2}\frac{\omega(r)}{\mu(r)}\left(1-\frac{1}{e}\right) + \frac{\gamma(r)}{2}\frac{r}{r-\epsilon}e^{\left(\frac{\epsilon-r}{r}\right)}\right]$$

$$\leq \mu(h)\left[\frac{1}{2}\frac{\omega(r)}{\mu(r)}\left(1-\frac{1}{e}\right) + \frac{1}{2e}\frac{r}{r-\epsilon}\frac{\omega(r)}{\mu(r)}\right]$$

$$\leq \frac{\mu(h)}{2}\frac{\omega(r)}{\mu(r)}\left[1+\frac{\epsilon}{e(r-\epsilon)}\right]. \tag{7.10}$$

Combining (7.7) and (7.10), we get

$$\sup_{h>0}\frac{||E(x+h,r)-E(x,r)||_p}{\mu(h)} \leq \frac{1}{2}\frac{\omega(r)}{\mu(r)}\left[1+\frac{\epsilon}{e(r-\epsilon)}\right]. \tag{7.11}$$

Finally from (7.6) and (7.11), we obtain our result

$$||E(x,r)||_{\mathcal{H}(p,\mu)} \leq \frac{5}{2}\omega_2(f,r) + \frac{1}{2}\frac{\omega(r)}{\mu(r)}\left[1+\frac{\epsilon}{e(r-\epsilon)}\right].$$

7.5 Corollaries

In this part, we prove some corollaries of Theorem 7.4.1

Corollary 7.5.1 *Let* $f \in \mathcal{H}(p, \omega)$, $1 \le p \le \infty$. *Then*

$$\|E(x,r)\|_{\mathcal{H}(p,\mu)} \le \frac{5}{2}\omega(r)\|f\|_{\mathcal{H}(p,\omega)} + \frac{1}{2}\frac{\omega(r)}{\mu(r)}\left[1 + \frac{\epsilon}{e(r-\epsilon)}\right], \tag{7.12}$$

for every $0 < r \le 1$.

Proof For $f \in \mathcal{H}(p, \omega)$,

$$\omega_2(f,t) \le \omega(f,t) \le \omega(t)\|f\|_{\mathcal{H}(p,\omega)}. \tag{7.13}$$

Using (7.5) and (7.13), we get (7.12).

Corollary 7.5.2 *Let* $f \in \overline{\mathcal{H}}(p, \omega)$, $1 \le p \le \infty$. *Then*

$$\|E(x,r)\|_{\mathcal{H}(p,\mu)} = O(\omega(r)) + \frac{1}{2}\frac{\omega(r)}{\mu(r)}\left[1 + \frac{\epsilon}{e(r-\epsilon)}\right], \tag{7.14}$$

for every $0 < r \le 1$.

Proof For $f \in \overline{\mathcal{H}}(p, \omega)$,

$$\omega_2(f,t) \le \omega(f,t) = o(\omega(t)). \tag{7.15}$$

Using (7.5) and (7.15), we get (7.14).

Corollary 7.5.3 *Let* $\omega(t) = t^\alpha$ *and* $\mu(t) = t^\beta$, $0 < \beta \le \alpha \le 1$, *for* $t > 0$. $f \in \mathcal{H}(p, \omega)$. *Then*

$$\|E(x,r)\|_{\mathcal{H}(p,\mu)} = O(r^{\alpha-\beta}).$$

Proof By putting the value of ω and μ in (7.12), we can easily get

$$\|E(x,r)\|_{\mathcal{H}(p,\mu)} \le cr^{\alpha-\beta},$$

where

$$c = \frac{1}{2}\max\left(5\|f\|_{\mathcal{H}(p,\omega)}, 1 + \frac{\epsilon}{e(r-\epsilon)}\right).$$

It is easy to observe that this result is also true if $f \in \overline{\mathcal{H}}(p, \omega, m)$, $1 \le p \le \infty$.

7.6 Inverse Approximation Theorem

In this part, we prove the result that is related to inverse approximation by means of Picard singular integral $\mathcal{P}_r(f, x)$.

Theorem 7.6.1 *Let $f \in \mathcal{H}(p, \omega, m)$, $1 \leq p \leq \infty$, such that*

$$||E(x, r)||_{\mathcal{H}(p,\omega,m)} \leq \omega(r), \quad r \to 0. \tag{7.16}$$

Then

$$\omega_2(f, \delta) \leq \frac{K\delta^2 + 12}{log2} \int_0^{\frac{1}{4}} \frac{\omega(x)}{x^3} dx,$$

where K is some constant and $\omega \in \mathcal{M}$.

Proof We write

$$S_n(x) = \begin{cases} \mathcal{P}_{\frac{1}{2^n}}(f, x), & n = 1, 2; \\ \mathcal{P}_{\frac{1}{2^n}}(f, x) - \mathcal{P}_{\frac{1}{2^{n-1}}}(f, x), & n = 3, 4, \ldots \end{cases}$$

Now we find

$$||S_n||_{\mathcal{H}(p,\omega,m)} \leq \left\| \mathcal{P}_{\frac{1}{2^n}}(f, x) - f(x) \right\|_{\mathcal{H}(p,\omega,m)} + \left\| \mathcal{P}_{\frac{1}{2^{n-1}}}(f, x) - f(x) \right\|_{\mathcal{H}(p,\omega,m)}.$$

Using (7.16), we get

$$||S_n||_{\mathcal{H}(p,\omega,m)} \leq \omega\left(\frac{1}{2^n}\right) + \omega\left(\frac{1}{2^{n-1}}\right) = 3\omega\left(\frac{1}{2^n}\right). \tag{7.17}$$

It is easy to observe from Theorem 7.4.1 that

$$\lim_{n\to\infty} \left\| f - \sum_{k=2}^{n} S_k \right\|_{\mathcal{H}(p,\omega,m)} = 0.$$

Now

$$\begin{aligned} &||f(x + h) + f(x - h) - 2f(x)||_{\mathcal{H}(p,\omega,m)} \\ &\leq \sum_{k=2}^{l} ||S_k(x + h) + S_k(x - h) - 2S_k(x)||_{\mathcal{H}(p,\omega,m)} \\ &\quad + 4 \sum_{k=l+1}^{\infty} ||S_k||_{\mathcal{H}(p,\omega,m)}. \end{aligned} \tag{7.18}$$

Now

$$S_k(x+h)+S_k(x-h)-2S_k(x)=\int_{-\frac{h}{2}}^{\frac{h}{2}}\int_{-\frac{h}{2}}^{\frac{h}{2}} S_k^{(2)}(x+t_1+t_2)dt_1dt_2.$$

By the Hölder-Minkowski inequality

$$||S_k(x+h)+S_k(x-h)-2S_k(x)||_{\mathcal{H}(p,\omega,m)}\leq h^2||S_k^{(2)}||_{\mathcal{H}(p,\omega,m)}. \tag{7.19}$$

Also

$$\begin{aligned} S_k(x)=&\mathcal{P}_{\frac{1}{2^k}}\left(f(x)-\mathcal{P}_{\frac{1}{2^{k-1}}}(f,x),x\right)\\ &-\mathcal{P}_{\frac{1}{2^{k-1}}}\left(f-\mathcal{P}_{\frac{1}{2^k}}(f,x),x\right),\ k=3,4,... \end{aligned} \tag{7.20}$$

From (7.20), we get

$$S_k^{(2)}(x)=\mathcal{P}^{(2)}_{\frac{1}{2^k}}\left(f(x)-\mathcal{P}_{\frac{1}{2^{k-1}}}(f,x),x\right)-\mathcal{P}^{(2)}_{\frac{1}{2^{k-1}}}\left(f-\mathcal{P}_{\frac{1}{2^k}}(f,x),x\right).$$

Now by taking $||.||_{\mathcal{H}(p,\omega,m)}$ on both sides, we get

$$\begin{aligned} ||S_k^{(2)}(x)||_{\mathcal{H}(p,\omega,m)}=&\left\|\mathcal{P}^{(2)}_{\frac{1}{2^k}}\left(f(x)-\mathcal{P}_{\frac{1}{2^{k-1}}}(f,x),x\right)\right\|_{\mathcal{H}(p,\omega,m)}\\ &+\left\|\mathcal{P}^{(2)}_{\frac{1}{2^{k-1}}}\left(f-\mathcal{P}_{\frac{1}{2^k}}(f,x),x\right)\right\|_{\mathcal{H}(p,\omega,m)}. \end{aligned}$$

Now by Lemma 7.3.3 and (7.16), we get

$$\begin{aligned} ||S_k^{(2)}(x)||_{\mathcal{H}(p,\omega,m)}\leq& K2^{2k}\left\|f(x)-\mathcal{P}_{\frac{1}{2^{k-1}}}(f,x)\right\|_{\mathcal{H}(p,\omega,m)}\\ &+K2^{2k-2}\left\|f-\mathcal{P}_{\frac{1}{2^k}}(f,x)\right\|_{\mathcal{H}(p,\omega,m)}\\ \leq& K\left(2^{2k}\omega\left(\frac{1}{2^{k-1}}\right)+2^{2k-2}\omega\left(\frac{1}{2^k}\right)\right)\\ \leq& K2^{2k}\omega\left(\frac{1}{2^k}\right). \end{aligned} \tag{7.21}$$

Finally, using (7.17)–(7.21), we get

$$\begin{aligned}
\omega_2(f,\delta) &\le K\delta^2 \sum_{k=2}^{l} 2^{2k}\omega\left(\frac{1}{2^k}\right) + 12 \sum_{k=l+1}^{\infty} \omega\left(\frac{1}{2^k}\right) \\
&\le K\delta^2 \sum_{k=2}^{\infty} 2^{2k}\omega\left(\frac{1}{2^k}\right) + 12 \sum_{k=2}^{\infty} 2^{2k}\omega\left(\frac{1}{2^k}\right) \\
&\le \frac{K\delta^2 + 12}{log2} \int_0^{\frac{1}{4}} \frac{\omega(x)}{x^3} dx.
\end{aligned}$$

This completes the proof.

7.7 Conclusion

In the literature, many results have been obtained on the rate of approximation of functions in the Hölder space of continuous functions by using the Picard singular integral. We extend this study for Lebesgue space with a generalized norm. We obtain the rate of approximation for the functions belonging to Lebesgue space by means of Picard singular integral and also proved the inverse approximation theorem. One can try to study the problem by using the Poisson-Cauchy singular integral [1, p. 117].

References

1. Mohapatra, R.N., Rodriguez, R.S.: On the rate of convergence of singular integrals for Hölder continuous functions. Math. Nachr. **149**(1), 117–124 (1990)
2. Butzer, P.L., Nessel, R.J.: Fourier Analysis and Approximation, vol. 1, Academic Press, Basel (1971)
3. Firlejy, B., Rempulska, L.: On some singular integrals in Hölder spaces. Math. Nachr. **170**(1), 93–100 (1994)
4. Gal, S.G.: Remark on the degree of approximation of continuous functions by singular integrals. Math. Nachr. **164**(1), 197–199 (1993)
5. Gal, S.G.: Degree of approximation of continuous functions by some singular integrals. Revue d'Analyse Numërique et de Théorie de l' approximation **27**(2), 251–261 (1998)
6. Rempulska, L., Walczak, Z.: On modified Picard and Gauss-Weierstrass singular integrals. Ukrain. Math. J. **57**(11), 1844–1852 (2005)
7. Lesniewicz, A., Rempulska, L., Wasiak, J.: Approximation properties of the Picard singular integral in exponential weighted spaces. Matematiques. **40**(2), 233–242 (1996)
8. Timan, A.F.: Theory of Approximation of Functions of Real Variable, Moscow (1960)

Chapter 8
Numerical Analysis on the Transport Process of Heat and Mass on MHD Nanofluid Flow Including Chemical Reaction and the Slip Effects at the Boundary of a Stretching Surface

Krishnandan Verma

Abstract A numerical inspection using Matlab solver bvp4c has been done to analyse the action of heat and mass transmission on hydromagnetic nanofluid flow on a surface that is stretching exponentially. Important effects like chemical reaction, viscous dissipation, radiation, and slip at the surface are added in the current work. The consequences of the important effects on the fluid motion, temperature, and concentration are displayed by graphs while numerical values are obtained through simulation for coefficient of skin-friction, Nusselt number, and Sherwood number to gain deep insight into the thermal and mass transmission activity as well as shear stress at the boundary of the surface. The investigation reveals that the action of chemical reaction leads to enhancement of mass diffusion that is responsible for the reduction of the volume fraction of nanoparticles near the surface. The action of thermophoresis, Brownian motion, and viscous dissipation reduce the rate of heat transmission, while thermophoresis enhances the mass transmission activity. The precision of the current investigation is checked by comparing our result with similar published work.

Keywords MHD · Exponentially stretching · Slip · Chemical reaction · bvp4c

8.1 Introduction

Thermal transport rate can be productively improved, if by some means the fluid's thermal conductivity can be augmented. Numerous techniques have been tested to raise the rate of heat transfer of fluids. Since solids have higher thermal conductivity than liquids, so researchers endeavour to boost the thermal efficiency

K. Verma (✉)
Department of Mathematics, Maryam Ajmal Women's College of Science and Technology, Hojai, India

A. Adhikari et al. (eds.), *Quantitative Frontiers*, Industrial and Applied Mathematics, https://doi.org/10.1007/978-981-96-9435-8_8

of base fluids by adding small or large sized solid materials. In comparison to pure liquids, nanofluids have added new dimension in the study of heat transfer as it can effectively enhance heat transfer rate of fluids. Due to the enhanced thermal properties of nanofluids, they are used as coolants in equipment's like heat exchangers, radiators, electronic cooling systems, etc., and different branch of engineering areas like continuous cooling process of fibre, condensation of liquid films, metallurgical process, production of plastic and rubber sheets, and solar collectors based on nanofluids, etc.

Over the years, a great number of researchers have reported many important works on the transfer of heat as well as mass past a stretching sheet. Saikiadis [1, 2] did the pioneering work on stretching sheet by developing a suitable similarity transformation of the flow field and obtained the numerical solution. Crane [3] studied fluid flow through a stretching surface with variable velocity. After Crane many notable researchers extended his work under various physical situations like Pavlov [4], Gupta and Gupta [5] and Chen and Char [6]. Banks [7] investigated both analytically and numerically stretching sheet problem with power law velocity variation and obtained similarity solution. Ali [8] examined the consequences of suction and injection on laminar, incompressible fluid flow through an elongating sheet with power law velocity variation and obtained a non-unique similarity solution. He found that the non-dimensional shear stress at the boundary surface increases for the increasing values of suction parameter for positive values of velocity exponent parameter. Salleh et al. [9] examined the thermal transmission on steady flow over an extending surface and observed that the conjugate parameters have notable effects on heat flux close to the sheet. Bachok et al. [10] investigated theoretically heat transmission on an unsteady nanofluid flow on a permeable extending/shrinking surface and obtained dual solutions for both extending and shrinking sheet. Bhattacharyya et al. [11] analysed numerically convective transfer of heat with thermal diffusion past a shrinking surface and obtained dual solutions. Devi et al. [12] investigated numerically heat transfer on MHD flow through an exponential extending sheet and found that heat transfer is enhanced for large values of magnetic field and Prandtl number. Zaimi et al. [13] examined numerically nanofluid flow through an extending/shrinking material having permeable surface by adopting two-phase model and found twin solutions for both extending and shrinking cases. Sharma and Aich [14] investigated numerically the actions of thermal diffusion through a non-isothermal stretching surface on steady MHD flow under the impact of heat generation parameter in porous medium. They found that Soret effect helps in restricting the temperature while Dufour effect decreases the fluid concentration near the surface of the sheet. Ali et al. [15] investigated analytically past an oscillating stretching sheet with heat transfer on MHD flow. They found that the rate of thermal transmission is enhanced for larger values of Prandtl number and smaller couple stress parameter. Babu and Sandeep [16] studied numerically MHD flow with nanoparticles over a sparse extending sheet with variable thickness in three-dimensional system considering the effects of slip and thermophoresis. Alavi et al. [17] examined the phenomenon of thermal transmission past a sheet near stagnation point, which is exponentially stretching, on hydromagnetic flow of a radiating fluid mixture. Mabood et al. [18] inquired into the consequences of thermal radiation on thermal and solutal transmission past a stretching/shrinking surface with second-order slip near stagnation point. Anuar et al. [19] studied numerically the impact of carbon nanotubes on an elongating/shrinking surface with the transfer of heat as well as mass in a homogeneous-heterogeneous reaction and also made the stability analysis. They found that a single solution exists for exponentially stretching sheet while the exponentially shrinking sheet has dual solutions where the first one is stable and the second one is unstable. Berrehal et al. [20] investigated the process of enhancement of heat transfer under the impacts of globular and non-globular shapes of silver nanoparticles on MHD water base nanofluid flow over a heated elongating sheet. Other important works are reported by Fenuga et al. [21], Tinker et al. [22], Govardhan et al. [23], Verma and Sharma [24], and Verma [25, 26].

The above literature reviews inspired us to study the heat and mass transmission characteristics of nanofluid flow past a vertical sheet that is elongating in an exponential manner. Many important effects like thermal radiation, chemical reaction, thermophoresis, and viscous dissipation are added to study the important changes due to these effects in the flow, thermal, and mass transmission characteristics. No slip condition is a universal condition but certain literature that includes the works of Haile and Awgichew [27] inspired us to investigate the flow characteristics incorporating the partial slip effects. Although slip can be observed at molecular level and it occurs at liquid-liquid interface in a very thin layer of liquid molecules that is attached to the solid boundary [28]. With the advancement of technology, the utilization of no-slip condition to some extent is being restricted. From the investigations of Thompson and Trojan [29], Barrat and Bocquet [30], Voronov et al. [31], Bocquet and Barrat [32], and Huang and Szlufarska [33], it is found that molecular dynamics simulation and linear response theory forms the base of theoretical studies of slip in liquid. Bocquet and Charlaix [34] modelled slip at solid-liquid interface as kinetic rate process and various research from molecular simulation reveals that at a very low shear stress, Navier slip condition appears which is highly dependent on the wettability of the interface. Investigations have also revealed that those interfaces which have excess wettability tends to produce slip whose length is of few nanometers or less while low wettability produces slip lengths in tens of nanometers [34]. In the current work, partial slip is considered at the solid-liquid interface and the changes in flow phenomenon are analysed numerically using Matlab solver bvp4c.

8.2 Construction of the Problem

Consider 2-D flow of nanofluid through a vertical flat permeable plate as shown in Fig. 8.1. The x-axis is considered along the plate while y-axis is perpendicular to it. The fluid is assumed to be incompressible and the flow to be steady. The surface is assumed to be stretching in an exponential manner. Let surface temperature of the

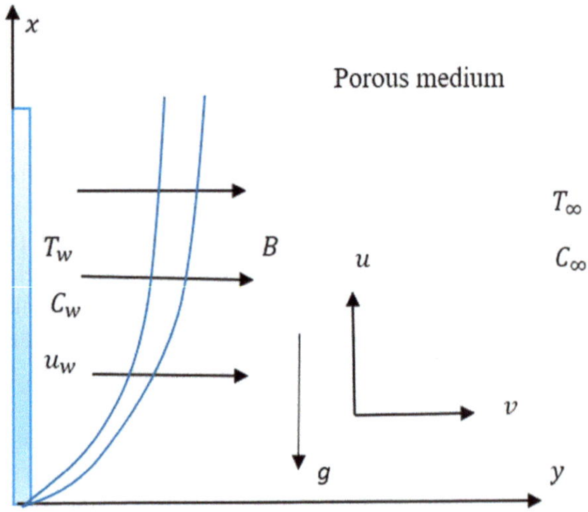

Fig. 8.1 Geometric model

plate be T_w and nanoparticles volume fraction at the surface be C_w. In the direction perpendicular to the plate, a varying magnetic field $B(x)$ is applied. The induced magnetic field possesses comparatively less strength than the applied magnetic field and so it is neglected. The acceleration due to gravity g is acting downwards in the direction parallel to the vertical plate as shown in Fig. 8.1. Further it assumed that there exists thermal equilibrium between the nanoparticles and the fluid. In accordance with the work of Haile and Awgichew [27], we have also incorporated the slip effects of velocity, thermal, and nanoparticles concentration at the solid boundary.

The mathematical equations in accordance with Haile and Awgichew [27] are

$$\frac{\partial u}{\partial x}+\frac{\partial v}{\partial y}=0, \tag{8.1}$$

$$u\frac{\partial u}{\partial x}+v\frac{\partial u}{\partial y}=\tilde{\upsilon}\frac{\partial^2 u}{\partial y^2}-\frac{\sigma B^2 u}{\rho}-\frac{\upsilon u}{K}+g\beta_T\left(T-T_\infty\right)+g\beta_C\left(C-C_\infty\right)-c^*\epsilon^2 u^2, \tag{8.2}$$

$$u\frac{\partial T}{\partial x}+v\frac{\partial T}{\partial y}=\alpha\frac{\partial^2 T}{\partial y^2}-\frac{1}{\left(\rho C_p\right)_f}\frac{\partial q_r}{\partial y}+\frac{\tilde{\upsilon}}{C_p}\left(\frac{\partial u}{\partial y}\right)^2+\tau\left(D_B\frac{\partial C}{\partial y}\frac{\partial T}{\partial y}+\frac{D_T}{T_\infty}\left(\frac{\partial T}{\partial y}\right)^2\right), \tag{8.3}$$

$$u\frac{\partial C}{\partial x}+v\frac{\partial C}{\partial y}=D_B\frac{\partial^2 C}{\partial y^2}+\frac{D_T}{T_\infty}\frac{\partial^2 T}{\partial y^2}-K_c\left(C-C_\infty\right), \tag{8.4}$$

where $\tau = \frac{(\rho C_p)_p}{(\rho C_p)_f}$, $B = B_0 e^{\left(\frac{x}{2L}\right)}$, $\alpha = \frac{k}{(\rho C_p)_f}$ and $K = K_0 e^{\left(-\frac{x}{L}\right)}$ represents ratio of nanoparticles heat capacity to base fluid heat capacity, varying magnetic field, thermal diffusivity and permeability of the porous medium, respectively. $\upsilon, \tilde{\upsilon}, c^*\epsilon^2, \beta_T, \beta_C, K_c$ represents kinematic viscosity, reference kinematic viscosity, inertia parameter, thermal expansion coefficient, concentration expansion coefficient, rate of reaction, respectively.

The restrictions constituted on the boundaries are

$$u = u_w + N\upsilon\frac{\partial u}{\partial y}, v = V_w(x), T = T_w + D_t\frac{\partial T}{\partial y}, C = C_w + D_m\frac{\partial C}{\partial y} \quad \text{at } y = 0, \tag{8.5}$$

$$u \to 0, T \to T_\infty, C \to C_\infty \quad \text{at } y \to \infty. \tag{8.6}$$

where

$$\begin{aligned} N = N_\upsilon^* e^{\left(\frac{-x}{2L}\right)}, D_t = D_1 e^{\left(\frac{-x}{2L}\right)}, D_m = D_2 e^{\left(\frac{-x}{2L}\right)}, u_w = u_0 e^{\left(\frac{x}{L}\right)}, \\ V_w(x) = -V_0 e^{\left(\frac{x}{2L}\right)}, C_w = C_\infty + C_0 e^{\left(\frac{x}{2L}\right)}, T_w = T_\infty + T_0 e^{\left(\frac{x}{2L}\right)}. \end{aligned} \tag{8.7}$$

Here N_υ^*, D_1, D_2, L, u_0, T_0, C_0, and V_0 are constants that represents slip factor of velocity, slip factor of temperature, slip factor of volume fraction of nanoparticles, reference length, reference velocity, reference temperature, reference concentration, and mass flux velocity, respectively.

Radiative heat flux,

$$q_r = -\frac{4\sigma^*}{3k^*} \cdot \frac{\partial T^4}{\partial y}, \tag{8.8}$$

where $T^4 \cong 4T_\infty^3 T - 3T_\infty^3$.

The following similarity transformations are introduced,

$$\begin{aligned} \eta = y\sqrt{\frac{u_0}{2\upsilon L}} e^{\left(\frac{x}{2L}\right)}, \psi = \sqrt{2\upsilon u_0 L}\, e^{\left(\frac{x}{2L}\right)} f(\eta), T = T_\infty + (T_w - T_\infty)\theta(\eta), \\ C = C_\infty + (C_w - C_\infty)\phi(\eta), \end{aligned} \tag{8.9}$$

The stream function ψ given by $u = \frac{\partial \psi}{\partial y}$ and $v = -\frac{\partial \psi}{\partial x}$ satisfies Eq. (8.1) while Eqs. (8.2)–(8.4) are transformed as:

$$V_t f''' + ff'' - 2(I_n + 1) f'^2 - 2Rf' + 2\lambda_1\theta + 2\lambda_2\phi = 0, \tag{8.10}$$

$$(1 + \text{Rd})\theta'' + \Pr\left(f\theta' - 4f'\theta + V_t Ec f''^2 + \text{Nb}\theta'\phi' + \text{Nt}\theta'^2\right) = 0, \tag{8.11}$$

$$\phi'' + \frac{\text{Nt}}{\text{Nb}}\theta'' + \text{Sc} f\phi' - \text{Sc}K_r\phi = 0. \tag{8.12}$$

Equations (8.5) and (8.6) transform to

$$f = S_0,\ f' = 1 + m_0 f'', \theta = 1 + t_0\theta', \phi = 1 + c_0\phi' \ \text{at}\ \eta = 0, \tag{8.13}$$

$$f' \to 0, \theta \to 0, \phi \to 0 \ \text{as}\ \eta \to \infty. \tag{8.14}$$

where

$$S_0 = -V_0\sqrt{\frac{2L}{\upsilon u_0}}, \mathrm{Pr} = \frac{\upsilon}{\alpha}, m_0 = N_\upsilon^*\sqrt{\frac{\upsilon u_0}{2L}}, Sc = \frac{\upsilon}{D_B}, t_0 = D_1\sqrt{\frac{u_0}{2\upsilon L}},$$

$$\mathrm{Nb} = \frac{\tau D_B\left(C_w - C_\infty\right)}{\upsilon}, c_0 = D_2\sqrt{\frac{u_0}{2\upsilon L}}, Ec = \frac{u_w^2}{C_p\left(T_w - T_\infty\right)},$$

$$\mathrm{Nt} = \frac{\tau D_T\left(T_w - T_\infty\right)}{\upsilon T_\infty},$$

$$I_n = c^*\epsilon^2 L, V_t = \frac{\tilde{\upsilon}}{\upsilon}, \lambda_1 = \frac{G_r}{\left(Re_x\right)^2} = \frac{g\beta_T L\left(T_w - T_\infty\right)}{u_w^2},$$

$$\lambda_2 = \frac{G_M}{\left(Re_x\right)^2} = \frac{g\beta_c L\left(C_w - C_\infty\right)}{u_w^2},$$

$$G_r = \frac{g\beta_T\left(T_w - T_\infty\right)}{\upsilon^2 L^3}, \mathrm{Rd} = \frac{16\sigma^* T_\infty^3}{3kk^*}, R = \frac{\upsilon L}{K_0 u_0} + \frac{\sigma B_0^2 L}{\rho u_0} \ \text{and}\ K_r = \frac{2LK_c}{u_0 e^{\frac{x}{L}}} \tag{8.15}$$

represents suction/injection parameter, Prandtl number, velocity slip parameter, Schmidt number, thermal slip parameter, Brownian motion parameter, concentration slip parameter, Eckert number, thermophoresis parameter, non-dimensional inertia parameter, viscosity ratio parameter, thermal convective parameter, mass convective parameter, Grashof number, radiation parameter, combine porous and magnetic parameter, and chemical reaction parameter, respectively.

In the problem coefficient of skin-friction, Nusselt number and Sherwood number are proportional to $-f''(0)$, $-\theta'(0)$ and $-\phi'(0)$, respectively.

8.3 Results and Discussion

The solution of the present problem has been procured numerically by using bvp4c, a solver in Matlab. The important parameters of the problem are given the following values: $V_t = 1, R = 0.5,\ \lambda_1 = 0.2, \lambda_2 = 0.01, \mathrm{Rd} = 1, \mathrm{Ec} = 0.2, \mathrm{Nb} = 0.1, \mathrm{Nt} = 0.1,$ $\mathrm{Sc} = 0.1,\ S_0 = 0.1,\ t_0 = 0.1,\ c_0 = 0.1,\ m_0 = 0.1,\ \mathrm{Pr} = 6.785$ and $K_r = 0.5$.

Table 8.1 Comparison of the current result with Haile and Awgichew [27] for some chosen values of Prandtl number of $-\phi'(0)$ when $\text{Nb} = \text{Nt} = S_0 = I_n = m_0 = t_0 = c_0 = 0.1$, $V_t = 1$, $\lambda_1 = 0.1$, $\lambda_2 = 0.2$, $R_d = 1$, $R = 0.5$, $\text{Ec} = 0.2$, and $K_r = 0$

Pr	Haile and Awgichew [27] $-\phi'(0)$	Present result $-\phi'(0)$
4	1.2391	1.2599
6.785	1.9193	1.9289
8	2.1723	2.0929
10	2.5508	2.5252

Table 8.2 Computed values of $-f''(0)$, $-\theta'(0)$, and $-\phi(0)$

Nt	Nb	K_r	E_c	$-f''(0)$	$-\theta'(0)$	$-\phi(0)$
0.1	0.1	0.5	0.2	1.4834	2.1937	1.5725
0.2				1.5159	2.1097	3.3638
0.3				1.5473	2.0397	5.0494
0.4				1.5769	1.9778	6.6414
0.1	0.1			1.4865	2.2152	1.4854
	0.2			1.4584	2.2033	0.5352
	0.5			1.4379	2.1066	0.0452
	1			1.4252	1.9373	0.0397
		0.1		1.4803	2.1741	1.8483
		0.2		1.4804	2.1804	1.7658
		0.5		1.4806	2.1918	1.5827
		1		1.4808	2.2029	1.3678
			0.1	1.4793	2.3157	1.4761
			0.2	1.4748	2.2212	1.3836
			0.3	1.4704	2.1279	1.2924
			0.4	1.4660	2.0359	1.2023

The precision of the current solution is verified by comparing our result with Haile and Awgichew [27] in Table 8.1. In order to analyse dimensionless shear stress on the boundary of the sheet, transmission of heat and mass phenomenon, the values of $-f''(0)$, $-\ \theta'(0)$ and $-\ \phi'(0)$ are computed for important parameters and it is represented in Table 8.2.

Thermophoresis is an important phenomenon that are used to separate tiny particles from their respective solvent in industries by a process called thermal field flow fractioning. Figure 8.2 illustrates the change in temperature distribution due to the influence of thermophoresis parameter. It is noted that the increase in thermophoresis parameter causes hike in fluid temperature. From (8.15) it is perceived that the thermophoresis parameter has direct dependence on the temperature gradient, i.e. $(T_w - T_\infty)$. So an increase in thermophoresis parameter increases the temperature gradient which accounts for the hike in fluid temperature. In Fig. 8.3, it is noticed that the nanoparticles volume fraction enhances with the increase in thermophoresis parameter. It is also noted that the $\phi(\eta)$ rises exponentially and reaches its peak value at $\eta \approx 0.5$ after which it declines. The enhancement in

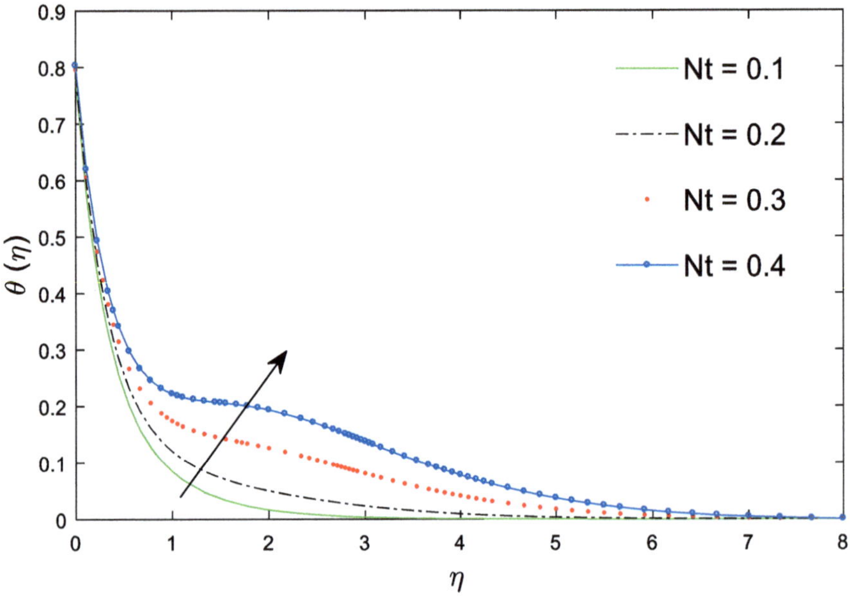

Fig. 8.2 Variations due to Nt on $\theta(\eta)$

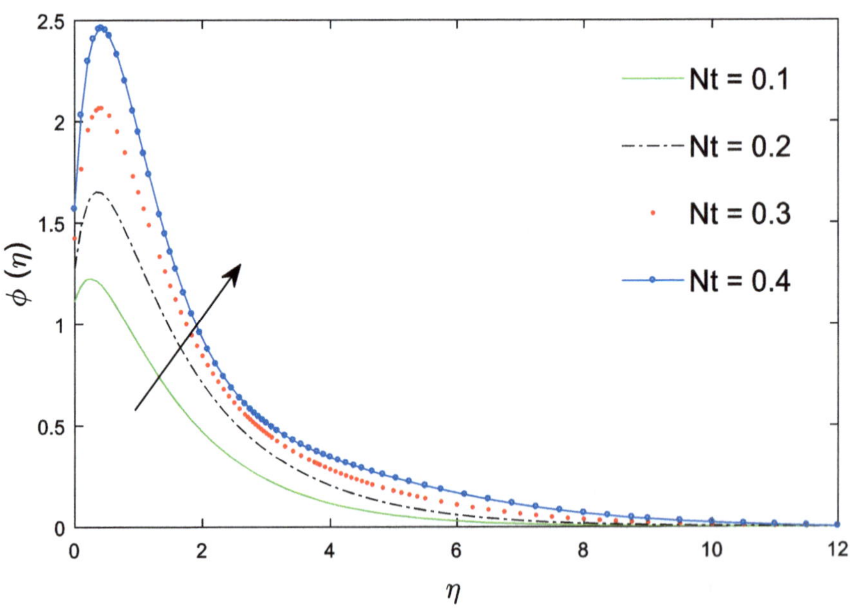

Fig. 8.3 Variations due to Nt on $\phi(\eta)$

$\phi(\eta)$ due to an increase in thermophoresis parameter also allows thickening of concentration boundary layer.

Brownian motion is an important phenomenon as it helps to differentiate a true solution from a colloid. The consequence of this motion on $\theta(\eta)$ and $\phi(\eta)$ are illustrated in Figs. 8.4 and 8.5 where we can see that the consequences of Brownian motion parameter on $\theta(\eta)$ and $\phi(\eta)$ are opposite of each other. The temperature of the fluid enhances with the rise in Brownian motion parameter while concentration

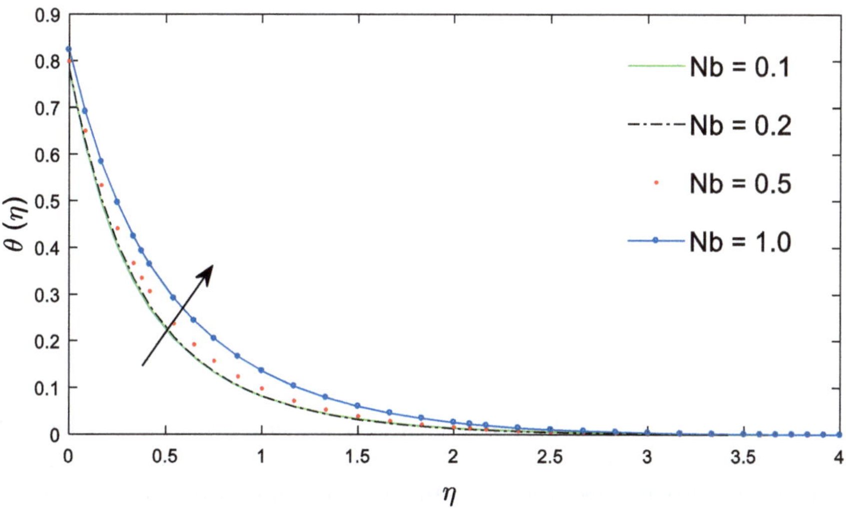

Fig. 8.4 Variations due to Nb on $\theta(\eta)$

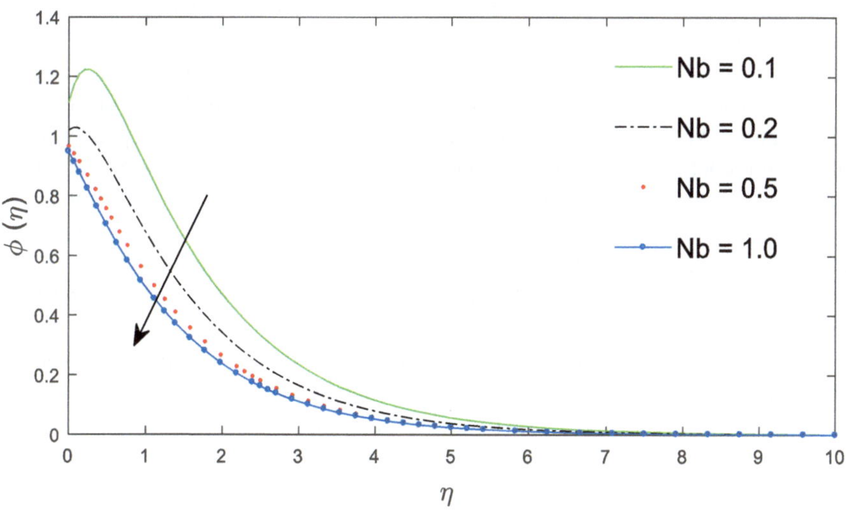

Fig. 8.5 Variations due to Nb on $\phi(\eta)$

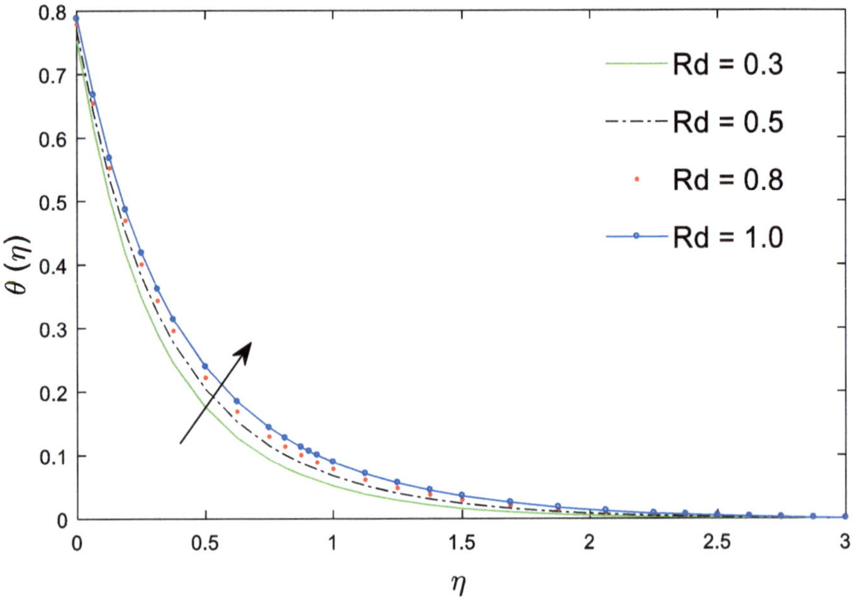

Fig. 8.6 Variations due to Rd on $\theta(\eta)$

of nanoparticles declines. Brownian motion is the unsystematically motion of particles suspended in the nanofluid. The rise in temperature due to N_b implies the increase in collision of the randomly suspended particles. Also, from Eq. (8.15) it is seen that Brownian motion parameter is directly proportional to concentration gradient, i.e. $(C_w - C_\infty)$. So, an increase in N_b increases the concentration which in turn enables mass diffusion away from the plate that causes decline in nanoparticles concentration.

The changes due to radiation on nanofluid temperature are demonstrated in Fig. 8.6. The result reveals that radiation parameter helps in the growth of fluid temperature. Mathematically, radiation parameter shows indirect dependence on absorption coefficient $k*$. So, it can be interpreted that hike in radiation parameter lowers the heat absorption effect which helps in the growth of temperature. In Fig. 8.7, it is seen that rise in Eckert number causes the enhancement of the fluid temperature. Eckert number depicts the consequence of viscous dissipation, and so it can be said that this important effect has the tendency to enhance the fluid temperature.

The changes due to chemical reaction parameter on nanoparticles volume fraction is shown in Fig. 8.8. It is spotted that nanoparticles volume fraction declines with the rise in chemical reaction parameter. The effect of chemical reaction boosts the mass diffusivity of the nanofluid, and the decrease in fluid concentration informs the diffusion of nanoparticles takes place away from the sheet.

From Fig. 8.9 it is spotted that fluid velocity decelerate with the hike in velocity slip parameter. Also Fig. 8.10 reveals that an increase in thermal slip parameter

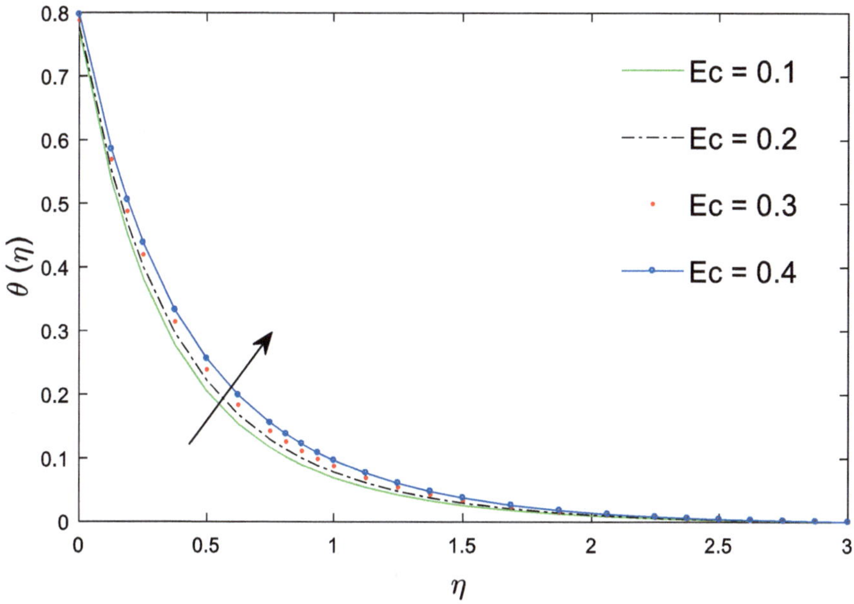

Fig. 8.7 Variations due to Ec on $\theta(\eta)$

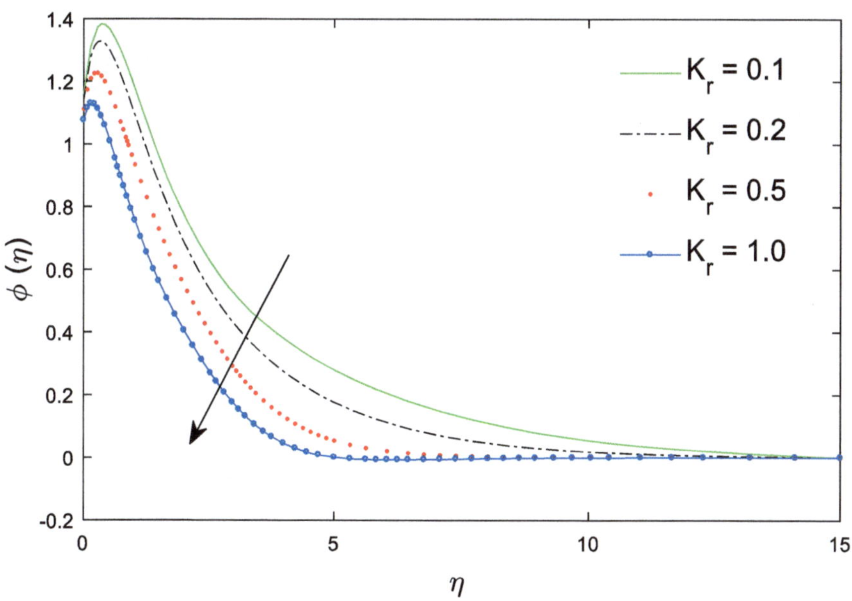

Fig. 8.8 Variations due to K_r on $\phi(\eta)$

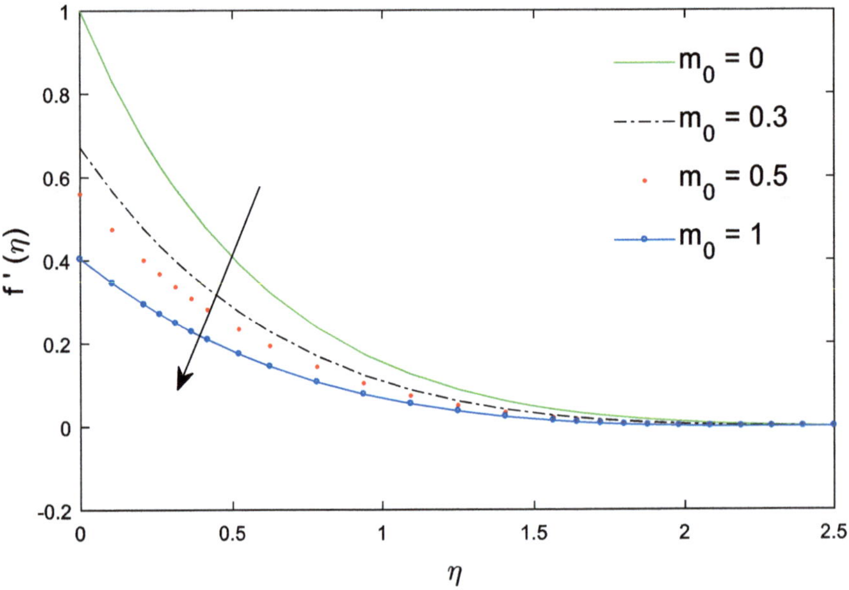

Fig. 8.9 Variations due to m_0 on $f'(\eta)$

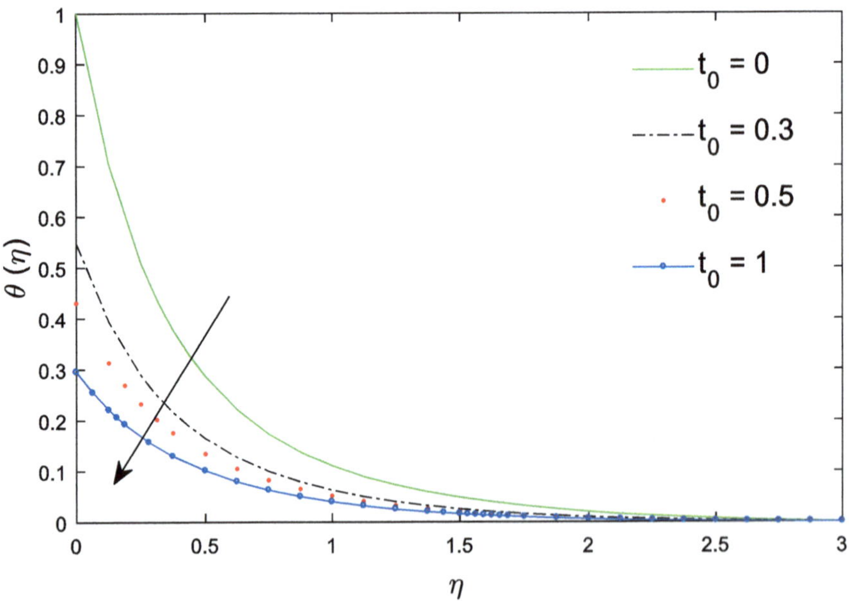

Fig. 8.10 Variations due to t_0 on $\theta(\eta)$

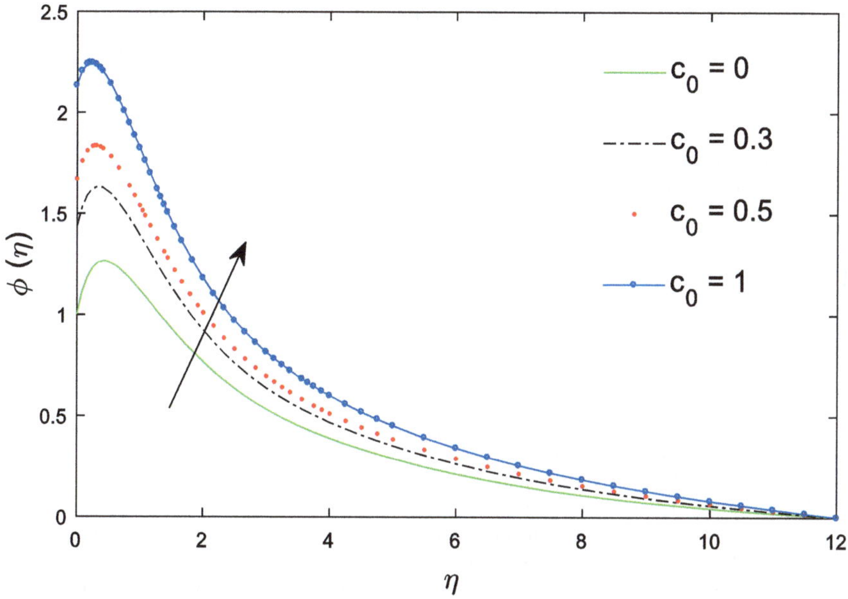

Fig. 8.11 Variations due to c_0 on $\phi(\eta)$

reduces the nanofluid temperature considerably. In Fig. 8.11, it is perceived that nanoparticles volume fraction enhances with the rise in concentration slip parameter.

8.4 Conclusions

The key outcomes from the present investigation are:

1. Velocity slip parameter decelerates the fluid flow.
2. The fluid temperature hikes with the rise in Nt, Nb, Ec, and Rd while temperature falls with the increase in thermal slip parameter.
3. Growth of nanoparticles volume fraction take place with the rise in thermophoresis parameter and concentration slip parameter while Brownian motion parameter and chemical reaction parameter decreases it.
4. Local skin-friction coefficient decreases due to Brownian motion parameter and Eckert number while it enhances with the rise in thermophoresis parameter and chemical reaction parameter.
5. The rate of heat transfer decreases due to thermophoresis parameter, Brownian motion parameter, and Eckert number while chemical reaction parameter boosts its growth.

6. The rate of mass transfer decreases with the rise in Eckert number, Brownian motion parameter, and chemical reaction parameter while thermophoresis parameter enhances it.

Acknowledgement I am extremely thankful to the reviewers for their important suggestions which helped to enhance the quality of this work.

References

1. Saikiadis, B.C.: Boundary layer behaviour on continuous solid surfaces: I, Boundary layer equations for two-dimensional and axisymmetric flows. AIChE J. **7**, 26–28 (1961)
2. Saikiadis, B.C.: Boundary layer behaviour on continuous solid surfaces: II, Boundary layer on continuous flat surface. AIChE J. **7**, 221–225 (1961)
3. Crane, L.: Flow past a stretching plate. Z. Angew. Math. Phys. **21**, 645–647 (1970)
4. Pavlov, K.B.: Magnetohydrodynamic flow of an incompressible viscous fluid caused by the deformation of a plane surface. Magn. Gidrodin. **4**, 146–147 (1974)
5. Gupta, P.S., Gupta, A.S.: Heat and mass transfer on a stretching sheet with suction or blowing. Can. J. Chem. Eng. **55**, 744–746 (1977)
6. Chen, C.K., Char, M.: Heat transfer of a continuous stretching surface with suction or blowing. J. Math. Anal. Appl. **135**, 568–580 (1988)
7. Banks, W.H.H.: Similarity solutions of the boundary layer equations for a stretching wall. J. de Mecanique Theorique et Appliquee. **2**(3), 375–392 (1983)
8. Ali, M.E.: The effect of suction or injection on laminar boundary layer development over a stretched surface. J. King Saud. Univ. **8**(1), 43–58 (1996)
9. Salleh, M.Z., Nazar, R., Pop, I.: Boundary layer flow and heat transfer over a stretching sheet with Newtonian heating. J. Taiwan Inst. Chem. Eng. **41**(6), 651–655 (2010)
10. Bachok, N., Ishak, A., Pop, I.: Unsteady boundary layer flow and heat transfer of a nanofluid over a permeable stretching/shrinking sheet. Int. J. Heat Mass Transf. **55**(7–8), 2102–2109 (2012)
11. Bhattacharyya, K., Layek, G.C., Seth, G.C.: Soret and Dufour effects on convective heat and mass transfer in stagnation point flow towards a shrinking surface. Phys Scr. **89**(9), 095203 (2014)
12. Devi, R.L.V.R., Poornima, T., Reddy, N.B., Venkataramana, S.: Radiation and mass transfer effects on MHD boundary layer flow due to an exponentially stretching sheet with heat source. Int. J. Eng. Innov. Technol. **3**(8), 33–38 (2014)
13. Zaimi, K., Ishak, A., Pop, I.: Flow past a permeable stretching/shrinking sheet in a nanofluid using two-phase model. PLoS One. **9**(11), e111743 (2014)
14. Sharma, B.R., Aich, A.: Influence of chemical reaction, magnetic field and radiation on heat and mass transfer by free convection flow near the lower stagnation point of an isothermal horizontal circular cylinder in a porous medium considering Soret and Dufour effects. Int. J. Comp. Appl. **116**(3), 5–11 (2015)
15. Ali, N., Khan, S.U., Sajid, M., Abbas, Z.: MHD flow and heat transfer of couple stress fluid over an oscillatory stretching sheet with heat source/sink in porous medium. Alex. Eng. J. **55**(2), 915–924 (2016)
16. Babu, J.M., Sandeep, N.: Three dimensional MHD slip flow of nanofluids over a slendering stretching sheet with thermophoresis and Brownian motion effects. Adv. Pow. Technol. **27**(5), 2039–2050 (2016)
17. Alavi, S.Q., Hussanan, A., Md. Kasim, A.R., Rosli, N., Md. Salleh, Z.: MHD stagnation point flow towards an exponentially stretching sheet with prescribed wall temperature and Heat flux. Int. J. Appl. Comput. Math. **3**, 3511–3523 (2017)

18. Mabood, F., Shafiq, A., Hayat, T., Abelman, S.: Radiation effects on stagnation point flow with melting heat transfer and second order slip. Results Phys. **7**, 31–42 (2017)
19. Anuar, N.S., Bachok, N., Arifin, N.M., Rosali, H.: Stagnation point flow and heat transfer over an exponentially stretching shrinking sheet in CNT with homogeneous-heterogeneous reaction: stability analysis. Symmetry. **11**(4), 522 (2019)
20. Berrehal, H., Sowmya, G., Makinde, O.D.: Shape effect of nanoparticles on MHD nanofluid flow over a stretching sheet in the presence of heat source/sink with entropy generation. Int. J. Num. Meth. Heat Fluid Flow. **32**(5), 1643–1663 (2022). https://doi.org/10.1108/HFF-03-2021-0225
21. Fenuga, O.J., Hassan, A.R., Olanrewaju, P.O.: Mixed convection in MHD flow and heat transfer rate near a stagnation-point on a non-linear vertical stretching sheet. Int. J. App. Mech. Eng. **25**(1), 37–51 (2020)
22. Tinker, S., Mishra, S.R., Sharma, R.P.: Influence of Soret and Dufour effect on MHD flow over an exponential stretching sheet: a numerical study. Indian J. Pure Appl. Phys. **58**, 558–568 (2020)
23. Govardhan, K., Narender, G., Sreedhar, G.S.: Heat and mass transfer in MHD nanofluid over a stretching surface along with viscous dissipation effect. Int. J. Math. Eng. Manag. Sci. **5**(2), 343–352 (2020)
24. Verma, K., Sharma, B.R.: Soret and Dufour effects on MHD flow with heat and mass transfer past an exponentially stretching sheet with viscous dissipation and Joule heating. Lat. Am. Appl. Res. **52**(2), 83–88 (2022)
25. Verma, K.: Heat and mass transport on a three-dimensional nanofluid flow past a non-linear stretching sheet under the influence of radiation and chemical reaction. Lat. Am. Appl. Res. **53**(3), 251–256 (2023)
26. Verma, K.: MHD flow in porous medium through an exponentially stretching sheet with surface heat and mass flux under the impact of chemical reaction. Lat. Am. Appl. Res. **53**(4), 295–302 (2023)
27. Haile, E., Awgichew, G.: Heat and mass transfer of magnetohydrodynamic nanofluid in a boundary layer with slip conditions along a permeable exponentially stretching sheet. J. Nanofluids. **8**, 1–12 (2019)
28. Rao, I.J., Rajagopal, K.R.: The effect of the slip boundary condition on the flow of fluids in a channel. Acta Mech. **135**, 113–126 (1999)
29. Thompson, P.A., Trojan, S.M.: A general boundary condition for liquid flow at solid surfaces. Nature. **389**, 360–362 (1997)
30. Barrat, J.L., Bocquet, L.: Large slip effect at a nonwetting fluid solid interface. Phys. Rev. Lett. **82**, 4671–4674 (1999)
31. Voronov, R.S., Papavassiliou, D.V., Lee, L.L.: Review of fluid slip over superhydrophobic surfaces and its dependence on the contact angle. Ind. Eng. Chem. Res. **47**, 2455–2477 (2008)
32. Bocquet, L., Barrat, J.L.: On the green-Kubo relationship for liquid-solid friction coefficient. J. Chem. Phys. **139**, 044704 (2013)
33. Huang, K., Szlufarska, I.: Green-Kubo relation for friction at solid-liquid interfaces. Phys. Rev. E Stat. Nonlin. Soft Matter. Phys. **89**, 032119 (2014)
34. Bocquet, L., Charlaix, E.: Nanofluids from bulk to interfaces. Chem. Soc. Rev. **39**, 1073–1095 (2010)

Chapter 9
Fibonacci Generation of Exact Distribution of Fibonacci Summands

Stein E. Johansen

Abstract The present contribution exhibits fundamental and exact structures in the distribution of Fibonacci summands as the structures themselves manifesting as strictly Fibonacci generated. Basically, the summands, both as (first) ordinal and as (second) cardinal numbers, are constituted from stepwise tandem between Fibonacci ordinal and cardinal numbers. The exhibition reveals that the distribution of summands is generated in formal congruence with Pascal's triangle, thus indicating a reinterpretation of Pascal's triangle as a Fibonacci-generated epistructure. This reinterpretation, as well as some other results from deeper contemplation of structures involved in Fibonacci generation of the distribution of summands, are shortly discussed in relation to some more recent advances in related mathematics and science, not excluding the previous attempt of the author to systematically reconstitute the foundations of number theory as generated from a deeper and more abstract conception of the Fibonacci algorithm.

Keywords Fibonacci algorithm · Ordinal numbers · Pascal's triangle · Epistructure · Distribution patterns

The proven *Zeckendorf theorem* [1, 2] stated:

> *Every natural number can be represented as a sum of distinct and non consecutive Fibonacci numbers (...). Using Fibonacci numbers, such a representation is always unique.*

The other way around Daykin [3] proved that Fibonacci numbers represent the *only* subset of natural numbers from which there is possible to construct sequences of natural numbers in such a way that said representation becomes unique.

Kologlu et al. [4] presented a study of the distribution of *summands* in Zeckendorf decompositions, i.e., of the Fibonacci numbers which when added represent

S. E. Johansen (✉)
Norwegian University of Science and Technology, Trondheim, Norway
e-mail: stein.e.johansen@ntnu.no

A. Adhikari et al. (eds.), *Quantitative Frontiers*, Industrial and Applied Mathematics, https://doi.org/10.1007/978-981-96-9435-8_9

sums uniquely corresponding to all natural numbers that are not Fibonacci numbers. Kologlu et al. claimed that "surprisingly, no one appears to have investigated the distribution of the number of summands," which appears somewhat exaggerated when considering the contribution by Stakhov [5]. The study by Kologlu et al. concluded that the distribution of summands converges to Gaussian. After 2010, the number of publications in the field has been growing, while still much focused on *probability* distributions (see, e.g., [6], pp. 88–89 for a list of references that far) or on various specialized mathematical *aspects* related to the summands.

In distinction to these lines of mathematical research, our focus will be to exhibit fundamental and exact structures in the distribution of summands and to present a deeper contemplation of these structures as themselves Fibonacci generated. We start out with displaying the exact composition of the Fibonacci summands when uniquely corresponding to the natural numbers in the correct succession:

In some analogy to chemistry, Fig. 9.1 describes Fibonacci numbers as *Fibonacci atoms*, and Fibonacci summands as *Fibonacci molecules* and applies the covering term *Fibonacci elements* to include both Fibonacci atoms and Fibonacci molecules. Column I describes the ordinal numbers of the Fibonacci atoms. Column II describes the ordinal numbers of Fibonacci elements, with the succession of the Fibonacci atoms as a subset of ordinal Fibonacci elements in column II-left, and with the succession of the Fibonacci molecules as a subset of ordinal Fibonacci elements in column II-right. We notice, from the ordinal numbers of Fibonacci elements in column II as a whole, that the cardinal numbers, or sizes, of the ordinal Fibonacci atoms correspond to their natural number. Column III describes the composition of ordinal Fibonacci molecules from the ordinal Fibonacci atoms which they contain. Column IV describes the cardinal numbers, or sizes, of the Fibonacci atoms contained in the respective ordinal Fibonacci molecules, and relates said numbers by the operator of addition. Column V describes the results of said addition, i.e., the cardinal numbers of Fibonacci molecules as determined by addition of the cardinal numbers of the contained Fibonacci atoms.

We notice, as a result from this successive procedure, that the cardinal numbers of Fibonacci molecules correspond exactly to their cardinal natural numbers, and in the same succession as between ordinal natural numbers. Thus every natural number represents one and only one corresponding Fibonacci element, either a Fibonacci atom or a Fibonacci molecule.

Figure 9.1 displays the *only* succession of Fibonacci molecules that is identical to the succession of natural numbers. This may indicate that it represents a deeper mathematical and ontological *truth* than other successions of summands for which Zeckendorf's theorem also hold. We can regard the composition and succession of Fibonacci elements as displayed in Fig. 9.1 as representing an *alternative* presentation of the very constitution of natural numbers *in general*.

When inspecting the succession of Fibonacci molecules and their composition of Fibonacci atoms as displayed in Fig. 9.1, we realize that the succession follows a rigorous structure. This structured succession can be represented by quite simple software. Such representation can rather obviously be formally achieved by *plural* alternative programs with entailed algorithms. However, if one assumes that this

I	II	III	IV	V	VI
F1	(1)				
F2	(1)				
F3	(2)				
F4	(3)				
	(4)	(2,4)	F2+F4=	1+3=	4
F5	(5)				
	(6)	(2,5)	F2+F5=	1+5=	6
	(7)	(3,5)	F3+F5=	2+5=	7
F6	(8)				
	(9)	(2,6)	F2+F6=	1+8=	9
	(10)	(3,6)	F3+F6=	2+8=	10
	(11)	(4,6)	F4+F6=	3+8=	11
	(12)	(2,4,6)	F2+F4+F6=	1+3+8=	12
F7	(13)				
	(14)	(2,7)	F2+F7=	1+13=	14
	(15)	(3,7)	F3+F7=	2+13=	15
	(16)	(4,7)	F4+F7=	3+13=	16
	(17)	(2,4,7)	F2+F4+F7=	1+3+13=	17
	(18)	(5,7)	F5+F7=	5+13=	18
	(19)	(2,5,7)	F2+F5+F7=	1+5+13=	19
	(20)	(3,5,7)	F3+F5+F7=	2+5+13=	20
F8	(21)				
	(22)	(2,8)	F2+F8=	1+21=	22
	(23)	(3,8)	F3+F8=	2+21=	23
	(24)	(4,8)	F4+F8=	3+21=	24
	(25)	(2,4,8)	F2+F4+F8=	1+3+21=	25
	(26)	(5,8)	F5+F8=	5+21=	26
	(27)	(2,5,8)	F2+F5+F8=	1+5+21=	27
	(28)	(3,5,8)	F3+F5+F8=	2+5+21=	28
	(29)	(6,8)	F6+F8=	8+21=	29
	(30)	(2,6,8)	F2+F6+F8=	1+8+21=	30
	(31)	(3,6,8)	F3+F6+F8=	2+8+21=	31
	(32)	(4,6,8)	F4+F6+F8=	3+8+21=	32
	(33)	(2,4,6,8)	F2+F4+F6+F8=	1+3+8+21=	33
F9	(34)				
	(35)	(2,9)	F2+F9=	1+34=	35
	(36)	(3,9)	F3+F9=	2+34=	36
	(37)	(4,9)	F4+F9=	3+34=	37
	(38)	(2,4,9)	F2+F4+F9=	1+3+34=	38
	(39)	(5,9)	F5+F9=	5+34=	39
	(40)	(2,5,9)	F2+F5+F9=	1+5+34=	40
	(41)	(3,5,9)	F3+F5+F9=	2+5+34=	41
	(42)	(6,9)	F6+F9=	8+34=	42
	(43)	(2,6,9)	F2+F6+F9=	1+8+34=	43
	(44)	(3,6,9)	F3+F6+F9=	2+8+34=	44

Fig. 9.1 Fibonacci generation of the natural numbers. *Columns:* (I) Ordinal Fibonacci atoms; (II) Ordinal Fibonacci elements (cardinal Fibonacci atoms to the left, ordinal Fibonacci molecules to the right); (III) Composition of ordinal Fibonacci molecules from ordinal Fibonacci atoms; (IV) Addition of ordinal Fibonacci atoms; (V) Addition of cardinal Fibonacci atoms; (VI) Cardinal Fibonacci molecules

(45)	(4,6,9)	F4+F6+F9=3+8+34=45
(46)	(2,4,6,9)	F2+F4+F6+F9=1+3+8+34=46
(47)	(7,9)	F7+F9=13+34=47
(48)	(2,7,9)	F2+F7+F9=1+13+34=48
(49)	(3,7,9)	F3+F7+F9=2+13+34=49
(50)	(4,7,9)	F4+F7+F9=3+13+34=50
(51)	(2,4,7,9)	F2+F4+F7+F9=1+3+13+34=51
(52)	(5,7,9)	F5+F7+F9=5+13+34=52
(53)	(2,5,7,9)	F2+F5+F7+F9=1+5+13+34=53
(54)	(3,5,7,9)	F3+F5+F7+F9=2+5+13+34=54
F10 **(55)**		
(56)	(2,10)	F2+F10=1+55=56
(57)	(3,10)	F3+F10=2+55=57
(58)	(4,10)	F4+F10=3+55=58
(59)	(2,4,10)	F2+F4+F10=1+3+55=59
(60)	(5,10)	F5+F10=5+55=60
(61)	(2,5,10)	F2+F5+F10=1+5+55=61
(62)	(3,5,10)	F3+F5+F10=2+5+55=62
(63)	(6,10)	F6+F10=8+55=63
(64)	(2,6,10)	F2+F6+F10=1+8+55=64
(65)	(3,6,10)	F3+F6+F10=2+8+55=65
(66)	(4,6,10)	F4+F6+F10=3+8+55=66
(67)	(2,4,6,10)	F2+F4+F6+F10=1+3+8+55=67
(68)	(7,10)	F7+F10=13+55=68
(69)	(2,7,10)	F2+F7+F10=1+13+55=69
(70)	(3,7,10)	F3+F7+F10=2+13+55=70
(71)	(4,7,10)	F4+F7+F10=3+13+55=71
(72)	(2,4,7,10)	F2+F4+F7+F10=1+3+13+55=72
(73)	(5,7,10)	F5+F7+F10=5+13+55=73
(74)	(2,5,7,10)	F2+F5+F7+F10=1+5+13+55=74
(75)	(3,5,7,10)	F3+F5+F7+F10=2+5+13+55=75
(76)	(8,10)	F8+F10=21+55=76
(77)	(2,8,10)	F2+F8+F10=1+21+55=77
(78)	(3,8,10)	F3+F8+F10=2+21+55=78
(79)	(4,8,10)	F4+F8+F10=3+21+55=79
(80)	(2,4,8,10)	F2+F4+F8+F10=1+3+21+55=80
(81)	(5,8,10)	F5+F8+F10=5+21+55=81
(82)	(2,5,8,10)	F2+F5+F8+F10=1+5+21+55=82
(83)	(3,5,8,10)	F3+F5+F8+F10=2+5+21+55=83
(84)	(6,8,10)	F6+F8+F10=8+21+55=84
(85)	(2,6,8,10)	F2+F6+F8+F10=1+8+21+55=85
(86)	(3,6,8,10)	F3+F6+F8+F10=2+8+21+55=86
(87)	(4,6,8,10)	F4+F6+F8+F10=3+8+21+55=87
(88)	(2,4,6,8,10)	F2+F4+F6+F8+F10=1+3+8+21+55=88
F11 **(89)**		
(90)	(2,11)	F2+F11=1+89=90
(91)	(3,11)	F3+F11=2+89=91
(92)	(4,11)	F4+F11=3+89=92

Fig. 9.1 (continued)

(93)	(2,4,11)	F2+F4+F11=1+3+89=93
(94)	(5,11)	F5+F11=5+89=94
(95)	(2,5,11)	F2+F5+F11=1+5+89=95
(96)	(3,5,11)	F3+F5+F11=2+5+89=96
(97)	(6,11)	F6+F11=8+89=97
(98)	(2,6,11)	F2+F6+F11=1+8+89=98
(99)	(3,6,11)	F3+F6+F11=2+8+89=99
(100)	(4,6,11)	F4+F6+F11=3+8+89=100
(101)	(2,4,6,11)	F2+F4+F6+F11=1+3+8+89= 101
(102)	(7,11)	F7+F11=13+89=102
(103)	(2,7,11)	F2+F7+F11=1+13+89=103
(104)	(3,7,11)	F3+F7+F11=2+13+89=104
(105)	(4,7,11)	F4+F7+F11=3+13+89=105
(106)	(2,4,7,11)	F2+F4+F7+F11=1+3+13+89=106
(107)	(5,7,11)	F5+F7+F11=5+13+89=107
(108)	(2,5,7,11)	F2+F5+F7+F11=1+5+13+89=108
(109)	(3,5,7,11)	F3+F5+F7+F11=2+5+13+89=109
(110)	(8,11)	F8+F11=21+89=110
(111)	(2,8,11)	F2+F8+F11=1+21+89=111
(112)	(3,8,11)	F3+F8+F11=2+21+89=112
(113)	(4,8,11)	F4+F8+F11= 3+21+89=113
(114)	(2,4,8,11)	F2+F4+F8+F11=1+3+21+89=114
(115)	(5,8,11)	F5+F8+F11=5+21+89=115
(116)	(2,5,8,11)	F2+F5+F8+F11=1+5+21+89=116
(117)	(3,5,8,11)	F3+F5+F8+F11=2+5+21+89=117
(118)	(6,8,11)	F6+F8+F11=8+21+89=118
(119)	(2,6,8,11)	F2+F6+F8+F11=1+8+21+89=119
(120)	(3,6,8,11)	F3+F6+F8+F11=2+8+21+89=120
(121)	(4,6,8,11)	F4+F6+F8+F11=3+8+21+89=121
(122)	(2,4,6,8,11)	F2+F4+F6+F8+F11=1+3+8+21+89=122
(123)	(9,11)	F9+F11=34+89=123
(124)	(2,9,11)	F2+F9+F11=1+34+89=124
(125)	(3,9,11)	F3+F9+F11=2+34+89=125
(126)	(4,9,11)	F4+F9+F11=3+34+89=126
(127)	(2,4,9,11)	F2+F4+F9+F11=1+3+34+89=127
(128)	(5,9,11)	F5+F9+F11=5+34+89=128
(129)	(2,5,9,11)	F2+F5+F9+F11=1+5+34+89=129
(130)	(3,5,9,11)	F3+F5+F9+F11=2+5+34+89=130
(131)	(6,9,11)	F6+F9+F11=8+34+89=131
(132)	(2,6,9,11)	F2+F6+F9+F11=1+8+34+89=132
(133)	(3,6,9,11)	F3+F6+F9+F11=2+8+34+89=133
(134)	(4,6,9,11)	F4+F6+F9+F11=3+8+34+89=134
(135)	(2,4,6,9,11)	F2+F4+F6+F9+F11=1+3+8+34+89=135
(136)	(7,9,11)	F7+F9+F11=13+34+89=136
(137)	(2,7,9,11)	F2+F7+F9+F11=1+13+34+89=137
(138)	(3,7,9,11)	F3+F7+F9+F11=2+13+34+89=138
(139)	(4,7,9,11)	F4+F7+F9+F11=3+13+34+89=139
(140)	(2,4,7,9,11)	F2+F4+F7+F9+F11=1+3+13+34+89=140
(141)	(5,7,9,11)	F5+F7+F9+F11=5+13+34+89=141
(142)	(2,5,7,9,11)	F2+F5+F7+F9+F11=1+5+13+34+89=142
(143)	(3,5,7,9,11)	F3+F5+F7+F9+F11=2+5+13+34+89=143

F12 (144)

...

Fig. 9.1 (continued)

unique structured succession is generated from some kind of *real* onto-mathematical constitutional logic, software programs will differ with respect to *how close* they reflect this generative logic. In the present context, we will not offer any discussion of this rather profound and—in our view—important topic, but merely refer to Johansen [7] for extensive treatment and reasoning.

In the present context, we regard the structured succession displayed in Fig. 9.1 as *already* established and seek to unfold some further mathematical structures relevant for the distribution of Fibonacci molecules, residing *enfolded* in this uniquely structured succession.

We apply symbol K_n to denote the *assembly* or *set* of Fibonacci molecules residing in the gap between Fibonacci atoms F_n and F_{n+1}. We describe F_n as the *Fibonacci root atom* in relation to K_n, due to the structured succession in Fig. 9.1 displaying that *all* Fibonacci molecules between F_n and F_{n+1} have F_n as rooting (last) digit or atom in their molecular decomposition.

We denote the *amount* of Fibonacci molecules contained in K_n with symbol S_n. Trivially, we have:

$$S_n = (F_{n+1} - F_n) - 1 \tag{9.1}$$

thus also:

$$S_{n-1} = (F_n - F_{n-1}) - 1 \tag{9.2}$$

and:

$$S_{n-2} = (F_{n-1} - F_{n-2}) - 1 \tag{9.3}$$

(9.1), (9.2), and (9.3) imply:

$$S_n = (F_{n-1} - 1) + F_{n-2} \tag{9.4}$$

Here we denote the first joint at r.h.s., i.e., $(F_{n-1} - 1)$, with symbol S_nA and the second joint at r.h.s., i.e., F_{n-2}, with symbol S_nB. Thus, by definition:

$$S_nA = F_{n-1} - 1 \tag{9.5}$$

$$S_nB = F_{n-2} \tag{9.6}$$

so that (9.4) is equivalent with:

$$S_n = S_nA + S_nB \tag{9.7}$$

This expresses a simple *numerical* relation with respect to how the *amount* of Fibonacci molecules contained in K_n is related to the amounts of Fibonacci molecules in the preceding assemblies or sets K_{n-1} and K_{n-2}.

We apply symbol K_nA to denote the set of Fibonacci molecules which in the amount of S_nA (= $F_{n-1} - 1$) is transported to K_n by undergoing a certain *transformation* of their molecular composition. We notice that this transformation is very simple, just *repeating* the atomic composition of all Fibonacci molecules from K_{n-1}, with the sole exception of changing the Fibonacci root atom (i.e., the last atom or digit) from the root atom of K_{n-1} into the Fibonacci root atom of K_n (which includes K_nA as a subset). As an example, all the seven Fibonacci molecules in K_7 (i.e., corresponding to natural numbers 14–20) in the gap between F_7 and F_8 reappear as Fibonacci molecules in K_8, i.e., in K_8A, with the sole modification that Fibonacci atom F_7 has become transformed to Fibonacci atom F_8. We also notice that the *succession* between the Fibonacci molecules remains the same after this transported transformation from K_{n-1} to K_nA as a subset of K_n.

We apply symbol K_nB to denote the set of Fibonacci molecules which in the amount of S_nB (= F_{n-2}) is transported to K_n by undergoing a certain *transformation* of their molecular composition. We notice that this transformation to K_nB is a bit less simple. *First*, different from the transformation of K_{n-1} into K_nA, the Fibonacci root atom F_{n-2} now becomes *included* as (the first) Fibonacci atom composing a new Fibonacci molecule inside K_nB. As an example, F_6 enters as atom in composing the (first) Fibonacci molecule (6,8) in K_8B. *Second*, different from the simple transformation of K_{n-1} into K_nA, *all* Fibonacci molecules in K_{n-2} are now *added* with one Fibonacci atom, namely the Fibonacci root atom F_n, in their transformation to Fibonacci molecules making up the set K_nB. Thus, *all* Fibonacci molecules transformed from K_{n-2} to K_nB reappear with *one more* Fibonacci atom, i.e., added with root atom F_n, as compared to their composition of Fibonacci atoms inside K_{n-2}.

Also in this case we notice that the *succession* between the Fibonacci molecules remains the same after this transported transformation from K_{n-2} to K_nB as a subset of K_n. Further, we notice that the succession between the *subsets* K_nA and K_nB always remains the same, with K_nA always representing the *first* subset and K_nB representing the *second* subset of K_n, i.e., that the Fibonacci molecules in set K_nA represent the *lowest* segment of natural numbers corresponding to the Fibonacci molecules in set K_n, and the Fibonacci molecules in set K_nB represent the *remaining, higher* segment of said natural numbers. As an example, the Fibonacci molecules in K_8, corresponding to the 12 natural numbers from 22 to 33, are composed by a transported transformation of K_7 to K_8A, corresponding to the 7 natural numbers from 22 to 28, and next by a transformation of K_6 (plus of F_6) to K_8B, corresponding to the natural numbers from 29 to 33.

With respect to the sets and subsets we always have:

$$K_n = K_nA \cup K_nB \tag{9.8}$$

and

$$K_nA \cap K_nB = \emptyset \tag{9.9}$$

This implies that from knowledge of sets K_{n-1} and K_{n-2} we can always calculate their respective transformations to subsets K_nA and K_nB, and thus the exact, unique, and complete make-up of set K_n of Fibonacci molecules in correct succession corresponding to the succession of natural numbers. Thus we realize that not only the succession of the Fibonacci numbers themselves, but also the succession and composition of the *Fibonacci molecules* rooted in their respective Fibonacci atoms, are generated by strict Fibonacci algorithms.

Trivially, (9.1) is equivalent with that the amount of Fibonacci molecules (i.e., summands) rooted in Fibonacci atom F_n is given by:

$$S_n = F_{n-1} - 1 \tag{9.10}$$

Evidently, the total amount of Fibonacci molecules before F_n is given as $F_n - n + 1$, denoted by symbol $\mathbf{S}_n$, so that the total amount of Fibonacci molecules *including* those rooted in F_n, is:

$$\mathbf{S}_n = F_n - n + 1 + S_n \tag{9.11}$$

(9.10) and (9.11) give:

$$\mathbf{S}_n = F_{n+1} - n \tag{9.12}$$

As an example of how to read the Table 9.1: The row starting with F12 displays the Fibonacci molecules rooted in Fibonacci atom F_{12}, i.e., the Fibonacci molecules residing in the gap between F_{12} and F_{13}, corresponding to the natural numbers between $F_{12} = 144$ and $F_{13} = 233$. In this gap, we observe that there occur

Table 9.1 Distribution of Fibonacci molecules with respect to their contained number of Fibonacci atoms, depending on which Fibonacci atom they are rooted in (© Stein E. Johansen)

Fib-root	Number of Fib atoms in molecules						
	2	**3**	**4**	**5**	**6**	S_n	$\mathbf{S}_n$
F4	1					1	1
F5	2					2	3
F6	3	1				4	7
F7	4	3				7	14
F8	5	6	1			12	26
F9	6	10	4			20	46
F10	7	15	10	1		33	79
F11	8	21	20	5		54	133
F12	9	28	35	15	1	88	221
F13	10	36	56	35	6	143	364

9 Fibonacci molecules composed of 2 Fibonacci atoms, 28 Fibonacci molecules composed of 3 atoms, 35 Fibonacci molecules composed of 4 atoms, 15 Fibonacci molecules composed of 5 atoms, and 1 Fibonacci molecule composed of 6 atoms.

We notice that the Fibonacci molecules are exactly distributed as *Pascal's triangle*, with the columns in the table being identical to the diagonals in Pascal's triangle. This rather startling observation becomes explainable when we approach the issue from our previous two transporting transformations of Fibonacci molecules, i.e., from K_{n-1} to K_nA and from K_{n-2} to K_nB.

As an example, we may look at the row for F10 where we observe that there reside 10 Fibonacci molecules that are composed of 4 Fibonacci atoms. By the transported transformation of K_9 to $K_{10}A$ the 4 Fibonacci molecules in K_9 containing 4 atoms, will reappear as 4 Fibonacci molecules in K_{10}, merely substituting the previous root atom F_9 in these molecules with the new root atom F_{10}. By the transported transformation of K_8 to $K_{10}B$ the 6 Fibonacci molecules in K_8 containing 3 atoms, will reappear in $K_{10}B$ as 6 Fibonacci molecules containing 4 atoms, due to the previous molecules now being extended by one new atom, namely the new root atom F_{10}.

We realize that this is just one case of the *general* implicate procedure, where any number in the table is found from two previous numbers from rows above, namely from picking the number one row above in the same column (consistent with the transformation from K_{n-1} to K_nA), and add the number one row above and one column to the left (consistent with the transformation from K_{n-2} to K_nB). This general procedure is formally equivalent with the addition procedure generating Pascal's triangle.

The emergence of the number "1" in the table, at the top of each column and at the far left for every second row, is always the result of the transformation from K_{n-2} to K_nB which creates one novel Fibonacci molecule by adding the new Fibonacci root atom to the *last* Fibonacci molecule in K_{n-2}, i.e., to the Fibonacci molecule in K_{n-2} corresponding to the *largest* natural number or "tail" in K_{n-2}. For this extension to result in only *one* new Fibonacci molecule in K_nB with one more atom than in K_{n-2}, the necessary and sufficient condition is that the last Fibonacci molecule in K_{n-2} is the *only* Fibonacci molecule in K_{n-2} that contains that many Fibonacci atoms as itself does. This can and must happen only for every *second* transformation from K_{n-2} to K_nB, more specifically for the transformation of (2,4) to (2,4,6) from K_4 to K_6B, of (2,4,6) to (2,4,6,8) from K_6 to K_8B, etc.

Next we will explore distribution of Fibonacci molecules that contain the *same* Fibonacci atom (Table 9.2).

We notice that this *subset* of Fibonacci molecules also is distributed according to Pascal's triangle, for reasons analogous to what we presented with respect to the Pascal distribution of the *whole* set of Fibonacci molecules displayed in Table 9.1. Now, the column of "1s" below **2** is added to the structure, which would be analogous to Table 9.1 if we had modified Table 9.1 to distribution of Fibonacci *elements* instead of Fibonacci *molecules*.

Table 9.2 Distribution of Fibonacci molecules containing F_2 as Fibonacci atom, depending on which Fibonacci atom they are rooted in

Fib-root	Number of F_2 atoms in molecules						
	2	**3**	**4**	**5**	**6**	$S_n(F_2)$	$\mathbf{S}_n(F_2)$
F4	1					1	1
F5	1					1	2
F6	1	1				2	4
F7	1	2				3	7
F8	1	3	1			5	12
F9	1	4	3			8	20
F10	1	5	6	1		13	33
F11	1	6	10	4		21	54
F12	1	7	15	10	1	34	88

F_2 must always be the *first* Fibonacci atom or digit in Fibonacci molecules containing it.

$$S_n\,(F_2) = F_{n-3} \tag{9.13}$$

and

$$\mathbf{S}_n\,(F_2) = F_{n-1}-1 \tag{9.14}$$

This subset of Fibonacci molecules containing F_3 as Fibonacci atom duplicates the distribution in the subset containing F_2, with the sole difference being that the structure in Table 9.2 is transported one row down. F_3 must always be the *first* Fibonacci atom or digit in Fibonacci molecules containing it (Table 9.3).

$$S_n\,(F_3) = F_{n-4} \tag{9.15}$$

and

$$\mathbf{S}_n\,(F_3) = F_{n-2}-1 \tag{9.16}$$

This subset of Fibonacci molecules containing F_4 as Fibonacci atom is also structured similar to Pascal's triangle, as indicated by the three numbers colored yellow, with some slight modification due to the initiating entering of (2,4) while there is no Fibonacci molecule rooted in F_5 that contains F_4 as Fibonacci atom.

Table 9.3 Distribution of Fibonacci molecules containing F_3 as Fibonacci atom, depending on which Fibonacci atom they are rooted in

Fib-root	Number of F_3 atoms in molecules						
	2	**3**	**4**	**5**	**6**	$S_n(F_3)$	$S_n(F_3)$
F5	1					1	1
F6	1					1	2
F7	1	1				2	4
F8	1	2				3	7
F9	1	3	1			5	12
F10	1	4	3			8	20
F11	1	5	6	1		13	33
F12	1	6	10	4		21	54
F13	1	7	15	10	1	34	88

Table 9.4 Distribution of Fibonacci molecules containing F_4 as Fibonacci atom, depending on which Fibonacci atom they are rooted in

Fib-root	Number of F_4 atoms in molecules						
	2	**3**	**4**	**5**	**6**	$S_n(F_4)$	$S_n(F_4)$
F4	1 0-1					1 0-1	1 0-1
F5	0					0	1 0-1
F6	1 1-0	1 0-1				2 1-1	3 1-2
F7	1 1-0	1 1-1				2 1-1	5 2-3
F8	1 1-0	2 1-1	1 0-1			4 2-2	9 4-5
F9	1 1-0	3 2-1	2 0-2			6 3-3	15 7-8
F10	1 1-0	4 3-1	4 1-3	1 0-1		10 5-5	25 12-13
F11	1 1-0	5 4-1	7 3-4	3 0-3		16 8-8	41 20-21
F12	1 1-0	6 5-1	11 6-5	7 1-6	1 0-1	26 13-13	67 33-34

F_4 can occur both as *first* atom/digit in Fibonacci molecules and as *second* atom/digit. In Table 9.4, these occurrences are indicated, respectively, as separated with dash sign "-". As an example: In the lowest row Fibonacci atom F_4 occurs 11 times in Fibonacci molecules rooted in F_{12}, 6 times as the *first* atom/digit in these Fibonacci molecules, and 5 times as *second* atom/digit in these Fibonacci molecules. We notice that also the distribution of first atoms/digits, indicated by color green, is structured similar to Pascal's triangle, and the same is the case for the distribution of second atoms/digits, indicated by color blue.

For rather obvious reasons, the *total* amount of occurrences as first atom/digit vs. as second atom/digit must be the *same* for Fibonacci molecules containing F_4

Table 9.5 Distribution of Fibonacci molecules containing F_5 as Fibonacci atom, depending on which Fibonacci atom they are rooted in

Fib-root	Number of F_5 atoms in molecules						
	2	**3**	**4**	**5**	**6**	$S_n(F_5)$	$\mathbf{S}_n(F_5)$
F4	0					0	0
F5	2 0-2					2 0-2	2 0-2
F6	0					0	0
F7	1 1-0	2 0-2				3 1-2	5 1-4
F8	1 1-0	2 0-2				3 1-2	8 2-6
F9	1 1-0	3 1-2	2 0-2			6 2-4	14 4-10
F10	1 1-0	4 2-2	4 0-4			9 3-6	23 7-16
F11	1 1-0	5 3-2	7 1-6	2 0-2		15 5-10	38 12-26
F12	1 1-0	6 4-2	11 3-8	6 0-6		24 8-16	62 20-42

as atom when the Fibonacci molecule is rooted in Fibonacci atoms larger than F_5. This is displayed in the column $S_n(F_4)$ when looking at the decompositions into first atoms/digits vs. second atoms/digits. When rooted in Fibonacci atoms larger than F_5, the amounts of both first atoms/digits and second atoms/digits progress according to the Fibonacci series, and we have:

$$S_n\,(F_4) = 2F_{n-5} \tag{9.17}$$

The two corresponding progressions when decomposing $\mathbf{S}_n(F_4)$ into first vs. second atoms/digits, give, respectively, $F_{n\text{-}3} - 1$ and $F_{n\text{-}3}$, so we have:

$$\mathbf{S}_n\,(F_4) = 2F_{n-3}{-}1 \tag{9.18}$$

This subset of Fibonacci molecules containing F_5 as Fibonacci atom is also structured similar to Pascal's triangle, as indicated by the three numbers colored yellow, with some slight modification due to the initiating entering of (3,5) while there is no Fibonacci molecule rooted in F_6 that contains F_5 as Fibonacci atom. And also in this subset, just as in the subset displayed in Table 9.4, we notice that the distribution of first atoms/digits, indicated by color green, is structured similar to Pascal's triangle, and the same is the case for the distribution of second atoms/digits, indicated by color blue (Table 9.5).

For rather obvious reasons, the *total* amount of occurrences as first atom/digit vs. as second atom/digit must happen in the proportion 1:2 for Fibonacci molecules containing F_5 as atom when the Fibonacci molecule is rooted in Fibonacci atoms larger than F_6. This is displayed in the column $S_n(F_5)$ when looking at the decompositions into first atoms/digits vs. second atoms/digits. When rooted in Fibonacci atoms larger than F_6, the amounts of first atoms/digits progress according to the Fibonacci series, and the amounts of second atoms/digits progress according

Table 9.6 Some resulting distribution of Fibonacci molecules containing F_a as Fibonacci atom, depending on which Fibonacci atom F_n they are rooted in

Atom F_a	$S_n = S_n$,1.atom + S_n,2.atom	$\mathbf{S}_n = \mathbf{S}_n$,1.atom + $\mathbf{S}_n$,2.atom
F_2	F_{n-3}	$F_{n-1} - 1$
F_3	F_{n-4}	$F_{n-2} - 1$
F_4	$2F_{n-5} = F_{n-5} + F_{n-5}$	$2F_{n-3} - 1 = F_{n-3} - 1 + F_{n-3}$
F_5	$3F_{n-6} = F_{n-6} + 2F_{n-6}$	$3F_{n-4} - 1 = F_{n-4} - 1 + 2F_{n-4}$

to *twice* the Fibonacci series. Thus we have:

$$S_n\ (F_5) = 3F_{n-6} \tag{9.19}$$

The two corresponding progressions when decomposing $\mathbf{S}_n(F_5)$ into first vs. second atoms/digits, give, respectively, $F_{n-4} - 1$ and $2F_{n-4}$, so we have:

$$\mathbf{S}_n\ (F_5) = 3F_{n-4}{-}1 \tag{9.20}$$

We compile the results with respect to S_n, $\mathbf{S}_n$, and their decomposition into related series of Fibonacci atoms/digits into the following table (Table 9.6):

The first row displays some results from investigating Table 9.2 which expressed the distribution of Fibonacci molecules containing Fibonacci atom F_2 (i.e., $a = 2$); the second row displays some results from investigating Table 9.3 which expressed the distribution of Fibonacci molecules containing Fibonacci atom F_3 (i.e., $a = 3$), etc.

Table 9.6 indicates the following *general formula* for the amounts of Fibonacci atom F_a occurring in Fibonacci molecules rooted in F_n, i.e., corresponding to the set of natural numbers occurring between F_n and F_{n+1}:

$$S_n\ (F_a) = F_{a-1}F_{n-a-1} \tag{9.21}$$

Table 9.6 indicates the following *general formula* for the aggregated amounts of Fibonacci atom F_a occurring in Fibonacci molecules rooted in Fibonacci atoms smaller than or equal to F_n:

$$\mathbf{S}_n\ (F_a) = F_{a-1}\ F_{n-a+1} \tag{9.22}$$

The mathematical *simplicity* of the formulas displayed in Table 9.6 and the generalized formulas (9.21) and (9.22) is rather startling, thus removing much of the mystery about the exact distribution and related composing of Fibonacci atoms into Fibonacci molecules/summands. We also notice that these formulas *themselves* are Fibonacci structured, thus indicating successive generation also of Fibonacci *molecules*, i.e., not merely of the Fibonacci numbers themselves, from strict and *self-referential* unfoldment of the basic Fibonacci algorithm.

With respect to the profound and far-reaching significance of self-referentiality D. Rapoport has presented extensive innovative studies in various regards and sciences, departing from the ontological significance of the self-referential *Klein-bottle* (and hyper-Klein-bottle) with related and crucial mathematical developments. Such significance will not be discussed in the present context, but at least it should be mentioned that Rapoport's presentation of Fibonacci generation from self-referential *Hadamard matrices* appears highly relevant as a link between Fibonacci generation and Klein-bottle dynamics [8].

Our results presented above give additional support to the novel mathematical *paradigm* presented in our treatise Johansen [7] which sought to reestablish number theory in its very foundations from some kind of Copernican turn, arguing that instead of regarding the Fibonacci series as a subset of natural numbers, the natural numbers themselves could be developed uniquely from a deeper and more abstract conception of the Fibonacci algorithm, in systematic and successive alternation between (first) *ordinal* Fibonacci numbers and *cardinal* Fibonacci numbers, as already indicated by the designations of the columns at top of Fig. 9.1.

The treatise was inspired by a further reflection ([9], pp. 124–127) on a certain result from our *philosophical* treatise Johansen [10] in qualitative informatics, indicating that the Fibonacci algorithm, considered in its most abstract sense, had to reside enfolded in the very "atom" of information and hence to represent the constitutive bridge between the qualitative and quantitative aspects of nature *in general*. (With respect to related philosophical issues, see also Johansen [11]; and Johansen [7, 12–15] for discussion of various implications to sciences outside mathematics.)

This treatise [7] also presented a development of the four basic arithmetic operators in a modified manner, consistent with the primacy to the Fibonacci algorithm, and some reasoning establishing a more fundamental and intimate connection between number theory and geometry when developed from the Fibonacci algorithm. In this last regard, there is an interesting parallel to the intimate relation between hadronic mathematics and hadronic mechanics, initiated by R.M. Santilli [16, 17], developed from detrivializing of "the unit." (See Johansen [14, 18] for related discussions.)

Also, the treatise presented detailed exhibition of and reflections on *how* the Fibonacci molecules become *self-referentially* composed in the succession that *resulted* in unique correspondence to the succession of natural numbers, i.e., on the *constitutional logic* unfolding from the Fibonacci algorithm and resulting in Fig. 9.1.

Further, the treatise argued that *Pascal's triangle* can be understood as a Fibonacci generated *epistructure*. The results above give further support for this line of reasoning. While it is trivial mathematical knowledge that the "shallow diagonals" in Pascal's triangle add up to the Fibonacci numbers, it is far less trivial that the distribution of Fibonacci *molecules generates* the *whole* of Pascal's triangle more or less exactly, as exhibited in the tables above with related formulas.

M.B. Ignatyev pioneered Russian robotics in the 1960's and developed later a universal formulation of cybernetics named *linguo-combinatorics* based on a certain set of differential equations applying only a distinction between units of meaning and their symbols [19–21]. Ignatyev anchors much of his treatment *exactly* in a certain structuring from *Pascal's triangle* which manifests the formula for "the basic law of cybernetics, informatics and synergetics for complex systems" [19]. The novelty, universality, and explanatory power represented by linguo-combinatorial cybernetics is indicated by, according to Ignatyev, proving able to develop exhaustive "models of all the known chemical elements, their isotopes, and molecular structures" ([21], p. 673). Thus, Ignatyev refers to the establishing of "cybernetic physics" ([20], p. 20) and states such cybernetic physics/chemistry as *superior* to the conventional method of linear combination of atomic orbitals, because "the linguo-combinatorial method considers all the combinations of interaction" ([21], p. 673).

When Pascal's triangle manifests in the over-all Fibonacci generation of Fibonacci molecules, this may suggest that Ignatyev's theory, having rather general relevance for analysis of natural systems, robotics and computation, can be given an even deeper formulation and anchoring from such Fibonacci generation. (See Johansen [13] for related further discussion.)

This may also have some relevance in relation to E. Trell's original line of research, successively building a new scientific theory bottom-up from the mathematical philosophy of Sophus Lie underlying his doctoral thesis *Over a Class of Geometric Transformations* [22]. From this Lie paradigm in geometry Trell started out with theoretically developing results in particle physics [23, 24], in amazing agreement with empirical measurements, next with theoretically developing the periodic system, and thereafter also proceeding to development of some molecular structures. There are some interesting similarities between Trell's theories and Ignatyev's insofar as they are built from deeper cybernetic laws and congruent with nanotechnological developments, in the case of Trell from stepwise generation of *space filling*. (See Trell [25] for some recent compressed presentation of his results, following a long record of published articles.)

Figure 9.2 performs a subdivision of K_{11} into Fibonacci-generated segments, applied as illustration in order to disclose an interesting general pattern *inside* K_n for distribution of Fibonacci molecules with respect to their contained number of atoms. While Table 9.1 presented a Fibonacci-generated Pascal structuring of the *total* number of Fibonacci molecules inside K_n containing, respectively, 2, 3, 4, 5,... Fibonacci atoms, Fig. 9.2 exhibits the internal *succession* inside K_n (illustrated by K_{11}) which *results* in the aggregated distribution of Fibonacci molecules for K_n as a whole.

The successive Fibonacci segments, separated by dashed lines, and the correspondence of their contained Fibonacci molecules to natural numbers, are displayed in the two columns to the left. As an example of how to read the figure: From natural number 102 to 109 we have 8 natural numbers which each correspond to a unique Fibonacci molecule, thus a segment consisting of 8 Fibonacci molecules. The first 5 of these Fibonacci molecules, corresponding to natural numbers 102 to 106 and displayed with color blue, occur in the succession of first one Fibonacci molecule containing 2 Fibonacci atoms, next 3 Fibonacci molecules containing 3 atoms, and next 1 Fibonacci molecule containing 4 atoms. After this, the 3 remaining

Amount of molecules in Fib segments		Amount of atoms in Fib elements 1	2	3	4	5	Amount of molecules in Fib sub-segments		Amount of diagonals in Fib segments
89 (=F11)	0	1							
90	1		1						
91	1		1						
92	2		1						
-93				1					
94	3		1						1 (1)
-96				2					
97	5		1						1 (1)
				3					
-101					1				
102	8		1				5		2 (1+1)
				3					
					1				
				1				3	
-109					2				
110	13		1				8		3 (1+2)
				3					
					1				
				1					
					2				
				1				5	
					3				
-122						1			
123	21		1				13		5 (2+3)
				3					
					1				
				1					
					2				
				1					
					3				
						1			
				1				8	
					3				
						1			
					1				
-143						2			
144 (=F12)									

Fig. 9.2 Fibonacci distribution of Fibonacci molecules with respect to their contained number of Fibonacci atoms, illustrated by the Fibonacci molecules between F11 and F12, i.e., the natural numbers between 89 and 144 (© Stein E. Johansen)

Fibonacci molecules in this segment, corresponding to natural numbers 107 to 109 and displayed with color red, occur in the succession of first a Fibonacci molecule containing 3 atoms, and next the 2 Fibonacci molecules completing the segment as containing 4 Fibonacci atoms.

We notice that numbers displayed in *red* constitute repeating diagonals in the form 1-2, while the numbers displayed in *blue* constitute repeating diagonals in the form 1-3-1. If we again look at the segment containing 8 Fibonacci molecules, corresponding to natural numbers from 102 to 109, we notice that the *blue diagonal 1-3-1* is just transported from the forerunning segment of 5 Fibonacci molecules, homologous to the transport from K_{n-1} to K_nA. We also notice that the *red diagonal 1-2* is transported from *two* segments back, while it at the same time is moved *one step to the right*, i.e., containing one more Fibonacci atom than the red diagonal two segments back. Thus, this transport is almost homologous to the transformation from K_{n-2} to K_nB.

For segments *after* the segment containing 8 Fibonacci molecules, the homologies to the transformations into K_nA and K_nB breaks down with respect to the *blue* vs. *red* diagonals, while we observe that these homologies still hold for the *larger* sub-segments containing *more* than only one blue and red diagonal. As an example, the sub-segment of 13 Fibonacci molecules, corresponding to natural numbers from 123 to 135, just repeats the structure from the forerunning segment from 110 to 122, while its added sub-segment of 8 Fibonacci molecules, corresponding to natural numbers from 136 to 143, repeats the structure from the segment from 102 to 109 while moving this segment one step to the right.

When we move to the right in Fig. 9.2 and inspect the two columns displaying sub-segments of the Fibonacci-generated segments, covering red and/or blue diagonals, we observe that the amount of Fibonacci molecules contained in the sub-segments also is Fibonacci generated.

When we move to the far right columns in Fig. 9.2, we observe that the number of red diagonals, the number of blue diagonals, and the number of red+blue diagonals covered by the successive Fibonacci segments, also are Fibonacci generated.

It is rather obvious from this illustration by K_{11} that the same structures of strict Fibonacci generation of Fibonacci molecules with related amounts of Fibonacci atoms, will manifest *internally* in any K_n. Thus, *both* the *aggregate* distribution of the number of atoms of Fibonacci molecules contained in K_n as a whole, as displayed in Table 9.1, *and* the *internal succession* of the number of atoms of Fibonacci molecules contained in K_n, manifest as *strictly Fibonacci generated.*

The succession of diagonal 1-2 vs. diagonal 1-3-1 might be of special interest as represented, respectively, by numbers 3 and 5. In the groundbreaking treatise of Rowlands [26], the fundamental and far-reaching significance of the symmetry break represented by the split code 3:5 is exhibited ([26], p. 72). Figure 9.2 displays, after segment 92-93, the succession between segments as 3/5/5-3/5-3-5/5-3-5-5-3, consistent with the Fibonacci succession of the last segment taking first the sequence of 3s and 5s from the preceding segment with it (while preserving their column positions) and next the sequence of 3s and 5s from the segment before this (while transporting their positions to the next column). When 5s occur as both the end of one segment and the beginning of the next segment, this might be interpreted as a Fibonacci-generated symmetry break, with possible interesting implications for further advances in Fibonacci anchored number theory.

In Johansen [27], we presented a deduction of the exact and complete pattern of prime numbers from generative elimination of composite numbers, which we later [28] named our *negative approach* to deduce the set of prime numbers. (See [29] for some elaborated application.) This deduction, with involved formulas, was illustrated as being correct by software developed by Jon M. Strand (cf. *appendix* in [28]), including software to exclusively pick twin primes inside segments high up in the number landscape. This negative approach was anchored in the split code 3:5 from the basic construction of the so-called *revolving chamber* exhibiting natural numbers as composed of 3s and 5s.

Our *positive approach* [30] in order to *attempt* to deduce the set of prime numbers uniquely and completely, is anchored in Fibonacci generation of the natural numbers. Fibonacci *molecules* can be *translated* to cover the whole field of composite natural numbers (cf. Figure 9.2 in [30]). However, this translation still does not explain the *cardinal* size of the prime numbers which the composite numbers are anchored in by analogy to the anchoring of Fibonacci molecules in Fibonacci root atoms. It is well known that the relation between Fibonacci numbers and prime numbers is intimate, while still much cloaked in mystery. For example, if n is prime, then n is either a factor in F_{n-1} or in F_{n+1}, while the opposite implication is not valid due to a few occurrences of the so-called Fibonacci pseudoprimes. In an analogous manner also Lucas numbers are intimately related to prime numbers [31].

Due to the significance of the split code 3:5, *both* in our *negative* approach deducing the prime numbers *and* in the generation of Fibonacci molecules, as illustrated by Fig. 9.2, it might be that this split code represents some key to achieve a complete deduction of prime numbers from the Fibonacci anchored *positive* approach.

And it might also be that the present exhibition and contemplation of Fibonacci summands in other regards than the 3:5 split code, represent some step in this direction, among possible other directions for further mathematical exploration.

References

1. Zeckendorf, E.: Représentation des nombres naturels par une somme de nombres de Fibonacci ou de nombres de Lucas. Bulletin de la Societe Royale des Sciences de Liege. **41**, 179–182 (1972a)
2. Zeckendorf, E.: A generalized Fibonacci numeration. Fibonacci Q. **10**, 365–372 (1972b)
3. Daykin, D.E.: Representation of natural numbers as sums of generalised Fibonacci numbers – II. J. Lond. Math. Soc. **35**, 143–160 (1960)
4. Kologlu, M., Kopp, G., Miller, S.J., Wang, Y.: On the number of summands in Zeckendorf decompositions. http://arxiv.org/abs/1008.3204v1 (2010)

5. Stakhov, A.P.: The Mathematics of Harmony. From Euclid to Contemporary Mathematics and Computer Science. Series of Knots and Everything, vol. 22. World Scientific, New Jersey (2009)
6. Catral, M., et al.: Generalizing Zeckendorf's theorem: the Kentucky sequence. Fibonacci Q. **52**(5), 68–90 (2014)
7. Johansen, S.E.: Fibonacci generation of natural numbers and prime numbers. In: Corda, C. (ed.) Proceedings of Third International Conference on Lie-Admissible Treatments of Irreversible Processes. Kathmandu University, Nepal, Jan. 3–7, 2011, pp. 305–410. Sankata Printing Press (2011)
8. Rapoport, D., Perez, J.C.: Golden ratio and Klein bottle logophysics: the keys of the codes of life and cognition. Quantum Biosyst. **9**(2), 8–76 (2018)
9. Johansen, S.E.: Initiation of 'hadronic philosophy', the philosophy underlying hadronic mechanics and chemistry. Hadron. J. **29**, 111–135 (2006)
10. Johansen, S.E.: Outline of Differential Epistemology. (In Norwegian: Grunnriss av en differensiell epistemologi. 2.ed.). Abstrakt, Oslo (2008a)
11. Johansen, S.E.: Systematic unfoldment of differential ontology from qualitative concept of information. In: Thomas, C. (ed.) Ch. 10: Ontology in Information Science, pp. 225–254. InTechOpen (2018)
12. Johansen, S.E.: Basic considerations about Kozyrev's theory of time from recent advances in specialist biology, mathematical physics and philosophical informatics. In: Shikhobalov, L.S. (ed.) Time and Stars: The Centenary of N.A. Kozyrev, pp. 652–703. Nestor-History, St. Petersburg (2008c)
13. Johansen, S.E.: Entering hi-time cybernetics – from recent liftings of science, philosophy, meaning theory and robotics onto the human condition. In: Vassiliev, J.S., Voronov, A.A., Ignatyev, M.B., Kozlov, V.N., Jusupov, R.M. (eds.) Time and Cybernetics – The Centenary of Academician A. A. Voronov, pp. 286–318. St. Petersburg State Polytechnic University, St. Petersburg (2010b)
14. Johansen, S.E.: Some ontological aspects of physics for matter and anti-matter. J. Comput. Methods Sci. Eng. **13**, 135–162 (2013)
15. Johansen, S.E.: Principle(s) of causality as *de facto* fundamental in mathematical physics, including for chance causality applied in quantum mechanics. In: Santilli, R.M. (ed.) Proceedings of International Teleconference on the Einstein-Podolsky-Rosen Argument that Quantum Mechanics is not a Complete Theory, pp. 607–644. Hadronic Press (2021)
16. Santilli, R.M.: Elements of iso-, geno-, hyper-mathematics for matter, their isoduals for antimatter, and their applications in physics, chemistry, and biology. Found. Phys. **33**, 1373–1416 (2003)
17. Santilli, R.M.: Hadronic Mathematics, Mechanics and Chemistry, vol. I–V. International Academic Press, US-Europe-Asia (2008) http://www.i-b-r.org/Hadronic-Mechanics.htm
18. Johansen, S.E.: Spacetimes of Santilli hypermechanics: from hadronic mechanics to hadronic biology and hadronic psychology. Hadron. J. **31**, 451–512 (2008b)
19. Ignatyev, M.B.: The study of the adaptational phenomenon in complex systems. AIP Conf. Proc. **839**, 322–330 (2006)
20. Ignatyev, M.B.: Semantics and self-organization in nanoscale physics. Int. J. Comput. Anticip. Syst. **22**, 17–23 (2008)
21. Ignatyev, M.B.: Necessary and sufficient conditions of nanorobot synthesis. Dokl. Math. **82**, 671–675 (2010)
22. Lie, S.: Over a class of geometrical transformations. (Translation from Norwegian by E. Trell.). Algebras Groups Geom. **15**, 395–445 (1998[1871])
23. Trell, E.: Real forms of the elementary particles with a report of the Σ resonances. Phys. Essays. **5**, 362–373 (1992)
24. Trell, E.: The eightfold eightfold way: application of Lie's true *geometriske transformationer* to elementary particles. Algebras Groups Geom. **15**, 447–471 (1998)
25. Trell, E.: A bottom-up 'Game of Lie' cellular automaton evolution in SO(3) root space both nucleating, crystallizing and space-filling the complete atomic realm. J. Phys. Conf. Ser. **2197**, 012025 (2022)

26. Rowlands, P.: Zero to Infinity. The Foundations of Physics. World Scientific, Singapore (2007)
27. Johansen, S.E.: Complete exposition of non-primes generated by a revolving approach from 8×8 sets of series and thereby *ad negativo* exposition of a systematic pattern for the totality of prime numbers. J. Dyn. Syst. Geom. Theor. **8**(2), 101–171 (2010a)
28. Johansen, S.E.: Negative approach: Implications for the relation between number theory and geometry, including connection to Santilli mathematics, from revolving geometric generation of natural numbers with recurrent outshooting leaving prime numbers. 10th International Conference on Mathematical Problems in Engineering, Aerospace and Sciences. AIP Conf. Proc. **1637**, 450–456 (2014a)
29. Johansen, S.E.: 2014 generation of exact and complete pattern of prime numbers vs. composite numbers as anticipating and computational prime numbering system. Int. J. Comput. Anticip. Syst. **26**, 239–256 (2014c)
30. Johansen, S.E.: Positive approach: Implications for the relation between number theory and geometry, including connection to Santilli mathematics, from Fibonacci reconstitution of natural numbers and of prime numbers. 10th International Conference on Mathematical Problems in Engineering, Aerospace and Sciences. AIP Conf. Proc. **1637**, 457–468 (2014b)
31. Rotkiewicz, A.: Lucas and Frobenius Pseudoprimes. Ann. Math. Silesiane. **17**, 17–39 (2003)

Chapter 10
Riemann Integral of Set-Valued Functions on Time Scales

Vikuozonuo Sekhose and Hemen Bharali

Abstract The Riemann integration for functions whose values are sets on time scales is introduced in this chapter. Efficient definitions of the set-valued Riemann Δ-integral and set-valued Riemann ∇-integral are put forward. The criterion of integrability is established, and some basic properties and results are formulated. The notion of convexity of the integral is discussed, and the relation between set-valued Riemann integrable functions and its support function is investigated. The relation between set-valued Riemann integral and the ordinary Riemann integral and the Hausdorff metric are also explored.

Keywords Set-valued Riemann Δ-integral · Set-valued Riemann ∇-integral · Time scales

10.1 Introduction

The analogy between discrete and continuous analyses and the aim to unify them have fascinated mathematicians throughout the decades. In 1988, S. Hilger introduced a theory of calculus named the time scale calculus, which in essence unified and extended discrete and continuous calculi [12]. S. Hilger, in [12], introduced the delta derivative. More than a decade after this so-called delta derivative was formulated, another derivative called the nabla derivative was introduced by F. M. Atici et al. [3], previously hinted in the works of C. D. Ahlbrandt et al. [1] who introduced a so-called alpha derivative which consisted both the delta and nabla derivatives as special cases.

For an excellent introduction to this subject with theoretical developmental summary and rich history, the reader is referred to the following [6, 12, 13, 15].

In literature, a number of integration concepts are treated in their constructive sense including the Riemann integral for real-valued functions on time scales, first

V. Sekhose (✉) · H. Bharali
Department of Mathematics, Assam Don Bosco University Kamarkuchi, Sonapur, Assam, India

A. Adhikari et al. (eds.), *Quantitative Frontiers*, Industrial and Applied Mathematics, https://doi.org/10.1007/978-981-96-9435-8_10

formulated by S. Sailer [6, 9] using the concept of Darboux sum definition of the integral and then by G. S. Guseinov et al. [9] using the concept of Riemann sum definition of the integral.

We investigate on the theory of Riemann integration for functions whose values are sets on time scales, and we will call these functions set-valued functions. In a recent paper [16, published 2024] the theory of Riemann-Stieltjes integration for functions whose values are sets on time scales has been introduced.

The integration theory for set-valued functions has been of great interest and fascination to mathematicians; some notable pioneers of this theory are R. J. Aumann [4], G. Debrue [8], P. M. Hukuhara [14], Z. Artstein et al. [2], etc.

For an excellent introduction to this subject, the reader is referred to the following [5].

As mentioned, we introduce the concept of Riemann integration for functions whose values are sets on time scales. We put forward efficient definitions of the set-valued Riemann Δ-integral and set-valued Riemann ∇-integral. The criterion of integrability is established, and some basic properties and results are formulated. The notion of convexity of the integral is discussed, and the relation between set-valued Riemann integrable functions and its support function is investigated. The relation between set-valued Riemann integral and the ordinary Riemann integral and the Hausdorff metric are also explored.

10.2 Preliminary

Here we recall a few notations, definitions, and results which will be used throughout the text.

Let $\mathrm{Com}(\mathbb{R}^n)$ represent the collection of all non-empty, compact, convex subsets of $\mathbb{R}^n$.

The scalar multiplication and Minkowski sum are defined as [5]- for $P, Q \in \mathrm{Com}(\mathbb{R}^n)$ and $\gamma \in \mathbb{R}$-

$$\gamma \cdot P = \{\gamma \cdot p | p \in P\}\ ; \text{ and } \ P + Q = \{p + q | p \in P, q \in Q\}.$$

For every $P, Q \in \mathrm{Com}(\mathbb{R}^n)$, the Hausdorff metric "$s$" is defined as [8]

$$s(P, Q) = \max\left\{\sup_{p\in P}\inf_{q\in Q}\|p - q\|, \sup_{q\in Q}\inf_{p\in P}\|q - p\|\right\};$$

here $\|\cdot\|$ stands for the norm of $\mathbb{R}^n$. The metric space $\big(\mathrm{Com}(\mathbb{R}^n), s\big)$ is complete.

The norm of set P, denoted by $\|P\|$, is defined as [17]

$$\|P\| = s\big(P, \{0\}\big) = \sup_{p\in P}\|p\|.$$

The diameter of set P, denoted by $D(P)$, is defined as [5]

$$D(P) = \sup_{p_1, p_2 \in P} \left\| p_1 - p_2 \right\|.$$

Definition 1 ([5] Convex Set) A set $P \subset \mathbb{R}^n$ is convex if $(1-\gamma)\cdot p_1 + \gamma \cdot p_2 \in P$ for all $p_1, p_2 \in P$ and for all $\gamma \in [0, 1]$.

Definition 2 ([5] Convex Hull) The convex hull of set P is defined as $co(P) = \bigcap_{P \subset Q} Q$, where Q is convex.

Proposition 10.2.1 ([5]) *Let $P \subset \mathbb{R}^n$; if P is convex, then $co(P) = P$.*

Proposition 10.2.2 ([5, 18]) *Let $P, Q, R, U \in \mathrm{Com}(\mathbb{R}^n)$ and $\gamma \in \mathbb{R}$. Then:*

1. $s(P, Q) = s(Q, P)$.
2. $s(co(P), co(Q)) \leq s(P, Q)$.
3. $s(\gamma \cdot P, \gamma \cdot Q) = |\gamma| \cdot s(P, Q)$.
4. $s(P, Q) \leq s(P, R) + s(R, Q)$.
5. $s(P + Q, R + U) \leq s(P, R) + s(Q, U)$.
6. $s(\{p\}, \{q\}) = \left\| p - q \right\|$.

Proposition 10.2.3 ([5]) *Let $P, Q \in \mathrm{Com}(\mathbb{R}^n)$ and $\gamma \in \mathbb{R}$. Then:*

1. $co(\gamma \cdot P) = \gamma \cdot co(P)$.
2. $co(P + Q) = co(P) + co(Q)$.

Theorem 10.2.1 ([5] Shapley-Folkman-Starr) *Let $P_h \in \mathrm{Com}(\mathbb{R}^n)$, where $h = 1, 2, \ldots, u$, $u \in \mathbb{N}$. Then,*

$$s\left(\sum_{h=1}^{u} P_h, \sum_{h=1}^{u} co(P_h)\right) \leq \frac{\sqrt{n}}{2} \max_{h=1,2,\ldots,u} D(P_h).$$

Definition 3 ([11] Support Function) Let set $P \in \mathrm{Com}(\mathbb{R}^n)$, $\eta \in \mathbb{R}^n \wedge \|\eta\| = 1$ (here "$\wedge$" stands for "and"). The function $\psi(\cdot, P)$ defined as $\psi(\eta, P) = \sup_{p \in P} \langle \eta, p \rangle$ is called the support function of P. Here $\langle \eta, p \rangle$ means the scalar product of η and p.

Set $P \in \mathrm{Com}(\mathbb{R}^n)$ (i.e., set P is convex and compact); the support function is characterized as [17]

$$P = \left\{ p \in \mathbb{R}^n | \langle \eta, p \rangle \leq \psi(\eta, P), \ \forall \, \eta \in \mathbb{R}^n \wedge \|\eta\| = 1 \right\}. \tag{10.1}$$

The relation between the Hausdorff metric and the support function for $P, Q \in \mathrm{Com}(\mathbb{R}^n)$ is defined as

$$s(P, Q) = \sup_{\eta \in \mathbb{R}^n \wedge \|\eta\| = 1} \left| \psi(\eta, P) - \psi(\eta, Q) \right|. \tag{10.2}$$

Proposition 10.2.4 ([2]) *Let* $P, Q \in \text{Com}(\mathbb{R}^n)$, $\eta \in \mathbb{R}^n \wedge \|\eta\| = 1$, *and* $\gamma \in \mathbb{R}$. *Then,*

1. $\psi(\eta, \gamma \cdot P) = \gamma \cdot \psi(\eta, P)$.
2. $\psi(\eta, P + Q) = \psi(\eta, P) + \psi(\eta, Q)$.

Function f is stated to be a set-valued function of t on $[y, z]$ if it assigns a non-empty set for every $\mathrm{t} \in [y, z]$.

Definition 4 ([5] Bounded) A set-valued function f is bounded if there exists a constant $C \geq 0$ such that $s\big(f(\mathrm{t}), \{0\}\big) \leq C$ for all $\mathrm{t} \in [y, z]$.

Definition 5 ([2] Integrably Bounded) A set-valued function f is integrably bounded if there exists an integrable real-valued function g such that $p \in f(\mathrm{t})$ implies $\sup_{p \in f(\mathrm{t})} \|p\| \leq g(\mathrm{t})$.

A set-valued function f is continuous if it is continuous with respect to the Hausdorff metrics "s" [8].

Definition 6 ([8] Continuity of Set-Valued Function) A function $f : [y, z] \to \text{Com}(\mathbb{R}^n)$ is continuous at $\mathrm{t}_0 \in [y, z]$, if for each $\varepsilon > 0$ there exists $\delta > 0$ such that $s\big(f(\mathrm{t}), f(\mathrm{t}_0)\big) < \varepsilon$ whenever $|\mathrm{t} - \mathrm{t}_0| < \delta$.

Definition 7 (Uniform Continuity of Set-Valued Function) A function $f : [y, z] \to \text{Com}(\mathbb{R}^n)$ is uniformly continuous on $[y, z]$, if for each $\varepsilon > 0$ there exists $\delta > 0$ such that for all $\mathrm{t}_0, \mathrm{t}_1 \in [y, z]$ we get $s\big(f(\mathrm{t}_0), f(\mathrm{t}_1)\big) < \varepsilon$ whenever $|\mathrm{t}_0 - \mathrm{t}_1| < \delta$.

A time scale $\mathcal{T}$ is any non-empty closed subset of $\mathbb{R}$ [12].

Definition 8 ([12] Forward Jump Operator) The forward jump operator denoted by σ is a mapping $\sigma : \mathcal{T} \to \mathcal{T}$ defined by $\sigma(\mathrm{t}) = \inf\{\tilde{t} \in \mathcal{T} : \tilde{t} > \mathrm{t}\}$.

Definition 9 ([12] Backward Jump Operator) The backward jump operator denoted by ρ is a mapping $\rho : \mathcal{T} \to \mathcal{T}$ defined by $\rho(\mathrm{t}) = \sup\{\tilde{t} \in \mathcal{T} : \tilde{t} < \mathrm{t}\}$.

Assuming $y \leq z$, intervals in $\mathcal{T}$ are defined as [6] $[y, z]_{\mathcal{T}} = \{\mathrm{t} \in \mathcal{T} : y \leq \mathrm{t} \leq z\}$; $(y, z)_{\mathcal{T}} = \{\mathrm{t} \in \mathcal{T} : y < \mathrm{t} < z\}$; $[y, z)_{\mathcal{T}} = \{\mathrm{t} \in \mathcal{T} : y \leq \mathrm{t} < z\}$; $(y, z]_{\mathcal{T}} = \{\mathrm{t} \in \mathcal{T} : y < \mathrm{t} \leq z\}$.

Let $[y, z]_{\mathcal{T}}$ be a closed interval on $\mathcal{T}$. Let $\mathfrak{P}$ represent the collection of all possible partitions of $[y, z]_{\mathcal{T}}$.

Definition 10 ([7] Convex Function) A function $g : [y, z]_{\mathcal{T}} \to \mathbb{R}$ is convex on $[y, z]_{\mathcal{T}}$ if $g\big(\gamma \cdot \mathrm{t}_1 + (1 - \gamma) \cdot \mathrm{t}_2\big) \leq \gamma \cdot g(\mathrm{t}_1) + (1 - \gamma) \cdot g(\mathrm{t}_2)$, for all $\mathrm{t}_1, \mathrm{t}_2 \in [y, z]_{\mathcal{T}}$ and all $\gamma \in [0, 1]$ such that $\gamma \cdot \mathrm{t}_1 + (1 - \gamma) \cdot \mathrm{t}_2 \in [y, z]_{\mathcal{T}}$.

Below we give the definition of the Riemann Δ-integral and Riemann ∇-integral for real-valued functions on time scales [9, 10]. For the sake of clarity, $\mathcal{Y} \in \mathfrak{P}$ will denote the partition for the Δ-integral and $\mathcal{Z} \in \mathfrak{P}$ will denote the partition for the ∇-integral.

Let $\mathcal{Y} = \{y = \mathrm{t}_0 < \mathrm{t}_1 < \ldots < \mathrm{t}_u = z\}$, $u \in \mathbb{N}$ with $\mathrm{t}_0, \mathrm{t}_1, \ldots, \mathrm{t}_u$ being the finite points of division. We consider Δ-subintervals of the form $[\mathrm{t}_{h-1}, \mathrm{t}_h)_{\mathcal{T}}$, for $1 \leq h \leq u$, and from each Δ-subinterval we choose ϑ_h arbitrarily defined as $\vartheta_h \in [\mathrm{t}_{h-1}, \mathrm{t}_h)_{\mathcal{T}}$ and call it the Δ-tag point of the respective Δ-subinterval. We define a point-interval collection as $\breve{\mathcal{Y}} = \{(\vartheta_h, [\mathrm{t}_{h-1}, \mathrm{t}_h)_{\mathcal{T}})\}_{h=1}^{u}$ and call it the Δ-tagged partition. The mesh of $\mathcal{Y}$ is defined as $\mathrm{mesh}(\mathcal{Y}) = \max_{1 \leq h \leq u}(\mathrm{t}_h - \mathrm{t}_{h-1}) > 0$. For some $\delta > 0$, $\mathcal{Y}_\delta$ will represent a partition of $[y, z]_{\mathcal{T}}$ with mesh δ satisfying the property: For each $h = 1, 2, \ldots, u$, we have either $(\mathrm{t}_h - \mathrm{t}_{h-1}) \leq \delta$ or $(\mathrm{t}_h - \mathrm{t}_{h-1}) > \delta \wedge \rho(\mathrm{t}_h) = \mathrm{t}_{h-1}$. Henceforth, $\breve{\mathcal{Y}}_\delta$ will mean a Δ-tagged partition with mesh δ.

We proceed to state the definition of Riemann Δ-integral on time scales according to G. S. Guseinov and B. Kaymakçalan [9, 10] for real-valued functions.

Definition 11 ([10]) A function $g : [y, z]_{\mathcal{T}} \to \mathbb{R}$ is Riemann Δ-integrable if there exists a real number $\overline{I} \in \mathbb{R}$ such that for any $\varepsilon > 0$ there exists $\delta > 0$; hence for any Δ-tagged partition $\breve{\mathcal{Y}}_\delta$, we have $\left| \sum_{h=1}^{u} g(\vartheta_h) \cdot (\mathrm{t}_h - \mathrm{t}_{h-1}) - \overline{I} \right| < \varepsilon$. Here $\overline{I} = \overline{\mathrm{R}} \int_y^z g(\mathrm{t}) \Delta \mathrm{t}$.

Now, let $\mathcal{Z} = \{y = \mathrm{t}_0 < \mathrm{t}_1 < \ldots < \mathrm{t}_u = z\}$, $u \in \mathbb{N}$ with $\mathrm{t}_0, \mathrm{t}_1, \ldots, \mathrm{t}_u$ being the finite points of division. We consider ∇-subintervals of the form $(\mathrm{t}_{h-1}, \mathrm{t}_h]_{\mathcal{T}}$, for $1 \leq h \leq u$, and from each ∇-subinterval, we choose ξ_h arbitrarily defined as $\xi_h \in (\mathrm{t}_{h-1}, \mathrm{t}_h]_{\mathcal{T}}$ and call it the ∇-tag point of the respective ∇-subinterval. We define a point-interval collection as $\breve{\mathcal{Z}} = \{(\xi_h, (\mathrm{t}_{h-1}, \mathrm{t}_h]_{\mathcal{T}})\}_{h=1}^{u}$ and call it the ∇-tagged partition. The mesh of $\mathcal{Z}$ is defined as $\mathrm{mesh}(\mathcal{Z}) = \max_{1 \leq h \leq u}(\mathrm{t}_h - \mathrm{t}_{h-1}) > 0$. For some $\delta > 0$, $\mathcal{Z}_\delta$ will represent a partition of $[y, z]_{\mathcal{T}}$ with mesh δ satisfying the property: for each $h = 1, 2, \ldots, u$, we have either $(\mathrm{t}_h - \mathrm{t}_{h-1}) \leq \delta$ or $(\mathrm{t}_h - \mathrm{t}_{h-1}) > \delta \wedge \sigma(\mathrm{t}_{h-1}) = \mathrm{t}_h$. Henceforth, $\breve{\mathcal{Z}}_\delta$ will mean a ∇-tagged partition with mesh δ.

We proceed to state the definition of Riemann ∇-integral on time scales according to G. S. Guseinov and B. Kaymakçalan [9, 10] for real-valued functions.

Definition 12 ([10]) A function $g : [y, z]_{\mathcal{T}} \to \mathbb{R}$ is Riemann ∇-integrable if there exists a real number $\underline{I} \in \mathbb{R}$ such that for any $\varepsilon > 0$ there exists $\delta > 0$; hence for any ∇-tagged partition $\breve{\mathcal{Z}}_\delta$ we have $\left| \sum_{h=1}^{u} g(\xi_h) \cdot (\mathrm{t}_h - \mathrm{t}_{h-1}) - \underline{I} \right| < \varepsilon$. Here $\underline{I} = \underline{\mathrm{R}} \int_y^z g(\mathrm{t}) \nabla \mathrm{t}$.

We proceed to give our definition of the Riemann integral for functions whose values are sets on time scales.

10.3 Riemann Integral for Set-Valued Functions

Let $[y, z]_{\mathcal{T}}$ be a closed interval on $\mathcal{T}$. Let $\mathfrak{P}$ represent the collection of all possible partitions of $[y, z]_{\mathcal{T}}$.

Before proceeding we first establish some preliminary information required for the definitions. For the sake of clarity, $\mathcal{Y} \in \mathfrak{P}$ will denote the partition for the Δ-integral and $\mathcal{Z} \in \mathfrak{P}$ will denote the partition for the ∇-integral.

10.3.1 Riemann Δ-Integral for Set-Valued Functions

Let $\mathcal{Y} = \{y = \mathrm{t}_0 < \mathrm{t}_1 < \ldots < \mathrm{t}_u = z\}$, $u \in \mathbb{N}$ with $\mathrm{t}_0, \mathrm{t}_1, \ldots, \mathrm{t}_u$ being the finite points of division. We consider Δ-subintervals of the form $[\mathrm{t}_{h-1}, \mathrm{t}_h)_{\mathcal{T}}$, for $1 \leq h \leq u$, and from each Δ-subinterval we choose ϑ_h arbitrarily defined as $\vartheta_h \in [\mathrm{t}_{h-1}, \mathrm{t}_h)_{\mathcal{T}}$ and call it the Δ-tag point of the respective Δ-subinterval. We define a point-interval collection as $\breve{\mathcal{Y}} = \{(\vartheta_h, [\mathrm{t}_{h-1}, \mathrm{t}_h)_{\mathcal{T}})\}_{h=1}^{u}$ and call it the Δ-tagged partition. The mesh of $\mathcal{Y}$ is defined as $\mathrm{mesh}(\mathcal{Y}) = \max_{1\leq h\leq u}(\mathrm{t}_h - \mathrm{t}_{h-1}) > 0$. For some $\delta > 0$, $\mathcal{Y}_\delta$ will represent a partition of $[y, z]_{\mathcal{T}}$ with mesh δ satisfying the property: For each $h = 1, 2, \ldots, u$, we have either $(\mathrm{t}_h - \mathrm{t}_{h-1}) \leq \delta$ or $(\mathrm{t}_h - \mathrm{t}_{h-1}) > \delta$ $\wedge$ $\rho(\mathrm{t}_h) = \mathrm{t}_{h-1}$. Henceforth, $\breve{\mathcal{Y}}_\delta$ will mean a Δ-tagged partition with mesh δ.

We proceed to state the definition of Riemann Δ-integral of set-valued functions on time scales, which we will call the set-valued Riemann Δ-integral.

We form the set-valued Riemann Δ-sum, $\overline{\mathrm{R}}_s(f; \breve{\mathcal{Y}}_\delta)$, of the set-valued function f evaluated at the Δ-tags as

$$\overline{\mathrm{R}}_s(f; \breve{\mathcal{Y}}_\delta) = \sum_{h=1}^{u} f(\vartheta_h) \cdot (\mathrm{t}_h - \mathrm{t}_{h-1}).$$

Definition 13 Set-Valued Riemann Δ-Integral: A function $f : [y, z]_{\mathcal{T}} \to \mathrm{Com}(\mathbb{R}^n)$ is set-valued Riemann Δ-integrable if there exists a set $\overline{I}_s \in \mathrm{Com}(\mathbb{R}^n)$ such that for any $\varepsilon > 0$ there exists $\delta > 0$; hence for any Δ-tagged partition $\breve{\mathcal{Y}}_\delta$, we have

$$s\Big(\overline{\mathrm{R}}_s(f; \breve{\mathcal{Y}}_\delta), \overline{I}_s\Big) = \max\left\{ \sup_{r\in\overline{\mathrm{R}}_s(f;\breve{\mathcal{Y}}_\delta)} \inf_{i\in\overline{I}_s} \|r - i\|, \sup_{i\in\overline{I}_s} \inf_{r\in\overline{\mathrm{R}}_s(f;\breve{\mathcal{Y}}_\delta)} \|i - r\| \right\} < \varepsilon.$$

Here $\overline{I}_s = \overline{\mathrm{R}}_s \int_y^z f(\mathrm{t})\Delta\mathrm{t}$, where $\overline{\mathrm{R}}_s \int_y^z f(\mathrm{t})\Delta\mathrm{t}$ is called the set-valued Riemann Δ-integral.

The set of all set-valued Riemann Δ-integrable functions on $[y, z]_{\mathcal{T}}$ will be denoted by $(\mathcal{R}_s)_\Delta[y, z]_{\mathcal{T}}$.

10.3.2 Riemann ∇-Integral for Set-Valued Functions

Let $\mathcal{Z} = \{y = \mathrm{t}_0 < \mathrm{t}_1 < \ldots < \mathrm{t}_u = z\}$, $u \in \mathbb{N}$ with $\mathrm{t}_0, \mathrm{t}_1, \ldots, \mathrm{t}_u$ being the finite points of division. We consider ∇-subintervals of the form $(\mathrm{t}_{h-1}, \mathrm{t}_h]_{\mathcal{T}}$, for $1 \leq h \leq u$, and from each ∇-subinterval we choose ξ_h arbitrarily defined as $\xi_h \in (\mathrm{t}_{h-1}, \mathrm{t}_h]_{\mathcal{T}}$ and call it the ∇-tag point of the respective ∇-subinterval. We define a point-interval collection as $\breve{\mathcal{Z}} = \{(\xi_h, (\mathrm{t}_{h-1}, \mathrm{t}_h]_{\mathcal{T}})\}_{h=1}^{u}$ and call it the ∇-tagged partition. The mesh of $\mathcal{Z}$ is defined as $\mathrm{mesh}(\mathcal{Z}) = \max_{1\leq h\leq u}(\mathrm{t}_h - \mathrm{t}_{h-1}) > 0$. For some $\delta > 0$, $\mathcal{Z}_\delta$ will represent a partition of $[y, z]_{\mathcal{T}}$ with mesh δ satisfying the property: For each $h = 1, 2, \ldots, u$ we have either $(\mathrm{t}_h - \mathrm{t}_{h-1}) \leq \delta$ or $(\mathrm{t}_h - \mathrm{t}_{h-1}) > \delta \wedge \rho(\mathrm{t}_h) = \mathrm{t}_{h-1}$. Henceforth, $\breve{\mathcal{Z}}_\delta$ will mean a ∇-tagged partition with mesh δ.

We proceed to state the definition of Riemann ∇-integral of set-valued functions on time scales, which we will call the set-valued Riemann ∇-integral.

We form the set-valued Riemann ∇-sum, $\underline{\mathrm{R}}_s(f; \breve{\mathcal{Z}}_\delta)$, of the set-valued function f evaluated at the ∇-tags as

$$\underline{\mathrm{R}}_s(f; \breve{\mathcal{Z}}_\delta) = \sum_{h=1}^{u} f(\xi_h) \cdot \big(\mathrm{t}_h - \mathrm{t}_{h-1}\big).$$

Definition 14 (Set-Valued Riemann ∇-Integral) A function $f : [y, z]_{\mathcal{T}} \to \mathrm{Com}(\mathbb{R}^n)$ is set-valued Riemann ∇-integrable if there exists a set $\underline{I}_s \in \mathrm{Com}(\mathbb{R}^n)$ such that for any $\varepsilon > 0$ there exists $\delta > 0$; hence for any ∇-tagged partition $\breve{\mathcal{Z}}_\delta$ we have

$$s\Big(\underline{\mathrm{R}}_s(f; \breve{\mathcal{Z}}_\delta), \underline{I}_s\Big) = \max\left\{ \sup_{r \in \underline{\mathrm{R}}_s(f; \breve{\mathcal{Z}}_\delta)} \inf_{i \in \underline{I}_s} \|r - i\|, \sup_{i \in \underline{I}_s} \inf_{r \in \underline{\mathrm{R}}_s(f; \breve{\mathcal{Z}}_\delta)} \|i - r\| \right\} < \varepsilon.$$

Here $\underline{I}_s = \underline{\mathrm{R}}_s \int_y^z f(\mathrm{t}) \nabla \mathrm{t}$, where $\underline{\mathrm{R}}_s \int_y^z f(\mathrm{t}) \nabla \mathrm{t}$ is called the set-valued Riemann ∇-integral.

The set of all set-valued Riemann ∇-integrable functions on $[y, z]_{\mathcal{T}}$ will be denoted by $(\mathcal{R}_s)_\nabla[y, z]_{\mathcal{T}}$.

10.4 Results

We now present a few basic results for set-valued Riemann Δ-integration, which will hold similarly for the set-valued Riemann ∇-integration hence not explicitly stated here.

Below we prove that if a set-valued function is given to be set-valued integrable, then its integral set is unique.

Theorem 10.4.1 *If $f \in (\mathcal{R}_s)_\Delta[y, z]_{\mathcal{T}}$, then the value of set integral, $\overline{I}_s$, is unique.*

Proof Let us assume that f on $[y, z]_{\mathcal{T}}$ has two set integrals, say $\overline{I}_s'$ and $\overline{I}_s''$, both satisfying the definition (Definition 13). Let $\varepsilon > 0$; there exists $\delta'_{\frac{\varepsilon}{2}} > 0$ such that for any Δ-tagged partition $\breve{\mathcal{Y}}_{\delta'_{\frac{\varepsilon}{2}}}$ of $[y, z]_{\mathcal{T}}$, the respective set-valued Riemann Δ-sum, $\overline{\mathrm{R}}_s'\Big(f; \breve{\mathcal{Y}}_{\delta'_{\frac{\varepsilon}{2}}}\Big)$, satisfies $s\Big(\overline{\mathrm{R}}_s'\Big(f; \breve{\mathcal{Y}}_{\delta'_{\frac{\varepsilon}{2}}}\Big), \overline{I}_s'\Big) < \frac{\varepsilon}{2}$. Also there exists $\delta''_{\frac{\varepsilon}{2}} > 0$ such that for any Δ-tagged partition $\breve{\mathcal{Y}}_{\delta''_{\frac{\varepsilon}{2}}}$ of $[y, z]_{\mathcal{T}}$, the respective set-valued Riemann Δ-sum, $\overline{\mathrm{R}}_s''\Big(f; \breve{\mathcal{Y}}_{\delta''_{\frac{\varepsilon}{2}}}\Big)$, satisfies $s\Big(\overline{\mathrm{R}}_s''\Big(f; \breve{\mathcal{Y}}_{\delta''_{\frac{\varepsilon}{2}}}\Big), \overline{I}_s''\Big) < \frac{\varepsilon}{2}$.

Now, let $\delta_\varepsilon = \min\left\{\delta'_{\frac{\varepsilon}{2}}, \delta''_{\frac{\varepsilon}{2}}\right\} > 0$, and let $\breve{\mathcal{Y}}_{\delta_\varepsilon}$ be the respective Δ-tagged partition of $[y,z]_\mathcal{T}$. Since mesh of $\breve{\mathcal{Y}}_{\delta_\varepsilon}$ is lesser than or equal to the mesh of the Δ-tagged partitions $\breve{\mathcal{Y}}_{\delta'_{\frac{\varepsilon}{2}}}$ and $\breve{\mathcal{Y}}_{\delta''_{\frac{\varepsilon}{2}}}$, hence taking $\overline{\mathrm{R}}_s(f;\breve{\mathcal{Y}}_{\delta_\varepsilon})$ to be the respective set-valued Riemann Δ-sum, we deduce

$$s\left(\overline{\mathrm{R}}_s(f;\breve{\mathcal{Y}}_{\delta_\varepsilon}), \overline{I}'_s\right) < \frac{\varepsilon}{2} \text{ and } s\left(\overline{\mathrm{R}}_s(f;\breve{\mathcal{Y}}_{\delta_\varepsilon}), \overline{I}''_s\right) < \frac{\varepsilon}{2},$$

whence, applying Proposition 10.2.2 *(4)*, we deduce

$$s(\overline{I}'_s, \overline{I}''_s) \le s\left(\overline{I}'_s, \overline{\mathrm{R}}_s(f;\breve{\mathcal{Y}}_{\delta_\varepsilon})\right) + s\left(\overline{\mathrm{R}}_s(f;\breve{\mathcal{Y}}_{\delta_\varepsilon}), \overline{I}''_s\right) < \varepsilon.$$

Since $\varepsilon > 0$ is arbitrary, thus we conclude that $\overline{I}'_s = \overline{I}''_s$.

We state and proof the criterion of integrability for set-valued Riemann Δ-integral on time scales.

Theorem 10.4.2 (Criterion of Integrability) *Function $f : [y,z]_\mathcal{T} \to \mathrm{Com}(\mathbb{R}^n)$ is set-valued Riemann Δ-integrable on $[y,z]_\mathcal{T}$ if and only if for each $\varepsilon > 0$ there exists $\delta > 0$ such that for Δ-partitions $\mathcal{Y}_1, \mathcal{Y}_2 \in \mathfrak{P}$ both having mesh δ implies*

$$s\left(\overline{\mathrm{R}}_s(f;\mathcal{Y}_1), \overline{\mathrm{R}}_s(f;\mathcal{Y}_2)\right) < \varepsilon. \tag{10.3}$$

Proof Suppose that $f \in (\mathcal{R}_s)_\Delta[y,z]_\mathcal{T}$; consider for $\varepsilon > 0$ there exists $\delta > 0$ such that for Δ-partitions $\mathcal{Y}_1$ and $\mathcal{Y}_2$ of $[y,z]_\mathcal{T}$ having $\mathrm{mesh}(\mathcal{Y}_1) < \delta$ and $\mathrm{mesh}(\mathcal{Y}_2) < \delta$ satisfies

$$s\left(\overline{\mathrm{R}}_s(f;\mathcal{Y}_1), \overline{I}_s\right) < \frac{\varepsilon}{2} \text{ and } s\left(\overline{\mathrm{R}}_s(f;\mathcal{Y}_2), \overline{I}_s\right) < \frac{\varepsilon}{2}.$$

Therefore applying Proposition 10.2.2 *(4)* we have

$$s\left(\overline{\mathrm{R}}_s(f;\mathcal{Y}_1), \overline{\mathrm{R}}_s(f;\mathcal{Y}_2)\right) \le s\left(\overline{\mathrm{R}}_s(f;\mathcal{Y}_1), \overline{I}_s\right) + s\left(\overline{\mathrm{R}}_s(f;\mathcal{Y}_2), \overline{I}_s\right) < \varepsilon.$$

Hence, given f is set-valued Riemann Δ-integrable implies Eq. (10.3) holds.

Conversely, for each $x \in \mathbb{N}$, let $\delta_x > 0$ be such that $\mathcal{Y}_1$ and $\mathcal{Y}_2$ are Δ-partitions both with mesh $< \delta_x$ and

$$s\left(\overline{\mathrm{R}}_s(f;\mathcal{Y}_1), \overline{\mathrm{R}}_s(f;\mathcal{Y}_2)\right) < \frac{1}{x}.$$

Evidently, we may assume that $\delta_x \geq \delta_{x+1}$ for $x \in \mathbb{N}$; otherwise we replace δ_x by $\delta'_x = \min\{\delta_1, \ldots, \delta_x\}$. For each $x \in \mathbb{N}$, let $\mathcal{Y}_x$ be a Δ-partition with mesh$(\mathcal{Y}_x) < \delta_x$. Clearly if $m > x$, then both $\mathcal{Y}_m$ and $\mathcal{Y}_x$ have mesh $< \delta_x$; hence

$$s\Big(\overline{\mathrm{R}}_s\big(f; \mathcal{Y}_x\big), \overline{\mathrm{R}}_s\big(f; \mathcal{Y}_m\big)\Big) < \frac{1}{x} \quad \text{for} \quad m > x. \tag{10.4}$$

Consequently, the sequence $\big\{\overline{\mathrm{R}}_s\big(f; \mathcal{Y}_m\big)\big\}_{m=1}^{\infty}$ is Cauchy; hence there exists set I such that $\lim_{m\to\infty} \overline{\mathrm{R}}_s\big(f; \mathcal{Y}_m\big) = I$. Fixing x and passing to the limit as $m \to \infty$ in Eq. (10.4), we obtain

$$s\Big(\overline{\mathrm{R}}_s\big(f; \mathcal{Y}_x\big), I\Big) < \frac{1}{x} \quad \text{for} \quad x \in \mathbb{N}. \tag{10.5}$$

To see that I is set-valued Riemann Δ-integral of f. Given $\varepsilon > 0$, let $K \in \mathbb{N}$ satisfy $K > \frac{2}{\varepsilon}$. If $\mathcal{Y}$ is any Δ-partition with mesh$(\mathcal{Y}) < \delta_K$, then applying Proposition 10.2.2 *(4)*, Eqs. (10.4), and (10.5) we obtain

$$s\Big(\overline{\mathrm{R}}_s\big(f; \mathcal{Y}\big), I\Big) \leq s\Big(\overline{\mathrm{R}}_s\big(f; \mathcal{Y}\big), \overline{\mathrm{R}}_s\big(f; \mathcal{Y}_K\big)\Big) + s\Big(\overline{\mathrm{R}}_s\big(f; \mathcal{Y}_K\big), I\Big) < \frac{1}{K} + \frac{1}{K} < \varepsilon.$$

Since $\varepsilon > 0$ is arbitrary, thus $f \in (\mathcal{R}_s)_\Delta[y, z]_\mathcal{T}$ with integral I.

Partition $\mathcal{Y}_1$ is said to be a refinement of partition $\mathcal{Y}_2$ if the points of $\mathcal{Y}_2$ form a subset of the points of $\mathcal{Y}_1$, i.e., $\mathcal{Y}_1$ refines $\mathcal{Y}_2$.

Corollary 10.4.1 *Let $f : [y, z]_\mathcal{T} \to \mathrm{Com}(\mathbb{R}^n)$ be set-valued Riemann Δ-integrable on $[y, z]_\mathcal{T}$ with Δ-partition $\mathcal{Y}_\varepsilon \in \mathfrak{P}$ for $\varepsilon > 0$. If Δ-partitions $\mathcal{Y}_1$ and $\mathcal{Y}_2$ of $[y, z]_\mathcal{T}$ refine $\mathcal{Y}_\varepsilon$, then $s\Big(\overline{\mathrm{R}}_s\big(f; \mathcal{Y}_1\big), \overline{\mathrm{R}}_s\big(f; \mathcal{Y}_2\big)\Big) < \varepsilon$.*

We now introduce the concept of set-valued convexity.

Definition 15 A function $f : [y, z]_\mathcal{T} \to \mathrm{Com}(\mathbb{R}^n)$ is a convex set-valued function if for all $\mathrm{t}_1, \mathrm{t}_2 \in [y, z]_\mathcal{T}$ and $\gamma \in [0, 1]$ we have $\gamma \cdot f(\mathrm{t}_1) + (1 - \gamma) \cdot f(\mathrm{t}_2) \subseteq f(\gamma \cdot \mathrm{t}_1 + (1 - \gamma) \cdot \mathrm{t}_2)$, for which $\gamma \cdot \mathrm{t}_1 + (1 - \gamma) \cdot \mathrm{t}_2 \in [y, z]_\mathcal{T}$.

Theorem 10.4.3 *If $f : [y, z]_\mathcal{T} \to \mathrm{Com}(\mathbb{R}^n)$ is set-valued Riemann Δ-integrable, then the set integral, $\overline{I}_s$, is a convex set.*

Theorem 10.4.4 *If function $f : [y, z]_\mathcal{T} \to \mathrm{Com}(\mathbb{R}^n)$ is set-valued Riemann Δ-integrable on $[y, z]_\mathcal{T}$, then $co(f)$ is also set-valued Riemann Δ-integrable on $[y, z]_\mathcal{T}$ with*

$$\overline{\mathrm{R}}_s \int_y^z co(f(\mathrm{t}))\Delta\mathrm{t} = co\bigg(\overline{\mathrm{R}}_s \int_y^z f(\mathrm{t})\Delta\mathrm{t}\bigg).$$

Proof Let $\varepsilon > 0$ and $\delta > 0$; then,

$$\sum_{h=1}^{u} co\big(f(\vartheta_h)\big) \cdot (\mathrm{t}_h - \mathrm{t}_{h-1}) = co\bigg(\sum_{h=1}^{u} f(\vartheta_h) \cdot (\mathrm{t}_h - \mathrm{t}_{h-1})\bigg),$$

following Proposition 10.2.3. Also Propositions 10.2.3 *(1)* and 10.2.2 *(2)* yield

$$s\bigg(\sum_{h=1}^{u} co\big(f(\vartheta_h)\big) \cdot (\mathrm{t}_h - \mathrm{t}_{h-1}), co\bigg(\overline{\mathrm{R}}_s \int_y^z f(\mathrm{t})\Delta\mathrm{t}\bigg)\bigg)$$

$$= s\bigg(co\bigg(\sum_{h=1}^{u} f(\vartheta_h) \cdot (\mathrm{t}_h - \mathrm{t}_{h-1})\bigg),$$

$$co\bigg(\overline{\mathrm{R}}_s \int_y^z f(\mathrm{t})\Delta\mathrm{t}\bigg)\bigg) \leq s\bigg(\sum_{h=1}^{u} f(\vartheta_h) \cdot (\mathrm{t}_h - \mathrm{t}_{h-1}), \overline{\mathrm{R}}_s \int_y^z f(\mathrm{t})\Delta\mathrm{t}\bigg) < \varepsilon,$$

following that $f \in (\mathcal{R}_s)_\Delta[y, z]_\mathcal{T}$.

Hence $s\Big(\sum_{h=1}^{u} co\big(f(\vartheta_h)\big) \cdot (t_h - t_{h-1}), \overline{\mathrm{R}}_s \int_y^z co\big(f(\mathrm{t})\big)\Delta\mathrm{t}\Big) < \varepsilon$.

Theorem 10.4.5 *If the convex hull of bounded set-valued function f is given to be set-valued Riemann Δ-integrable on $[y, z]_\mathcal{T}$, then f is also set-valued Riemann Δ-integrable on $[y, z]_\mathcal{T}$ with*

$$\overline{\mathrm{R}}_s \int_y^z co\big(f(\mathrm{t})\big)\Delta\mathrm{t} = \overline{\mathrm{R}}_s \int_y^z f(\mathrm{t})\Delta\mathrm{t}.$$

Proof Let $\varepsilon > 0$ and $\delta \leq \frac{2\varepsilon}{\sqrt{n}C} > 0$; here $C > 0$ is the bound of function f. Also given $co(f) \in (\mathcal{R}_s)_\Delta[y, z]_\mathcal{T}$, we have

$$s\bigg(\sum_{h=1}^{u} co\big(f(\vartheta_h)\big) \cdot (\mathrm{t}_h - \mathrm{t}_{h-1}), \overline{\mathrm{R}}_s \int_y^z co\big(f(\mathrm{t})\big)\Delta\mathrm{t}\bigg) < \varepsilon.$$

Now, Proposition 10.2.2 *(4)* and Theorem 10.2.1 yield

$$s\bigg(\sum_{h=1}^{u} f(\vartheta_h) \cdot (\mathrm{t}_h - \mathrm{t}_{h-1}), \overline{\mathrm{R}}_s \int_y^z co\big(f(\mathrm{t})\big)\Delta\mathrm{t}\bigg)$$

$$\leq s\bigg(\sum_{h=1}^{u} f(\vartheta_h) \cdot (\mathrm{t}_h - \mathrm{t}_{h-1}), \sum_{h=1}^{u} co\big(f(\vartheta_h)\big) \cdot$$

$$(\mathrm{t}_h - \mathrm{t}_{h-1})\bigg) + s\bigg(\sum_{h=1}^{u} co\big(f(\vartheta_h)\big) \cdot (\mathrm{t}_h - \mathrm{t}_{h-1}), \overline{\mathrm{R}}_s \int_y^z co\big(f(\mathrm{t})\big)\Delta\mathrm{t}\bigg)$$

$$\leq \frac{\sqrt{n}}{2} \max_{1 \leq h \leq u} D\big(f(\vartheta_h) \cdot (\mathrm{t}_h - \mathrm{t}_{h-1})\big) + \varepsilon$$

$$\leq \frac{\sqrt{n}}{2} \cdot \frac{2\varepsilon}{\sqrt{n}C} \cdot C + \varepsilon = 2\varepsilon.$$

Since $\varepsilon > 0$ is arbitrary, hence we conclude.

Corollary 10.4.2 *If function $f : [y, z]_{\mathcal{T}} \to \mathrm{Com}(\mathbb{R}^n)$ is set-valued Riemann Δ-integrable on $[y, z]_{\mathcal{T}}$, then*

$$\overline{\mathrm{R}}_s \int_y^z f(\mathrm{t})\Delta\mathrm{t} = co\left(\overline{\mathrm{R}}_s \int_y^z f(\mathrm{t})\Delta\mathrm{t}\right).$$

Below we investigate the relation between set-valued Riemann Δ-integral and the ordinary Riemann Δ-integral.

Theorem 10.4.6 *Let $f : [y, z]_{\mathcal{T}} \to \mathrm{Com}(\mathbb{R}^n)$ and $g : [y, z]_{\mathcal{T}} \to \mathbb{R}^n$ with $f(\mathrm{t}) = \{g(\mathrm{t})\}$ for all $\mathrm{t} \in [y, z]_{\mathcal{T}}$. Then f is set-valued Riemann Δ-integrable on $[y, z]_{\mathcal{T}}$ if and only if g is Riemann Δ-integrable on $[y, z]_{\mathcal{T}}$, and $\overline{\mathrm{R}}_s \int_y^z f(\mathrm{t})\Delta\mathrm{t} = \left\{\overline{\mathrm{R}} \int_y^z g(\mathrm{t})\Delta\mathrm{t}\right\}$.*

Proof "$\Rightarrow$" Let $\varepsilon > 0$ and $\delta > 0$; then

$$s\left(\overline{\mathrm{R}}_s(f; \breve{\mathcal{Y}}_\delta), \overline{I}_s\right) = s\left(\sum_{h=1}^{u} f(\vartheta_h) \cdot \left(\mathrm{t}_h - \mathrm{t}_{h-1}\right), \overline{I}_s\right) < \varepsilon.$$

Given $f(\mathrm{t}) = \{g(\mathrm{t})\}$ and applying Proposition 10.2.2 *(6)*, we get

$$\left\| \sum_{h=1}^{u} g(\vartheta_h) \cdot \left(\mathrm{t}_h - \mathrm{t}_{h-1}\right) - \overline{I}_s \right\| < \varepsilon.$$

"$\Leftarrow$" holds similarly in reverse.

Theorem 10.4.7 *Function $f : [y, z]_{\mathcal{T}} \to \mathrm{Com}(\mathbb{R}^n)$ is set-valued Riemann Δ-integrable on $[y, z]_{\mathcal{T}}$ if and only if its support function $\psi(\eta, f)$ is uniformly Riemann Δ-integrable on $[y, z]_{\mathcal{T}}$, given $\eta \in \mathbb{R}^n \wedge \|\eta\| = 1$, and*

$$\overline{\mathrm{R}} \int_y^z \psi(\eta, f(\mathrm{t}))\Delta\mathrm{t} = \psi\left(\eta, \overline{\mathrm{R}}_s \int_y^z f(\mathrm{t})\Delta\mathrm{t}\right).$$

Proof " $\Rightarrow$" Since $f \in (\mathcal{R}_s)_\Delta[y,z]_\mathcal{T}$, for $\varepsilon > 0$ there exists $\delta > 0$ such that $s\Big(\overline{\mathrm{R}}_s(f;\breve{\mathcal{Y}}_\delta), \overline{I}_s\Big) < \varepsilon$. For $\eta \in \mathbb{R}^n \wedge \|\eta\| = 1$, we consider

$$\begin{aligned}
&\left|\sum_{h=1}^{u}\psi\Big(\eta, f(\vartheta_h)\cdot(\mathrm{t}_h-\mathrm{t}_{h-1})\Big)-\psi\big(\eta,\overline{I}_s\big)\right| \\
&\leq \sup_{\eta\in\mathbb{R}^n\wedge\|\eta\|=1}\left|\sum_{h=1}^{u}\psi\Big(\eta, f(\vartheta_h)\Big)\cdot(\mathrm{t}_h-\mathrm{t}_{h-1})-\psi\big(\eta,\overline{I}_s\big)\right| \\
&= \sup_{\eta\in\mathbb{R}^n\wedge\|\eta\|=1}\left|\psi\Big(\eta, \sum_{h=1}^{n} f(\vartheta_h)\cdot(\mathrm{t}_h-\mathrm{t}_{h-1})\Big)-\psi\big(\eta,\overline{I}_s\big)\right|;
\end{aligned}$$

following Proposition 10.2.4 and applying Eq. (10.2), we get

$$s\left(\sum_{h=1}^{u} f(\vartheta_h)\cdot(\mathrm{t}_h-\mathrm{t}_{h-1}), \overline{I}_s\right) < \varepsilon.$$

Hence $\psi(\eta, f)$ is Riemann Δ-integrable on $[y,z]_\mathcal{T}$ uniformly for all $\eta \in \mathbb{R}^n \wedge \|\eta\| = 1$.

" $\Leftarrow$" Since $\psi(\eta, f)$ is uniformly Riemann Δ-integrable on $[y,z]_\mathcal{T}$ for all $\eta \in \mathbb{R}^n \wedge \|\eta\| \leq 1$, hence, there exists $\overline{I}_\eta \in \mathbb{R}$ such that given $\varepsilon > 0$ there exists $\delta > 0$ such that

$$\left|\sum_{h=1}^{u}\psi\Big(\eta, f(\vartheta_h)\Big)\cdot(\mathrm{t}_h-\mathrm{t}_{h-1})-\overline{I}_\eta\right| < \varepsilon.$$

To prove $\big\{p \in \mathbb{R}^n | \langle\eta, p\rangle \leq \overline{I}_\eta,\ \eta \in \mathbb{R}^n \wedge \|\eta\| \leq 1\big\} \in \mathrm{Com}(\mathbb{R}^n)$.

Here $\big\{p \in \mathbb{R}^n | \langle\eta, p\rangle \leq \overline{I}_\eta,\ \eta \in \mathbb{R}^n \wedge \|\eta\| \leq 1\big\}$ is non-empty. Let $p_1, p_2 \in \big\{p \in \mathbb{R}^n | \langle\eta, p\rangle \leq \overline{I}_\eta,\ \eta \in \mathbb{R}^n \wedge \|\eta\| \leq 1\big\}$; then $\langle\eta, p_1\rangle \leq \overline{I}_\eta$, $\langle\eta, p_2\rangle \leq \overline{I}_\eta$, so for any $\gamma \in [0,1]$,

$$\langle\eta, \gamma\cdot p_1+(1-\gamma)\cdot p_2\rangle = \gamma\cdot\langle\eta, p_1\rangle+(1-\gamma)\cdot\langle\eta, p_2\rangle \leq \gamma\cdot\overline{I}_\eta+(1-\gamma)\cdot\overline{I}_\eta = \overline{I}_\eta,$$

following Proposition 10.2.4; hence set $\big\{p \in \mathbb{R}^n | \langle\eta, p\rangle \leq \overline{I}_\eta,\ \eta \in \mathbb{R}^n \wedge \|\eta\| \leq 1\big\}$ is a convex subset of $\mathbb{R}^n$.

Let $\{p_m\}_{m=1}^{\infty} \subset \big\{p \in \mathbb{R}^n | \langle\eta, p\rangle \leq \overline{I}_\eta,\ \eta \in \mathbb{R}^n \wedge \|\eta\| \leq 1\big\}$; then for each $\eta \in \mathbb{R}^n \wedge \|\eta\| = 1$, $\langle p_m, \eta\rangle \leq \overline{I}_\eta = \overline{\mathrm{R}}\int_y^z \psi(\eta, f)(\mathrm{t})\Delta\mathrm{t}$, i.e., $< p_m, \eta >$ is bounded. So for any $\eta \in \mathbb{R}^n \wedge \|\eta\| = 1$, $\langle p_m, \eta\rangle$ there exists convergent subsequence $\langle p_{m_k}, \eta\rangle$.

Let $\lim_{k\to\infty}\langle p_{m_k}, \eta\rangle = b_\eta$; then $b_\eta \le \overline{I}_\eta$, apparently,

$$\left\langle \lim_{k\to\infty} p_{m_k}, \eta \right\rangle = \lim_{k\to\infty} \langle p_{m_k}, \eta\rangle = b_\eta \le \overline{I}_\eta,$$

i.e., $\lim_{k\to\infty} p_{m_k} \in \left\{p \in \mathbb{R}^n | \langle \eta, p\rangle \le \overline{I}_\eta,\ \eta \in \mathbb{R}^n \wedge \|\eta\| \le 1\right\}$. So $\left\{p \in \mathbb{R}^n | \langle \eta, p\rangle \le \overline{I}_\eta,\ \eta \in \mathbb{R}^n \wedge \|\eta\| \le 1\right\}$ is a compact set on $\mathbb{R}^n$. Let $\lim_{k\to\infty} p_{m_k} \in \left\{p \in \mathbb{R}^n | \langle \eta, p\rangle \le \overline{I}_\eta,\ \eta \in \mathbb{R}^n \wedge \|\eta\| \le 1\right\} = \overline{I} \in \mathrm{Com}(\mathbb{R}^n)$. By Eq. (10.1) $\psi(\eta, I) = \overline{I}_\eta$. So, by Eq. (10.2) and Proposition 10.2.4, we deduce

$$\begin{aligned} &s\left(\sum_{h=1}^{u} f(\vartheta_h)\cdot(\mathrm{t}_h - \mathrm{t}_{h-1}), \overline{I}\right) \\ &= \sup_{\eta\in\mathbb{R}^n \wedge \|\eta\|=1} \left|\psi\left(\eta, \sum_{h=1}^{u} f(\vartheta_h)\cdot(\mathrm{t}_h - \mathrm{t}_{h-1})\right) - \psi(\eta, \overline{I})\right| \\ &= \sup_{\eta\in\mathbb{R}^n \wedge \|\eta\|=1} \left|\sum_{h=1}^{u} \psi(\eta, f(\vartheta_h))\cdot(\mathrm{t}_h - \mathrm{t}_{h-1}) - \overline{I}_\eta\right| \le \varepsilon. \end{aligned}$$

We will now formulate a few basic results which will hold true, given a set-valued function is set-valued Riemann Δ-integrable.

Theorem 10.4.8 (Linearity) *Let* $f : [y, z]_{\mathcal{T}} \to \mathrm{Com}(\mathbb{R}^n)$, $f_1 : [y, z]_{\mathcal{T}} \to \mathrm{Com}(\mathbb{R}^n)$, *and* $f_2 : [y, z]_{\mathcal{T}} \to \mathrm{Com}(\mathbb{R}^n)$ *be set-valued Riemann* Δ*-integrable functions, and let* $\gamma \in \mathbb{R}$. *Then:*

1. $\gamma \cdot f$ *is set-valued Riemann* Δ*-integrable on* $[y, z]_{\mathcal{T}}$ *and*

$$\overline{\mathrm{R}}_s \int_y^z (\gamma\cdot f)(\mathrm{t})\Delta\mathrm{t} = \gamma\cdot \overline{\mathrm{R}}_s \int_y^z f(\mathrm{t})\Delta\mathrm{t}.$$

2. $f_1 + f_2$ *is set-valued Riemann* Δ*-integrable on* $[y, z]_{\mathcal{T}}$ *and*

$$\overline{\mathrm{R}}_s \int_y^z (f_1 + f_2)(\mathrm{t})\Delta\mathrm{t} = \overline{\mathrm{R}}_s \int_y^z f_1(\mathrm{t})\Delta\mathrm{t} + \overline{\mathrm{R}}_s \int_y^z f_2(\mathrm{t})\Delta\mathrm{t}.$$

3. $\gamma \cdot co(f)$ *is set-valued Riemann* Δ*-integrable on* $[y, z]_{\mathcal{T}}$ *and*

$$\overline{\mathrm{R}}_s \int_y^z (\gamma\cdot co(f))(\mathrm{t})\Delta\mathrm{t} = \gamma\cdot \overline{\mathrm{R}}_s \int_y^z co(f(\mathrm{t}))\Delta\mathrm{t},$$

also

$$co\left(\overline{\mathrm{R}}_s\int_y^z(\gamma\cdot f)(\mathrm{t})\Delta\mathrm{t}\right)=\gamma\cdot co\left(\overline{\mathrm{R}}_s\int_y^z f(\mathrm{t})\Delta\mathrm{t}\right).$$

4. *$co(f_1+f_2)$ is set-valued Riemann Δ-integrable on $[y,z]_{\mathcal{T}}$ and*

$$\overline{\mathrm{R}}_s\int_y^z co(f_1+f_2)(\mathrm{t})\Delta\mathrm{t}=\overline{\mathrm{R}}_s\int_y^z co(f_1(\mathrm{t}))\Delta\mathrm{t}+\overline{\mathrm{R}}_s\int_y^z co(f_2(\mathrm{t}))\Delta\mathrm{t},$$

also

$$co\left(\overline{\mathrm{R}}_s\int_y^z(f_1+f_2)(\mathrm{t})\Delta\mathrm{t}\right)=co\left(\overline{\mathrm{R}}_s\int_y^z f_1(\mathrm{t})\Delta\mathrm{t}\right)+co\left(\overline{\mathrm{R}}_s\int_y^z f_2(\mathrm{t})\Delta\mathrm{t}\right).$$

Remark 10.4.1 Taking $\gamma=-1$,

$$\overline{\mathrm{R}}_s\int_y^z(-f)(\mathrm{t})\Delta\mathrm{t}=-\overline{\mathrm{R}}_s\int_y^z f(\mathrm{t})\Delta\mathrm{t}.$$

Remark 10.4.2 In general, if $f_1, f_2, \ldots, f_u\in(\mathcal{R}_s)_{\Delta}[y,z]_{\mathcal{T}}$ then $(f_1+f_2+\ldots+f_u)\in(\mathcal{R}_s)_{\Delta}[y,z]_{\mathcal{T}}$.

Theorem 10.4.9 *Let $f:[y,z]_{\mathcal{T}}\to\mathrm{Com}(\mathbb{R}^n)$, $f_1:[y,z]_{\mathcal{T}}\to\mathrm{Com}(\mathbb{R}^n)$, and $f_2:[y,z]_{\mathcal{T}}\to\mathrm{Com}(\mathbb{R}^n)$ be set-valued Riemann Δ-integrable functions. Then:*

1. *If $f\in(\mathcal{R}_s)_{\Delta}[y,z]_{\mathcal{T}}$, then f is set-valued Riemann Δ-integrable on any subinterval of $[y,z]_{\mathcal{T}}$, and if there exist mutually exclusive intervals I_1, I_2, …, I_m so as to $[y,z]_{\mathcal{T}}=\bigcup_{h=1}^{m}I_h$, then*

$$\overline{\mathrm{R}}_s\int_y^z f(\mathrm{t})\Delta\mathrm{t}=\sum_{h=1}^{m}\overline{\mathrm{R}}_s\int_{I_h}f(\mathrm{t})\Delta\mathrm{t}.$$

2. *If $f_1(\mathrm{t})\subseteq f_2(\mathrm{t})$ for all $\mathrm{t}\in[y,z]_{\mathcal{T}}$, then*

$$\overline{\mathrm{R}}_s\int_y^z f_1(\mathrm{t})\Delta\mathrm{t}\subseteq\overline{\mathrm{R}}_s\int_y^z f_2(\mathrm{t})\Delta\mathrm{t}.$$

Theorem 10.4.10 *Suppose that $f:[y,z]_{\mathcal{T}}\to\mathrm{Com}(\mathbb{R}^n)$ and $y<w<z$. Then f is set-valued Riemann Δ-integrable on $[y,z]_{\mathcal{T}}$ if and only if it is set-valued Riemann Δ-integrable on $[y,w]_{\mathcal{T}}$ and $[w,z]_{\mathcal{T}}$*

$$\overline{\mathrm{R}}_s\int_y^z f(\mathrm{t})\Delta\mathrm{t}=\overline{\mathrm{R}}_s\int_y^w f(\mathrm{t})\Delta\mathrm{t}+\overline{\mathrm{R}}_s\int_w^z f(\mathrm{t})\Delta\mathrm{t}.$$

Theorem 10.4.11 *Let* $f : [y, z]_{\mathcal{T}} \to \mathrm{Com}(\mathbb{R}^n)$ *be set-valued Riemann* Δ*-integrable on* $[y, z]_{\mathcal{T}}$*; then* f *is bounded.*

Proof By contradiction, suppose there exists a set-valued Riemann Δ-integrable function f on $[y, z]_{\mathcal{T}}$ that is unbounded. Let $\overline{I}_s$ denote the value $\overline{\mathrm{R}}_s \int_y^z f(\mathrm{t})\Delta\mathrm{t}$.

Taking $\varepsilon := 1$, there exists $\delta > 0$ such that $s\big(\overline{\mathrm{R}}_s(f; \breve{\mathcal{Y}}), \overline{I}_s\big) < 1$, for all Δ-tagged partition $\breve{\mathcal{Y}}$ of $[y, z]_{\mathcal{T}}$ having mesh strictly less than δ. Proposition 10.2.2 *(4)* yields

$$s\Big(\overline{\mathrm{R}}_s(f; \breve{\mathcal{Y}}), \{0\}\Big) \leq s\Big(\overline{\mathrm{R}}_s(f; \breve{\mathcal{Y}}), \overline{I}_s\Big) + s\Big(\overline{I}_s, \{0\}\Big) < 1 + s\Big(\overline{I}_s, \{0\}\Big) \tag{10.6}$$

for all such $\breve{\mathcal{Y}}$.

We now construct a special Δ-tagged partition $\breve{\mathcal{W}}$ which will contradict the above.

Let $\mathcal{W} = \big\{[\mathrm{t}_{h-1}, \mathrm{t}_h)_{\mathcal{T}}\big\}_{h=1}^{u}$ be any Δ-partition of $[y, z]_{\mathcal{T}}$ whose mesh is strictly less than δ. Since f is unbounded on $[y, z]_{\mathcal{T}}$, it is unbounded on at least one Δ-subinterval of $\mathcal{W}$. Hence, there is an index $j = 1, 2, \ldots, u$ and $\vartheta_j \in [\mathrm{t}_{j-1}, \mathrm{t}_j)_{\mathcal{T}}$ such that

$$s\Big(f(\vartheta_j) \cdot (\mathrm{t}_j - \mathrm{t}_{j-1}), \{0\}\Big) > 1 + s\Big(\overline{I}_s, \{0\}\Big) - s\Bigg(\sum_{h \neq j} f(\vartheta_h) \cdot (\mathrm{t}_h - \mathrm{t}_{h-1}), \{0\}\Bigg),$$

applying Proposition 10.2.2 *(5)*. Hence

$$s\Big(\overline{\mathrm{R}}_s(f; \breve{\mathcal{W}}), \{0\}\Big) = s\Bigg(f(\vartheta_j) \cdot (\mathrm{t}_j - \mathrm{t}_{j-1}) + \sum_{h \neq j} f(\vartheta_h) \cdot (\mathrm{t}_h - \mathrm{t}_{h-1}), \{0\}\Bigg),$$

applying Proposition 10.2.2 *(5)* we get

$$\begin{aligned} &> 1 + s\Big(\overline{I}_s, \{0\}\Big) - s\Bigg(\sum_{h \neq j} f(\vartheta_h) \cdot (\mathrm{t}_h - \mathrm{t}_{h-1}), \{0\}\Bigg) \\ &+ s\Bigg(\sum_{h \neq j} f(\vartheta_h) \cdot (\mathrm{t}_h - \mathrm{t}_{h-1}), \{0\}\Bigg) \\ &> 1 + s\Big(\overline{I}_s, \{0\}\Big), \end{aligned}$$

which contradicts Eq. (10.6), and hence we conclude the statement that if f is given to be set-valued Riemann Δ-integrable on $[y, z]_{\mathcal{T}}$, then f is set-valued bounded on $[y, z]_{\mathcal{T}}$.

Theorem 10.4.12 *If* $f : [y, z]_{\mathcal{T}} \to \mathrm{Com}(\mathbb{R}^n)$ *is set-valued Riemann* Δ*-integrable on* $[y, z]_{\mathcal{T}}$*, then* f *is integrably bounded.*

Theorem 10.4.13 *Let $f : [y, z]_{\mathcal{T}} \to \mathrm{Com}(\mathbb{R}^n)$ be a continuous set-valued function; then for each $\varepsilon > 0$ there exists $\delta > 0$ such that for any Δ-tagged partition $\breve{\mathcal{Y}}$ of $[y, z]_{\mathcal{T}}$ whose mesh is strictly less than δ and any refinement $\breve{\mathcal{W}}$ of $\breve{\mathcal{Y}}$, the following holds:*

$$s\Big(\overline{\mathrm{R}}_s(f; \breve{\mathcal{Y}}), \overline{\mathrm{R}}_s(f; \breve{\mathcal{W}})\Big) < \varepsilon.$$

Proof Since f is a continuous set-valued function and $[y, z]_{\mathcal{T}}$ is compact, thus the set-valued function f is uniformly continuous. Therefore, given $\varepsilon > 0$, there exists $\delta > 0$ such that

$$s\big(f(\tilde{t}_0), f(\tilde{t}_1)\big) < \frac{\varepsilon}{z - y},$$

for all $\tilde{t}_0, \tilde{t}_1 \in [y, z]_{\mathcal{T}}$ whenever $|\tilde{t}_0 - \tilde{t}_1| < \delta$.

Let $\breve{\mathcal{Y}}$ be a Δ-tagged partition of $[y, z]_{\mathcal{T}}$ with $\mathrm{mesh}(\mathcal{Y}) < \delta$.

Suppose that $\breve{\mathcal{W}}$ is any refinement of $\breve{\mathcal{Y}}$ with its own set of Δ-tags. Any Δ-subinterval $[\mathrm{t}_{h-1}, \mathrm{t}_h)_{\mathcal{T}}$ of $\breve{\mathcal{Y}}$ may then be decomposed into a finite union of Δ-subintervals from $\breve{\mathcal{W}}$ as

$$[\mathrm{t}_{h-1}, \mathrm{t}_h)_{\mathcal{T}} = \bigcup_{j=1}^{k_h} \big[\hat{\mathrm{t}}_{h_j-1}, \hat{\mathrm{t}}_{h_j}\big)_{\mathcal{T}}.$$

Therefore,

$$\begin{aligned} & s\Big(\overline{\mathrm{R}}_s(f; \breve{\mathcal{Y}}), \overline{\mathrm{R}}_s(f; \breve{\mathcal{W}})\Big) \\ &= s\Bigg(\sum_{h=1}^{u} f(\vartheta_h) \cdot (\mathrm{t}_h - \mathrm{t}_{h-1}), \sum_{h=1}^{u}\sum_{j=1}^{k_h} f(\hat{\vartheta}_{h_j}) \cdot (\hat{\mathrm{t}}_{h_j} - \hat{\mathrm{t}}_{h_j-1})\Bigg) \\ &\leq \sum_{h=1}^{u}\sum_{j=1}^{k_h} (\hat{\mathrm{t}}_{h_j} - \hat{\mathrm{t}}_{h_j-1}) \cdot s\Big(f(\vartheta_h), f(\hat{\vartheta}_{h_j})\Big), \end{aligned}$$

applying Proposition 10.2.2 (2 and 4); here ϑ_h are Δ-tags from $\breve{\mathcal{Y}}$ and $\hat{\vartheta}_h$ are Δ-tags from $\breve{\mathcal{W}}$. Since the $\mathrm{mesh}(\breve{\mathcal{Y}}) < \delta$ implies,

$$s\Big(\overline{\mathrm{R}}_s(f; \breve{\mathcal{Y}}), \overline{\mathrm{R}}_s(f; \breve{\mathcal{W}})\Big) < \frac{\varepsilon}{z - y} \cdot \sum_{h=1}^{u}\sum_{j=1}^{k_h} (\hat{\mathrm{t}}_{h_j} - \hat{\mathrm{t}}_{h_j-1}) = \varepsilon.$$

Theorem 10.4.14 *Let $f : [y, z]_{\mathcal{T}} \to \mathrm{Com}(\mathbb{R}^n)$ be a continuous set-valued function; then f is set-valued Riemann Δ-integrable on $[y, z]_{\mathcal{T}}$.*

Below we investigate the relation between set-valued Riemann Δ-integral and the Hausdorff metric "s."

Theorem 10.4.15 *Let* $f_1 : [y, z]_{\mathcal{T}} \to \mathrm{Com}(\mathbb{R}^n)$, $f_2 : [y, z]_{\mathcal{T}} \to \mathrm{Com}(\mathbb{R}^n)$, *and a real function* $s(f_1, f_2) : [y, z]_{\mathcal{T}} \to \mathbb{R}^+$ *be* Δ*-integrable in the Riemann sense on* $[y, z]_{\mathcal{T}}$. *Then*

$$s\left(\overline{\mathrm{R}}_s \int_y^z f_1(\mathrm{t})\Delta\mathrm{t}, \overline{\mathrm{R}}_s \int_y^z f_2(\mathrm{t})\Delta\mathrm{t}\right) \leq \overline{\mathrm{R}} \int_y^z s\big(f_1(\mathrm{t}), f_2(\mathrm{t})\big)\Delta\mathrm{t}.$$

Proof For an arbitrary natural number u, let us take Δ-partition $\mathcal{Y} = \{\mathrm{t}_0, \mathrm{t}_1, \ldots, \mathrm{t}_u\}$ such that $\mathrm{t}_0 = y$, $\mathrm{t}_u = z$, and $\mathrm{t}_h - \mathrm{t}_{h-1} = \frac{z-y}{u}$ for all $h \in \{1, \ldots, u\}$ and take $\vartheta_h \in \{\mathrm{t}_0, \ldots, \mathrm{t}_{u-1}\}$. We consider the set-valued Riemann Δ-sums of f_1 and f_2 as

$$\overline{\mathrm{R}}_s(f_1; \breve{\mathcal{Y}}) = f_1(\mathrm{t}_0) \cdot (\mathrm{t}_1 - \mathrm{t}_0) + \ldots + f_1(\mathrm{t}_{u-1}) \cdot (\mathrm{t}_u - \mathrm{t}_{u-1}),$$

and

$$\overline{\mathrm{R}}_s(f_2; \breve{\mathcal{Y}}) = f_2(\mathrm{t}_0) \cdot (\mathrm{t}_1 - \mathrm{t}_0) + \ldots + f_2(\mathrm{t}_{u-1}) \cdot (\mathrm{t}_u - \mathrm{t}_{u-1}).$$

$$\begin{aligned} &s\left(\overline{\mathrm{R}}_s(f_1; \breve{\mathcal{Y}}), \overline{\mathrm{R}}_s(f_2; \breve{\mathcal{Y}})\right) \\ &\quad = s\Big(f_1(\mathrm{t}_0) \cdot (\mathrm{t}_1 - \mathrm{t}_0) + \ldots + f_1(\mathrm{t}_{u-1}) \cdot (\mathrm{t}_u - \mathrm{t}_{u-1}), f_2(\mathrm{t}_0) \cdot (\mathrm{t}_1 - \mathrm{t}_0) + \\ &\qquad \ldots + f_2(\mathrm{t}_{u-1}) \cdot (\mathrm{t}_u - \mathrm{t}_{u-1})\Big); \end{aligned}$$

applying Proposition 10.2.2 (*2* and *4*) we obtain

$$s\left(\overline{\mathrm{R}}_s(f_1; \breve{\mathcal{Y}}), \overline{\mathrm{R}}_s(f_2; \breve{\mathcal{Y}})\right) \leq \sum_{h=1}^{u} (\mathrm{t}_h - \mathrm{t}_{h-1}) \cdot s\Big(f_1(\mathrm{t}_{h-1}), f_2(\mathrm{t}_{h-1})\Big). \tag{10.7}$$

Since $\lim_{u \to \infty} \max \left\{\mathrm{t}_h - \mathrm{t}_{h-1} : h \in \{1, \ldots, u\}\right\} = 0$ as $u \to \infty$, Eq. (10.7) implies our assertion.

10.5 Conclusion

We investigated on the theory of Riemann integral for functions whose values are sets on time scales and established some basic properties and results.

References

1. Ahlbrandt, C.D., Bohner, M., Ridenhour, J.: Hamiltonian systems on time scales. Aust. J. Math. Anal. Appl. **250**(2), 561–578 (2000)
2. Artstein, Z., Burns, J.A.: Integration of compact set-valued functions. Pac. J. Math. **58**(2), 297–307 (1975)
3. Atici, F.M., Guseinov, G.S.: On Green's functions and positive solutions for boundary value problems on time scales. J. Comput. Appl. Math. **141**(1), 75–99 (2002)
4. Aumann, R.J.: Integrals of Set-Valued Functions. J. Math. Anal. Appl. **12**, 1–12 (1965)
5. Baier, R., Gerdts, M.: Intensive Course on Set-Valued Numerical Analysis and Optimal Control. Borovets and Sofia, Bulgaria (2005)
6. Bohner, M., Georgiev, S.G.: Multivariable Dynamic Calculus on Time Scales. Springer International Publishing, Switzerland (2016)
7. Dinu, C.: Convex functions on time scales. In: Annals of the University of Craiova, Mathematics and Computer Science Series, vol. 35, pp. 87–96 (2008)
8. Debrue, G.: Integration of correspondences. In: Proceedings of the Fifth Berkeley Symposium on Mathematical Statistics and Probability, vol. II, pp. 351–372. University of California Press, California (1967)
9. Guseinov, G.S., Kaymakçalan, B.: Basics of Riemann Delta and Nabla Integration on time scales. J. Differ. Equations Appl. **8**(11), 1001–1017 (2002)
10. Guseinov, G.S.: Integration on time scales. J. Math. Anal. Appl. **285**, 107–127 (2003)
11. Hai, S.X., Kong, F.D., Chen, J.S.: The Henstock integral of set-valued functions. Int. J. Pure Appl. Math. **37**(4), 507–511 (2007)
12. Hilger, S.: Analysis on Measure Chains- A unified approach to continuous and discrete calculus. Results Math. **18**, 18–56 (1990)
13. Hilger, S.: Differential and Difference Calculus- Unified!. Nonlinear Anal. Theory Methods Appl. **30**(5), 2683–2694 (1997)
14. Hukuhara, P.M.: Integration des applications measurables dont la valeur est un compact convexe. Funkcialaj Ekvacioj **10**, 205–223 (1967)
15. Lakshmikantham, V., Sivasundaram, S., Kaymakcalan, B.: Dynamic Systems on Measure Chains. Springer, Dordrecht (1996)
16. Sekhose, V., Bharali, H.: Set-Valued Riemann-Stieltjes integration on time scales. Kyungpook Math. J. **64**, 649–667 (2024)
17. Stefanini, L.: A generalization of Hukuhara difference and division for interval and fuzzy arithmetic. Fuzzy Sets Syst. **161**, 1564–1584 (2010)
18. Wibowo, R.B.E., Muslikh, M.: The Henstock-Kurzweil integral of set-valued function. Int. J. Math. Anal. **8**(55), 2741–2755 (2014)

Chapter 11
Real Projective Space with Topological Surgery Theory

Tanmay Saha

Abstract The real projective space in dimension 3, i.e., $\mathbb{RP}^3$, has been embedded in various manifolds and has a large impact on Physics and Mathematics. In this chapter, we describe $\mathbb{RP}^m$ in two formal ways and analyze it with the help of topological surgery theory over time to give a clear visualization of $\mathbb{RP}^m$, in dimension $m = 1, 2, 3$.

Keywords Real projective space · Quotient space · Space-time · Topological Surgery theory

11.1 Introduction

Revaluation has great significance in scientific research, and it opens up new paths of scientific discovery. Starting with ancient Greeks, who were always fascinated with the geometry of line and plane, the idea of a projective plane arose, and later scientists established a formal definition of real projective space. Although it has a wide range of applications in topology, geometry, differential geometry, and quantum mechanics, till now scientists have believe it is hard to visualize it in dimension 3. This chapter, by the technique of topological surgery theory, gives a nice visualization of real projective space in dimension 3.

11.2 Real Projective Space

The real projective space of dimension m is a topological manifold of all lines passing through the origin in the real Euclidean space $\mathbb{R}^{m+1}$. It is denoted by $\mathbb{RP}^m$.

T. Saha (✉)
Department of Mathematics, Central University of Kerala, Periye, India

A. Adhikari et al. (eds.), *Quantitative Frontiers*, Industrial and Applied Mathematics, https://doi.org/10.1007/978-981-96-9435-8_11

The $\mathbb{RP}^m$ can be formed by taking the quotient of $\mathbb{R}^{m+1}\backslash\{0\}$ under the equivalence relation $x \sim \lambda x$ for $\lambda \in \mathbb{R}\backslash\{0\}$. This gives the equivalence class [1]

$$[x] = [x_1, x_2, ..., x_{m+1}] = \{\lambda x | \lambda \in \mathbb{R}\backslash\{0\}, x \in \mathbb{R}^{m+1}\backslash\{0\}\},$$

which can be identified with the antipodal points on the sphere S^m in $\mathbb{R}^{m+1}$.

We can construct real projective space $\mathbb{RP}^m$ in another by considering the closed m-dimensional disk D^m, by identifying antipodal points on its boundary, i.e., $\partial D^m = S^{m-1}$.

The quotient space $D^m/\sim$ is defined as the set of equivalence classes of points in the closed m-dimensional disk D^m under the relation that identifies antipodal points on the boundary $\partial D^m = S^{m-1}$, where

$$[x] = \{x | x = y \text{ or x antipodal to y}; x, y \in D^m\}.$$

The above two definitions are equivalent.

Throughout this study, we will consider $\mathbb{RP}^m$ as the quotient space of $D^m/\sim$.

11.3 Topological Surgery Theory

Space-Time By a space-time we mean a paracompact Hausdorff space $\mathcal{M}$ together with a continuous function $\mathbf{t} : \mathcal{M} \to \mathbb{R}$, called time. The image of $\mathbf{t}$ can be finite or infinite intervals, called time interval of the definition of space-time. So, according to that the starting point of the interval is called the initial time, and the final point is called the final time. The level sets of $\mathbf{t}$ are called the time slice of space-time, and we denote by M_t the manifold at time t.

We will perform our surgery process over the space with the flow of time.

Surgery Theory In topology, surgery theory is a procedure to transform one manifold to another manifold in a controlled way.

There are different kinds of surgery on a manifold.

The formal way is to identify an embedded structure X on the manifold M that we want to cut and then the embedding

$$\varphi : X \to M$$

To exclude it, do the cutting operation

$$M\backslash(int(\varphi(X)),$$

where *int* is the interior of a set.

For gluing back (if $\partial X = \partial N$), use the operation

$$M \cup_{\varphi|\partial N} N.$$

Condition: $dim\ X = dim\ M = dim\ N = dim\ (\partial N + 1)$ [2].

11.4 Analysis of $\mathbb{R}P^m$

Surgery Process Our initial manifold is D^m, $m = 1, 2, 3$, and to get the quotient space $D^m/\sim$, we have to join the antipodal points on the boundary of D^m, that is, S^{m-1}. As we tried to join the antipodal points in a higher dimensional manifold, it started to form a collapsing shape, which was difficult to visualize and study, so we take help of topological surgery theory to complete the joining of antipodal points more easily.

For better understanding let us divide the time interval into subintervals like t_0 to t_1 and t_1 to t_2.

At the initial time t_0 we have $M_{t_0} = D^m$. Consider the embedding

$$\varphi : D^{m-1} \times D^1 \to D^m.$$

After t_1 time let us do the cut operation on D^m, i.e.,

$$D^m \backslash (int(\varphi(D^{m-1} \times D^1)).$$

According to the dimension m, we will get the two half-balls H_1^m (upper) and H_2^m (down) with respective boundaries, it is our M_{t_1}, and we get the cutting region $D^{m-1} \times D^1$; in cutting region we take the length of D^1 is very small quantity $\epsilon > 0$.

Now if we imagine an imaginary Y-axis in between these two half-balls, we can see that if we take any one point on H_1^m, then we have to join this with H_2^m at the point which is in the opposite side of Y-axis.

Let us put the H_2^m same and turn H_1^m around 180°, and put it under H_2^m. Now if we imagine at t_2 time antipodal points come on the same side of the imaginary Y-axis except the cutting region, but because of 180° rotation opposite of one half-ball came to the side of another half-ball, since they are antipodal points, so we can join them and get M_{t_2}.

So, as we join the antipodal points without any difficulty, after joining we will get again a connected manifold, but the problem occurs in the cutting region that we have to rejoin.

Now we use the gluing operation on M_{t_2}, i.e.,

$$M_{t_2} \cup_{\varphi_{|\partial(D^{m-1} \times D^1)}} (D^{m-1} \times D^1)$$

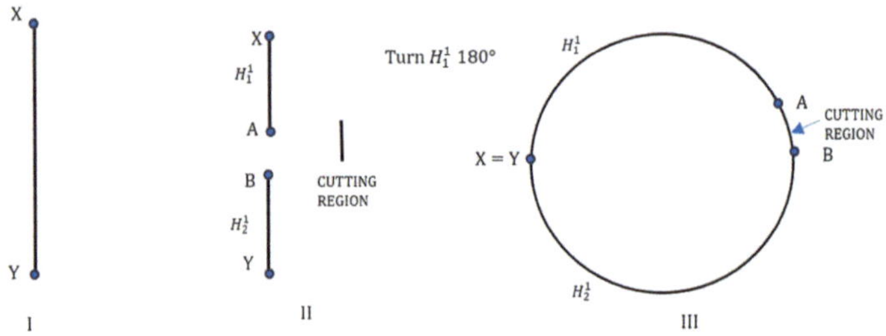

I. At the initial time t_0, D^1 and X, Y are antipodal points; it is our M_{t_0} **II.** At the time t_1, after applying cut operation H_1^1 (upper) and H_2^1 (lower), it is our M_{t_1} and cutting region D^1 **III.** At t_2 time join antipodal points X and Y, to get M_{t_2}, and gluing operation on M_{t_2} to get $\mathbb{RP}^1$

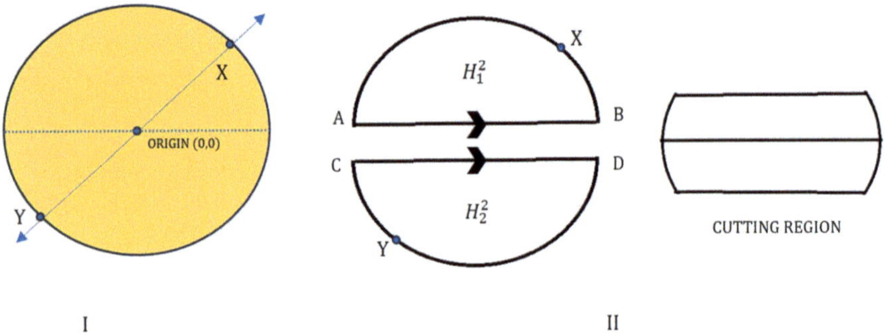

I. At the Initial time t_0, D^2 and X, Y are antipodal points joined by a line passing through the origin; it is our M_{t_0} **II.** At the time t_1, after applying cut operation H_1^2(upper) and H_2^2(lower), it is our M_{t_1} and cutting region $D^1 \times D^1$

to get our desire $\mathbb{RP}^m$, for $m = 1, 2, 3$. Here we cannot do this step in dimension m, we have to go in dimension greater than m, and by Whitney embedding theorem an m-dimensional manifold can be embedded into $\mathbb{R}^{2m+1}$.

Let us look $\mathbb{RP}^m$**, for** $m = 1, 2, 3$.

1. $\mathbb{RP}^1$: **Embedding,** $D^0 \times D^1 = D^1$, $\varphi : D^1 \to D^1$, **and cut operation,** $D^1 \setminus (int(\varphi(D^1))$ **andgluing operation on**M_{t_2}, i.e., $M_{t_2} \cup_{\varphi_{|\partial D^1}} D^1 = \mathbb{RP}^1$ (Fig. 11.1).

2. $\mathbb{RP}^2$: **Embedding,** $\varphi : D^1 \times D^1 \to D^2$, **and cut operation,** $D^2 \setminus (int(\varphi(D^1 \times D^1))$ (Figs. 11.2, 11.3, and 11.4).

Gluing operation on M_{t_2}, i.e., $M_{t_2} \cup_{\varphi_{|\partial(D^1 \times D^1)}} (D^1 \times D^1) = \mathbb{RP}^2$ (Fig. 11.5).

3. $\mathbb{RP}^3$: **Embedding,** $\varphi : D^2 \times D^1 \to D^3$, **and cut operation,** $D^3 \setminus (int(\varphi(D^2 \times D^1))$ (Figs. 11.6, 11.7, and 11.8).

Gluing operation on M_{t_2}, i.e., $M_{t_2} \cup_{\varphi_{|\partial(D^2 \times D^1)}} (D^2 \times D^1) = \mathbb{RP}^3$ (Fig. 11.9).

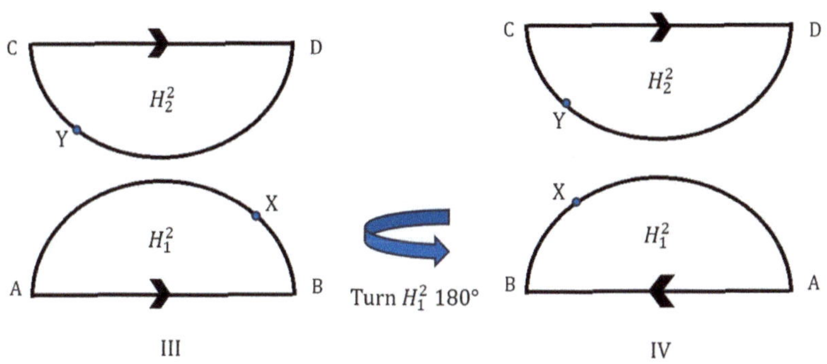

III. Put H_1^2 below H_2^2 **IV.** Turn H_1^2 around 180° (In IV since we turn H_1^2, AB direction changes)

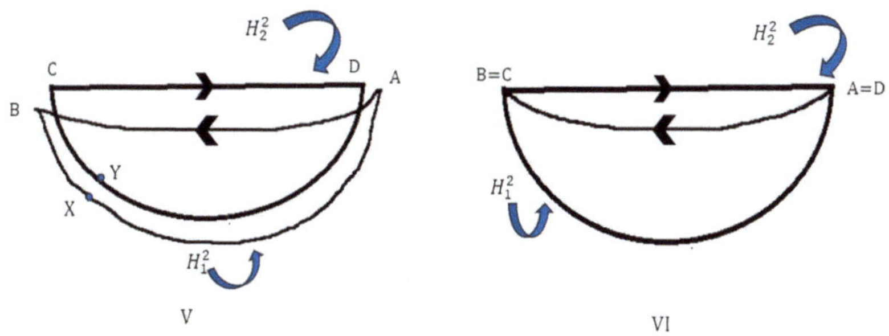

V. Here join the antipodal points X and Y and other all antipodal points in H_1^2 with H_2^2, since in step IV we turn H_1^2 180°, so we can join the antipodal points of the boundary of cutting region in H_1^2 and H_2^2, so $B = C$ and $A = D$ **VI.** At time t_2, all antipodal points are joined; it is our M_{t_2}

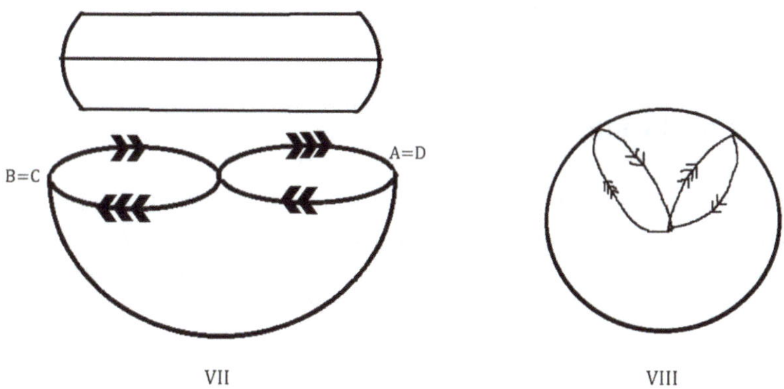

VII. Gluing operation on M_{t_2} **VIII.** $\mathbb{RP}^2$

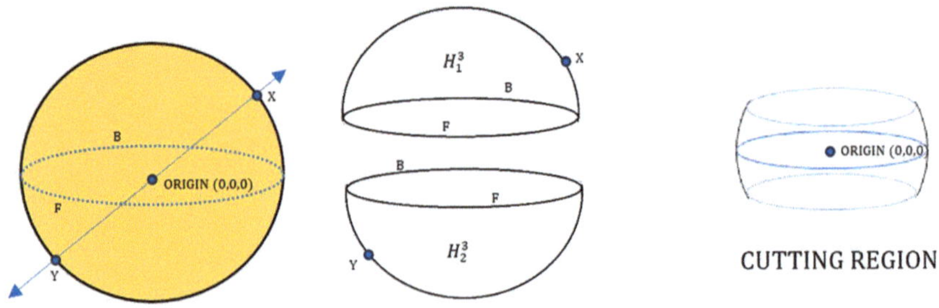

I II

I. At the initial time t_0, D^3 and X, Y are antipodal points joined by a line passing through the origin; it is our M_{t_0} **II.** At the time t_1, after applying cut operation H_1^3(upper) and H_2^3(lower) and cutting region $D^2 \times D^1$, it is our M_{t_1} B = back side and F = front side

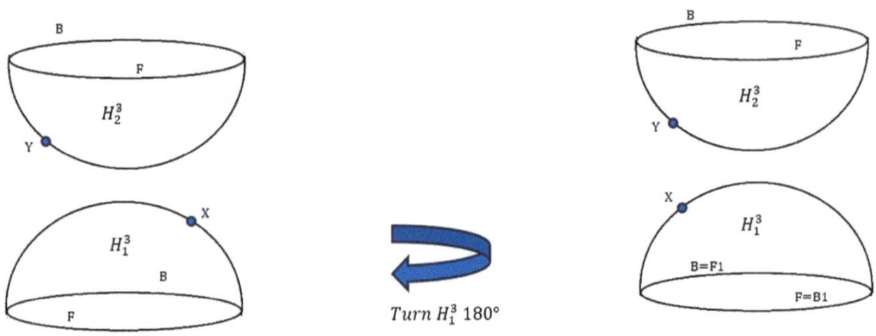

III IV

III. Put H_1^m below of H_2^m **IV.** Turn H_1^m around 180° (In IV since we turn H_1^m so original came front and the original front went back) B= Back side and F= Front side B1= Original Back side and F= Original Front side

11.5 Conclusion

In this chapter, we put forward a construction of real projective space, $\mathbb{RP}^m$, in dimension 1, 2, 3 by a controlled topological surgery process on D^m. This gives a clear visualization of $\mathbb{RP}^m$ mainly in dimension 3, which will help the whole science community work with the embedded $\mathbb{RP}^3$ structure in manifolds.

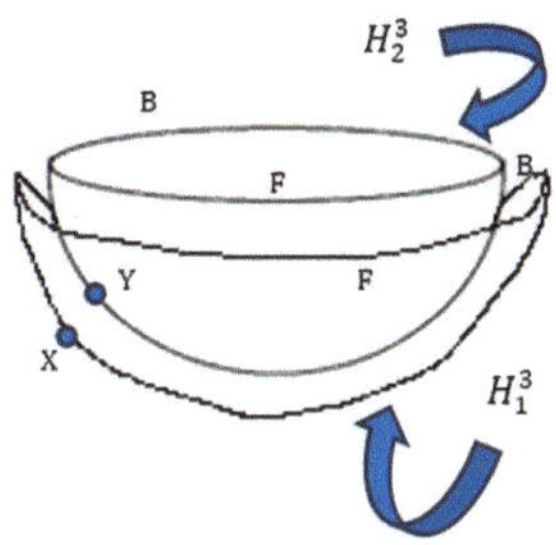

V

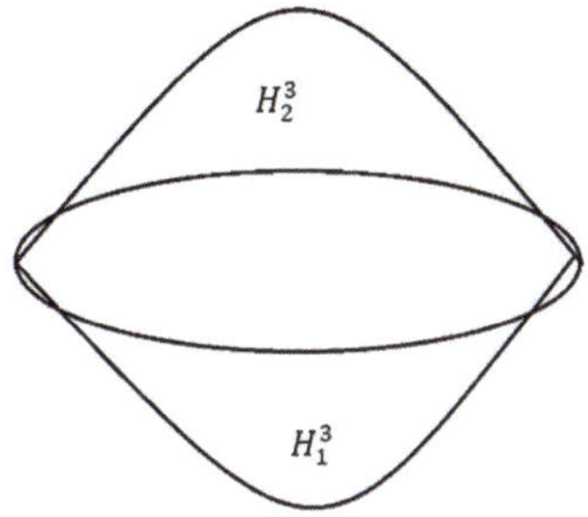

VI

V. Here join the antipodal points X and Y and other all antipodal points in H_1^m with H_2^m, since in step IV we turn H_1^m 180°, so we can join the antipodal points of the boundary of cutting region in H_1^m and H_2^m B = back side and F = front side **VI.** At time t_2, join all antipodal points; it is our M_{t_2}

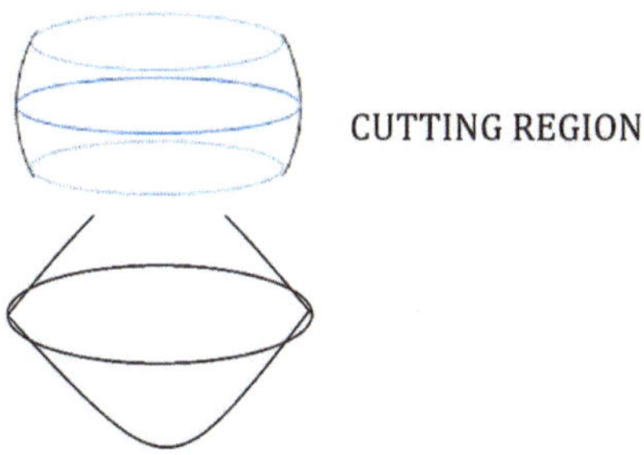

VII

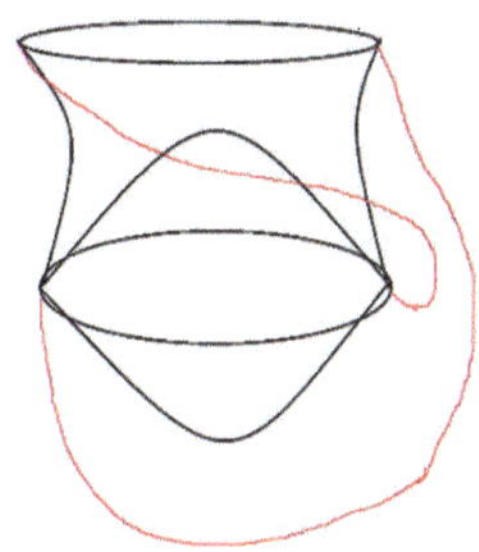

VIII

VII. Gluing operation on M_{t_2} **VIII.** $\mathbb{RP}^3$

References

1. Bhattacharya, A., Bhattacharya, R.: Nonparametric inference on manifolds with applications to shape space, chapter-12. In: Real Projective Spaces and Projective Shape Spaces, pp. 147–155. Cambridge University, Cambridge (2012)
2. Lück, W.: A basic introduction to surgery theory. In: Lecture Notes presented at summer school "High-dimensional Manifold Theory" in Trieste in May/June 2001 (2004)

Chapter 12
Fixed Point Analysis of a Homogeneous Isotropic Turbulent Flow

Amit Kumar Laha

Abstract Analyses of fixed point and numerical integration are commonly carried out in obtaining asymptotically consistent high Reynolds number solution in case of energy decay equation of homogeneous isotopic turbulence. In this chapter we have solved the *Von-Kármán-Howarth* equation vis *á* vis the energy spectrum equation and carried out fixed point analysis for a class of self-preserving solutions of homogeneous isotopic turbulence at high Reynolds number. It is shown that the decay laws obtained by similarity method and fixed point method due to the mathematical applications are compatible.

Keywords Fixed point · Isotopic turbulence · Energy spectrum · Self-preserving solution

12.1 Introduction

The chapter focuses on applying the mathematical analysis of the concept of fixed points to the physical turbulence problems of applied interest and, in particular, concentrating the attention to the roles of fixed points to the behavior of the solutions of the problems of homogeneous isotropic turbulence in fluid mechanics. Here in the present case, we would deal with the consequences of the applications of fixed points to the well-known equation of *Von-Kármán-Howarth* [1] in homogeneous isotopic turbulence.

In 1948, Sir G.K. Batchelor [2] studied the self-preserving features of the decaying-isotopic turbulence based on the assumption of invariance of Loitsiansky [3] integral. It was shown later on such invariance of Loitsiansky [3] integral, which is not completely valid. So our analysis should concern primarily with the development of isotopic turbulence under different conditions.

A. K. Laha (✉)
Amity University Kolkata, Newtown, West Bengal, India

A. Adhikari et al. (eds.), *Quantitative Frontiers*, Industrial and Applied Mathematics, https://doi.org/10.1007/978-981-96-9435-8_12

Speziale and Bernard [4, 5] pointed out that generally two types of asymptotic solutions are obtainable for isotopic turbulence, e.g., the turbulent kinetic energy decay as $K \sim t^{-1}$ and $K \sim t^{-\alpha}$, where $\alpha > 1$ with explicit dependence on the initial conditions. It is to be mentioned that by numerical integration and analysis of fixed point of the exact one-point equations, the turbulent kinetic energy is $K \sim t^{-1}$, which is consistent with asymptotic validity of high Reynolds number solutions.

In a recent paper Chasnov [6] commented that in three dimensions, high Reynolds number decay laws can be obtained by assuming that these laws continue to depend on invariant or near-invariant low wave number spectral coefficients but become independent of viscosity.

Further, the closure problems at high Reynolds numbers even are yet unknown. Nevertheless Chasnov [6] wrote that asymptotic decay laws of energy have been obtained by working in analogy to the final period of decay solutions. We discuss this point later on.

Accordingly, one assumes that scaling of the energy now depends nonlinearly on the low wave number spectral coefficient and the time $'t'$ alone. Viscosity no longer enters the scaling law as an independent parameter. Chasnov [6] mentioned by dimension analysis that with the usual turbulence phenomenology, the energy dissipation rate is a nonzero limit with vanishing viscosity, e.g., [6]

$$\langle u^2 \rangle = c_0 (4\pi B_0)^{\frac{2}{5}} t^{-\frac{6}{5}}, l \propto B_0^{\frac{1}{5}} t^{\frac{2}{5}} \tag{12.1}$$

$$\langle u^2 \rangle = c_2 (4\pi B_2)^{\frac{2}{7}} t^{-\frac{10}{7}}, l \propto B_0^{\frac{1}{7}} t^{\frac{2}{7}} \tag{12.2}$$

The result (12.1) is due to Saffman [7], and the second result (12.2) was forwarded by Kolmogorov [8]. It is to be noted that the time dependence in (12.1) is expected to be exact because of invariance of B_0 and the explicit time dependence is modified by the time dependence of B_2.

In the present chapter our aim will be to include some additional decay laws of homogeneous and isotopic turbulence. That is, we will discuss in the process following Sen's [9] a class of self-preserving solutions of homogeneous and isotopic turbulence in respect of their relevance to fixed point analogy.

12.2 Historical Background

12.2.1 Von-Kármán-Howarth Equation and Isotopic Turbulence

To start with we consider two points x and x' at time t inside the homogeneous and isotopic turbulent flow field and accordingly

$$\xi = x' - x, \qquad \xi_i \xi_i = r^2$$

$$\overline{u_i u'_j} = R_i^j = u'^2 \left[-\frac{1}{2r}\frac{\partial f}{\partial r}\xi_i \xi_j + \left(f + \frac{1}{2}\frac{\partial f}{\partial r}\delta_{ij} \right) \right]$$

where

$$\delta_{ij} = \begin{cases} 1, & \text{if } i = j, \\ 0, & \text{otherwise.} \end{cases} \tag{12.3}$$

$$u'^2 = \overline{u_1}^2 = \overline{u_2}^2 = \overline{u_3}^2,$$

$$u'^2 f(r) = (\overline{u_i u'_j})_{i=j, \xi_i = r}$$

$$u'^2 k(r) = (\overline{u_i u_j u'_k})_{i=j=k, \xi_i = r}$$

12.2.2 Navier-Stokes Equation

The most important equation [10] of homogeneous isotopic turbulence at a point (x, t) can be written as

$$\frac{\partial u_i}{\partial t} + u_j \frac{\partial u_i}{\partial x_j} = -\frac{\partial p}{\partial x_i} + \nu \nabla_x^2 u_i \tag{12.4}$$

$$\frac{\partial u_i}{\partial x_i} = 0 \tag{12.5}$$

where u_i is the turbulent velocity vector component, p is the fluctuating pressure, and ν is the kinematic viscosity. Two point double and triple longitudinal velocity correlations are denoted by $f(r, t)$ and $k(r, t)$, respectively, and are defined in the standard way

$$f(r, t) = \frac{\overline{u(x, t) u(x + r, t)}}{\overline{u}^2} \tag{12.6}$$

$$k(r, t) = \frac{\overline{u^2(x, t) u(x + r, t)}}{(\overline{u}^2)^{3/2}} \tag{12.7}$$

where u is any component of turbulent velocity and x and $x + r$ are any two spatial points separated by a distance $r = \| \overrightarrow{r} \|$ in the direction of u.

Von-Kármán-Howarth [1] derived the most important governing equation of homogeneous and isotopic turbulence as

$$\frac{\partial(\overline{u^2 f})}{\partial t} = (\overline{u}^2)^{3/2}\left(\frac{\partial k}{\partial r} + \frac{4}{r}k\right) + 2\nu\overline{u}^2\left(\frac{\partial^2 f}{\partial r^2} + \frac{4}{r}\frac{\partial f}{\partial r}\right) \tag{12.8}$$

12.2.3 Some Interesting Results of Turbulence

Now, we shall discuss about the solutions of this equation under different conditions and carricd out thc typical review of the analysis following Speziale and Bernard [5].

We define turbulent kinetic energy as $K = \frac{1}{2}\overline{u_i u_i}$ is a solution of the differential equation

$$\dot{K} = -\epsilon \tag{12.9}$$

where $\epsilon = \nu\overline{\omega_i\omega_i} \equiv \nu\omega^2$ is the turbulent dissipation rate, ω_i is the velocity vector, and ω^2 is the entropy.

In addition to specifying K and ϵ we would consider one-point statistic at the initial time and specify $f(r)$ and $k(r)$ at any instant of time which satisfies the *Von-Kármán-Howarth* equation.

We shall now consider several important relations that were derived in the decay process of turbulence, e.g.,

$$\frac{d\epsilon}{dt} = -2\nu\frac{\partial^3 S_{il,i}}{\partial r_l \partial r_j^2}(0) - 2\nu^2\frac{\partial^4 \mathcal{R}_{ii}}{\partial r_j^2 \partial r_l^2}(0) \tag{12.10}$$

where $\mathcal{R}_{ii} = \overline{u}^2(3f + rf')$.

$$\frac{\partial^4 \mathcal{R}_{ii}}{\partial r_j^2 \partial r_l^2}(r) = \overline{u}^2\left[\frac{24}{r}f'''(r) + 11f^{iv}(r) + rf^{v}(r)\right] \tag{12.11}$$

The Taylor series expansion of $f'''(r)$ yields

$$f'''(r) = rf^{iv}(0) + \frac{r^3}{3!}f^{vi}(0) + \ldots, \tag{12.12}$$

$$\frac{\partial^4 \mathcal{R}_{ii}}{\partial r_j^2 \partial r_l^2}(0) = 35\overline{u}^2 f^{iv}(0) \tag{12.13}$$

and

$$\frac{\partial^3 S_{il,i}}{\partial r_l \partial r_j^2}(0) = \frac{35}{2} u_{rms}^3 k'''(0) \tag{12.14}$$

From Eq. (12.10) we get

$$\frac{d\epsilon}{dt} = -35\nu u_{rms}^3 k'''(0) - 70\nu^2 \overline{u}^2 f^{iv}(0) \tag{12.15}$$

where $u_{rms} = \sqrt{2/3K}$. A form of (12.15) that is more useful for later analysis can be had by replacing $k'''(0)$ and $f^{iv}(0)$ by nondimensional parameters.

The skewness of the velocity derivative field is

$$S_k = -\frac{\overline{(\frac{\partial u}{\partial x})^3}}{\overline{(\frac{\partial u}{\partial x})^2}^{3/2}} = -\lambda^3 \left[\frac{\partial^3 K}{\partial r^3}\right]_{r=0} \tag{12.16}$$

and the palenstrophy coefficient is given by

$$G = \frac{\overline{(\frac{\partial^2 u}{\partial x^2})^2}}{\overline{(\frac{\partial u}{\partial x})^2}^2} = \lambda^4 \left[\frac{\partial^4 f}{\partial r^4}\right]_{r=0} \tag{12.17}$$

where $R_T = \frac{K^2}{\nu\epsilon}$ and $\lambda = [\frac{10\nu K}{\epsilon}]^{\frac{1}{2}}$, and we get $\epsilon = -15\nu\overline{u^2} f''(0) = 15\nu\overline{(\frac{\partial u}{\partial x})^2}$.

The standard form of the ϵ-equation for homogeneous and isotopic turbulence can be derived as

$$\frac{d\epsilon}{dt} = S_K^* R_T^{\frac{1}{2}} \frac{\epsilon^2}{K} - G^* \frac{\epsilon^2}{K} \tag{12.18}$$

where $S_K^* = \frac{7}{3\sqrt{15}} S_K$, $G^* = \frac{7}{15} G$, and $R_T = \frac{K^2}{\nu\epsilon}$. R_T is the dimensionless parameter that may be interpreted as a Reynolds number.

12.2.4 Similarity Scales

It is important to mention that for isotopic turbulence to be self-preserving in the sense of *Von-Kármán-Howarth* and Batchelor [2], we must express $f(r, t)$ and $k(r, t)$ as

$$f(r,t) = \tilde{f}(r/L), \qquad k(r,t) = \tilde{k}(r/L)$$

where $L = L(t)$ is the unique specified similarity length scale.

For complete self-preserving of all scales of turbulence, we define full range of $0 \le r \le \infty$ and restricted range of $0 \le r \le r_{max}$.

Following Batchelor [2], we identify the relations:

$$\epsilon = -10\nu K \left[\frac{\partial^2 f}{\partial r^2}\right]_{r=0}, \qquad \lambda^2 \left[\frac{\partial^2 f}{\partial r^2}\right]_{r=0} = -1$$

Now, for complete self-preserving isotopic turbulence

$$\frac{\lambda^2}{L^2}\tilde{f}''(0) = -1$$

from which it can be concluded that $L \propto \lambda$ (since $\tilde{f}''(0)$ is constant). For complete self-preserving of isotopic turbulence, we may set $L = \lambda$, and we get that $S_K = -\tilde{K}'''(0)$=constant and $G = f^{iv}(0)$=constant, where a prime denotes derivative with respect to the variable of $\frac{r}{\lambda}$. Consequently, $S_K = S_{K_0}$, $G = G_0$. This is a closed system for determining K and ϵ once K_0, ϵ_0, S_{K_0}, and G_0 are provided. The assumption of complete self-preservation is seen to lead to the closure assumption, e.g., "if initial conditions for the skewness and the palinstrophy coefficient are provided in addition to initial conditions for K and ϵ then the energy decay can be calculated explicitly for all later times."

The *Von-Kármán-Howarth* equation may now be put in the form

$$\begin{aligned} 10\tilde{f} + 2\eta^{-4}\frac{d}{d\eta}\left(\eta^4\frac{d\tilde{f}}{d\eta}\right) + \eta\frac{d\tilde{f}}{d\eta}\left(\frac{7}{3}G_0 - 5\right) \\ = R_\lambda(\frac{7}{6}S_{K_0}\eta\frac{d\tilde{f}}{d\eta} - \eta^{-4}\frac{d(\eta^4\tilde{K})}{d\eta} \end{aligned} \tag{12.19}$$

where

$$R_\lambda = (\overline{u^2})^{1/2}\frac{\lambda}{\nu} = \left(\frac{20}{3}\right)^{1/2} R_t^{1/2}$$

Speziale and Bernard [5] revisited the decay of turbulence problem and discussed that the similarity singularity scale ensures complete self-preservation of the full Von-Kármán-Howarth equation [1]. Using a fixed point analysis they demonstrated rigorously that the two asymptotic fully self-preserving regimes of decay are possible.

12.3 Fixed Point Analysis

Zheng Ran [11] presented in their report "large scale dynamics of isotopic turbulence" provided a table for comparison of self-preserving regimes with fixed point analysis.

For carrying out fixed point analysis of

$$\frac{\partial(\overline{u^2 f})}{\partial t} = (\overline{u}^2)^{3/2}\left(\frac{\partial k}{\partial r} + \frac{4}{r}k\right) + 2\nu\overline{u}^2\left(\frac{\partial^2 f}{\partial r^2} + \frac{4}{r}\frac{\partial f}{\partial r}\right) \tag{12.20}$$

$$\dot{\epsilon} = \frac{7}{3\sqrt{15}} S_K \sqrt{R_T \frac{\epsilon^2}{K}} - \frac{7}{15} G \frac{\epsilon^2}{K} \tag{12.21}$$

we combine them into a single transformed equation for the turbulence Reynolds number R_t. Since

$$\dot{R_t} = \frac{2k\dot{k}}{\nu\epsilon} - \frac{K^2}{\nu\epsilon^2}\dot{\epsilon} \tag{12.22}$$

it follows that

$$\dot{R_t} = -\frac{2K}{\nu} - \frac{7}{3\sqrt{15}} S_{K_0}\sqrt{R_t}\frac{K}{\nu} + \frac{7}{15} G_0 \frac{K}{\nu} \tag{12.23}$$

If the transformed dimensionless of time τ-defined by the relation

$$d\tau = \frac{\epsilon}{K} dt \tag{12.24}$$

is introduced into (12.23), we obtain the equation

$$\frac{dR_t}{d\tau} = R_t\left(\frac{7}{15}G_0 - 2 - \frac{7}{3\sqrt{15}} S_{K_0}\sqrt{R_t}\right) \tag{12.25}$$

The fixed of (12.25) is obtained by setting $\frac{dR_t}{d\tau} = 0$ which yields the equation

$$R_{t\infty}\left(\frac{7}{15}G_0 - 2 - \frac{7}{3\sqrt{15}} S_{K_0}\sqrt{R_{t\infty}}\right) = 0 \tag{12.26}$$

where $(.)_\infty$ denotes the equilibrium value in the limit as $\tau \to \infty$.

The Eq. (12.26) has the solutions:

$$R_{t\infty} = 0, \qquad \text{for} \frac{7}{15} G_0 \leq 2; \tag{12.27}$$

$$R_{t\infty} = \left(\frac{\frac{7}{15}G_0 - 2}{\frac{7}{3\sqrt{15}S_{K0}}} \right)^2, \qquad \text{for} \frac{7}{15}G_0 > 2 \tag{12.28}$$

It is also evident for (12.28) that in order to have an equilibrium high Reynolds number isotopic flow field under self-preserving condition, it is necessary that $G_0 \sim \sqrt{R_{t\infty}}$.

By high Reynolds number isotopic turbulence we mean the case where $R_t >> 1$; for a low Reynolds number isotopic turbulence $R_t = O(1)$.

The fixed point $R_{t_\infty} = 0$ is associated with asymptotic solutions of K and ϵ that satisfy the differential equations

$$\dot{K} = -\epsilon \tag{12.29}$$

$$\dot{\epsilon} = -\frac{7}{15}G_0\frac{\epsilon^2}{K} \tag{12.30}$$

Equations (12.29) and (12.30) have the exact solution

$$K = K_0 \left(1 + \frac{1}{\alpha}\frac{\epsilon_0 t}{K_0}\right)^{-\alpha} \tag{12.31}$$

$$\epsilon = \epsilon_0 \left(1 + \frac{1}{\alpha}\frac{\epsilon_0 t}{K_0}\right)^{-\alpha-1} \tag{12.32}$$

where $\alpha = \frac{1}{(\frac{7}{15}G_0 - 1)} > 1$ an exact asymptotic solution for K and ϵ may now be explained as

$$K \sim t^{-\alpha}, \epsilon \sim t^{-\alpha+1} \tag{12.33}$$

We may now thought of a physically realizable range for $\frac{7}{15}G_0$ as less than 2.

12.4 Spectrum of Homogeneous Isotopic Turbulence Flow

Now we embark upon the spectral analysis of homogeneous isotopic turbulence with

$$E(k, t) = u^2(t)l(t)F(kl(t))$$

where $u^2(t)$ is the standard derivation of the characteristic velocity of the flow and it is related to turbulent kinetic energy $K(t)$ as $u^2(t) = \frac{2}{3}K(t)$, $l(t)$ is a single length scale corresponding to velocity scale $u(t)$, and F is a time independent spectrum shape function.

We have the dynamical version of the energy spectrum $E(k, t)$ for isotopic turbulence

$$\frac{\partial E(k,t)}{\partial t} + 2\nu k^2 E(k,t) = T(k,t) \tag{12.34}$$

where $T(k, t)$ is the nonlinear transfer term. This equation may also be put in the form

$$\frac{\partial}{\partial t}\int_0^k E(k)dk = -2(\nu + \nu'(k))\int_0^k k^2 E(k)dk \tag{12.35}$$

where $\nu'(k) = \mathcal{K}_H \int_k^\infty (\frac{E(k)}{k^3})^{1/2} dk$. We must mention that Eq. (12.34) is equivalent to the Von-Kármán-Howarth equation [1] ($Stanisić$).

We shall now consider a most general type of self-similar solution of Eq. (12.34) as forwarded by Sen [9] with Heisenberg's form the energy transfer function $T(k, t)$.

Then we get that

$$E(k,t) = \frac{1}{k^2}\frac{1}{k_0^3 t_0^2}\left(\frac{t_0}{t}\right)^{(2-3c)} f\left(\frac{k}{k_0}\frac{t^c}{t_0^c}\right) \tag{12.36}$$

where c is a constant with $0 < c < \frac{2}{3}$. Let $x = (\frac{k}{k_0}\frac{t^c}{t_0^c})$; we find important properties of the asymptotic solution $f(x)$ as

$$\begin{aligned} f(x) &\sim x^{(\frac{2-3c}{c})} && (x \to 0) \\ f(x) &\sim x^{(-\frac{5}{3})} && (x \to \infty) \end{aligned} \tag{12.37}$$

We now calculate the turbulent energy by the formula

$$u'^2 = \int_0^\infty E(k,t)dk \tag{12.38}$$

by the relation

$$u'^2 = \frac{1}{k^2}\frac{1}{(k_0 t_0^c)^2} t^{2(c-1)} \int_0^\infty f(x)dx \tag{12.39}$$

We can write this as

$$u'^2 \sim t^{2(c-1)} \tag{12.40}$$

From Taylor's scale of turbulence [12], we obtain that

$$\lambda^2 = \left(\frac{5}{1-c}\right)\nu t \tag{12.41}$$

12.5 Discussion of Results

We now construct the following two tables and put the numerical estimates of various characters of isotopic turbulent flows in it.

Two additional slower decay laws, e.g., $u'^2 \sim t^{-1.36}$ and $u'^2 \sim t^{-1.43}$ that have been included in the present analysis, satisfy all the criteria like the other power laws of tables and the cases have also been dealt by Chasnov [6]. It is to notice that the calculations carried out with $c = .285$ are very close to the case that corresponds to the case with $c = \frac{2}{7}$ (Tables 12.1 and 12.2).

Heisenberg's turbulence energy spectrum $f(x)$ of one parameter family of functions, the parameter being Reynolds number R_λ, is given by

$$\int_0^\infty f(x)dx - \frac{1}{2}xf(x) = 2\left[\frac{1}{\alpha} + \int_x^\infty \left(\frac{f(x)}{x^5}\right)^{\frac{\lambda}{2}} dx\right]\int_0^x x^2 f(x)dx \tag{12.42}$$

$$R_\lambda^2 = \frac{u'^2\lambda^2}{\nu^2} = \frac{2}{3}\frac{\lambda^2}{\nu^2}\int_0^\infty E(k)dk = \frac{2}{3}\frac{\lambda^2}{\nu\epsilon}\alpha k^{-2} I_0 \tag{12.43}$$

Table 12.1 Analysis

Parameter c	Power law $k \to 0$	Taylor's scale law	Law for u'^2	Power law $k \to \infty$
1/2	$f(x) \sim x$	$\lambda^2 = 10\nu t$	$u'^2 \sim t^{-1}$	$E(k,t) \sim x^{-\frac{5}{3}}$
2/5	$f(x) \sim x^2$	$\lambda^2 = \frac{25}{3}\nu t$	$u'^2 \sim t^{-\frac{6}{5}}$	$E(k,t) \sim x^{-\frac{5}{3}}$
1/3	$f(x) \sim x^3$	$\lambda^2 = \frac{15}{2}\nu t$	$u'^2 \sim t^{-\frac{4}{3}}$	$E(k,t) \sim x^{-\frac{5}{3}}$
.32	$f(x) \sim x^{3.25}$	$\lambda^2 = 7.35\nu t$	$u'^2 \sim t^{-1.36}$	$E(k,t) \sim x^{-\frac{5}{3}}$
2/7	$f(x) \sim x^4$	$\lambda^2 = 7\nu t$	$u'^2 \sim t^{-\frac{10}{7}}$	$E(k,t) \sim x^{-\frac{5}{3}}$
.285	$f(x) \sim x^{4.02}$	$\lambda^2 = 6.99\nu t$	$u'^2 \sim t^{-1.43}$	$E(k,t) \sim x^{-\frac{5}{3}}$

Table 12.2 Analysis

Parameter c	Value of α	Value of $\frac{7}{15}G_0$
1/2	1	2
2/5	$6/5 = 1.2$	$11/6 = 1.83$
1/3	$4/3 = 1.3$	$7/4 = 1.75$
.32	1.36	$17/10 = 1.735$
2/7	$10/7 = 1.428$	$17/10 = 1.70$
.285	1.43	$17/10 = 1.699$

where $I_0 = \int_0^x f(x)dx$, $\lambda^2 = 10\gamma t$

For self-preserving $\lambda^2 R_\lambda^2 = \frac{20}{3}\alpha I_0$.

On Loitsiansky's assumption of the rapidly vanishing of the correlation function, e.g., $\lim_{r\to\infty} r^4 f' = 0$ and $\lim_{r\to\infty} r^4 \lambda = 0$, the following relation holds:

$$u'^2 \int_0^\infty f(r) r^4 dr = J_0 \tag{12.44}$$

where J_0 is a constant. Batchelor [2] also showed that

$$\lim_{k\to 0} \frac{E(k,\epsilon)}{k^4} = J_0 \tag{12.45}$$

This indicates that spectral energy function $E(k,t)$ behaves at small k, $E(k,t) \sim k^4$. Thus we may conclude that this point of the energy spectrum does not belong to the similar spectrum of Heisenberg, e.g., $u'^2 \sim t^{-1}$, $\lambda^2 = 10\nu t$, $E(k,t) \sim k(k \to 0)$, $E(k,t) \sim k^{\frac{-5}{3}}(k \to \infty)$.

In a recent research article by Meldi and Segant [13] discussed measures that "for arbitrary values of decay exponent of Kinetic energy, e.g., $E(t) \sim t^{n_k}$ with $n_k \leq -1$ can be obtained arbitrarily by neglecting some terms in the equation for prescribing some invariant quantities as mentioned by *Von-Kármán-Howarth*, or similar way according to Noether's theorem, enforcing a symmetry in the problem."

Also with the consideration of single velocity and a single length scale, George's theory [14] leads to a general decay regime with $l(t) \sim t^{\frac{1}{2}}$ and $K(t) \sim t^{n_k}$, where n_k is fixed by the initial conditions or the existence of an invariant quantity. The classical case $n_k = -1$ can be recovered easily.

12.6 Concluding Remarks

Keeping the same theoretical framework, any departure from this regime needs to be interpreted as a breakdown of exact self-preservation. In spite of that, small departures from exact self-preservation are expected to yield $K(t) \sim t^{-1}$ as exact self-preservation does.

In the present analysis, we have been able to discuss that fixed point method may be applied well to the turbulence problems when it is characterized by a large Reynolds number and the flow permits comparatively for a large time. It is shown that the energy spectrum $E(k,t)$ should exist as a self-preserving solution for the entire spectrum (Heisenberg's case).

During the early period decay the nonlinear inertia term may dominate over viscous dissipation terms characterized by the parameter c having values between 0 and 2/3. That is, the *Kármán-Howarth* equation may correspond to the case when ν is negligible as in Sen's treatment. In Heisenberg's case for $c = \frac{1}{2}$, the energy decay law is $u'^2 \sim t^{-1}$, while for $c = \frac{2}{7}$ the decay law applies even up to $u'^2 \sim t^{-\frac{10}{7}}$ due to Kolmogorov or for $c = .285$ where $u'^2 \sim t^{-1.43}$ that mentioned by Chasnov [6]. The fact that ought to be mentioned is Sen's solution may be applied to the large Reynolds number cases completely, while in the last case, for example, the case $c = \frac{2}{7}$ is valid for the final period decay. To avoid discrepancy [15] prescribed that $u'^2 = \alpha t^{-1} + \beta$, where α, β are constant. Lin [16] showed further that Heisenberg's case is very probably stable when viscosity term remains present in the energy decay equation, while all the cases are unstable when viscous dissipation is negligible as shown by Mazumdar [17]. We may encompass the idea of George [14]; the energy spectrum approach may also be applied suitably to analyze the decay of energy spectrum in homogeneous isotopic turbulence. Finally, we may conclude that the present fixed point approach may be applied in a straightforward manner to all the cases discussed here above.

Acknowledgments I would like to thank to Late Professor Himadri Pai Mazumdar, PAMU, Indian Statistical Institute, Kolkata, West Bengal, India-700108 for his guidance and active involvement in the preparation of this manuscript. I would also like to thank the Honorable reviewers for the comments and helpful suggestions.

References

1. Von Karman, Howarth L.: On Statistical theory of isotopic turbulence. Proc. R. Soc. Lond. **A-164**, 192–215 (1938)
2. Batchelor, G.K.: Energy decay and self-preserving correlation functions in isotopic turbulence. Quart. Applied Math. **6**, 207–216 (1948)
3. Loitsianski, L.G.: Some basic laws for isotopic turbulent flow. Trudy Tsentr. Aero.-Giedrodin. Inst. **31**, 440 (1939)
4. Speziale, C.G., Bernard, P.S.: The energy decay in self preserving isotopic turbulence revisited. J. Fluid Mech. **241**, 645–667 (1992)
5. Speziale, C.G., Bernard, P.S.: The energy decay in self preserving isotopic turbulence revisited. NASA Contract. Rep. **187640**, 645–667 (1991)
6. Chasnov, J.R., Decaying turbulence in two and three dimensions. In: Advances in DNS/LES. Greyden Press, Columbus, Ohio (1997)
7. Saffman, P.G.: Note on decay of homogeneous turbulence. Phys. Fluids **10**, 1349 (1967)
8. Kolmogoroff, A.N.: C. R. Acad. Sci. U.R.S.S. **30** (1941).
9. Sen, N.R.: On Heisenberg spectrum of turbulence. Bull. Cal. Math. Soc. **43**, 1–7 (1951)
10. Pope, S.B.: Turbulent flows. Cambridge University, Cambridge (2009)
11. Ran, Z.: Multiscales and cascading in isotropic turbulence. Letter Fluid Mechanics **56**(27), 2889–2892 (2011)
12. Taylor, G.I.: Statistical theory of turbulence. In: Proceedings of the Royal Society of London, vol. 151, pp. 421–444 (1935)
13. Meldi, M., Sagaut, P.: Further insights into self-similarity and self-preservation in freely decaying isotropic turbulence. J. Turbul., 14(8), 24–53 (2013)
14. George, W.K.: A theory of the decay of homogeneous isotropic turbulence. Tech. Rep.(Turbulence research laboratory), vol. 116 (1987)

15. Lin, C.C.: Note on the law of decay of isotropic turbulence. Proc. Natl. Acad. Sci. USA **34**(11), 540–543 (1948). https://doi.org/10.1073/pnas.34.11.540. PMID: 16588831; PMCID: PMC1079163
16. Lin, C.C.: On law of decay of isotropic turbulence. Proc. VII Int. Congr. Appl. Mech. **2**(1), 127–139 (1948)
17. Mazumdar, H.P.: Some aspect of Heisenberg spectrum of homogeneous and isotopic turbulence. Rev. Roum. Sci. Techn-Mec. Appl. Tome **22**, 863–869 (1977)

Chapter 13
Symmetric and Asymmetric Axial and Directional Models and Bivariate Extensions Thereof

Barry C. Arnold and Ashis SenGupta

Abstract Some well-known and not so well-known circular and axial models are reviewed. Attention is paid to the necessary modifications to adapt to frequently encountered asymmetric. The development of bivariate extensions of the models, which are needed to analyze dependent variables, is discussed in some detail. Special attention is paid to the models involving mixtures of bivariate models with independent coordinate variables.

Keywords Additive components · Cardioid · Conditional specification · Mixture · von Mises · Wrapped Cauchy

13.1 Introduction

Florence Nightingale's introduction of certain graphical representations of periodic data configurations set in motion a major effort to identify suitable models for analyzing such data. The literature on such topics is now vast. More than 2,000 papers have published with "circular distribution" either in the title or in the list of key words. Add another 600 dealing with axial distributions, and one realizes what a challenge it would be to attempt a survey of the area. The present chapter focuses on densities related to two of the most popular circular densities, the circular normal and the cardioid model and its generalizations. Even subject to that restriction, it will be seen that there are plenty of alternatives to consider.

Models for directional and axial data typically are symmetric in character. In real-world datasets, such symmetry is frequently absent. Several authors have suggested

B. C. Arnold
University of California, Riverside, CA, USA
e-mail: barnold@ucr.edu

A. SenGupta (✉)
Department of Population Health Sciences, MCG, Augusta University, Augusta, GA, USA

Department of Statistics, Middle East Technical University, Ankara, Turkey

A. Adhikari et al. (eds.), *Quantitative Frontiers*, Industrial and Applied Mathematics, https://doi.org/10.1007/978-981-96-9435-8_13

variations of the usual univariate directional and/or axial models in order to allow fitting of asymmetric data configurations. In this chapter we will review several of these proposed models. In addition, discussion of bivariate and multivariate symmetric and asymmetric models will be provided. We begin with a brief review of univariate symmetric models.

13.2 Popular One-Dimensional Circular Models

Undoubtedly, the most frequently encountered circular model is the *circular normal* or *von Mises* distribution. A random variable Θ will be said to have circular normal distribution with parameters $\mu \in [0, 2\pi)$ and $\kappa \in (0, \infty)$ if its density is of the form

$$f_\Theta(\theta; \mu, \kappa) \propto \exp[\cos\{\kappa(\theta - \mu)\}]I(0 \leq \theta < 2\pi). \tag{13.2.1}$$

In this representation, the parameter μ corresponds to the mode of the density, while κ is a precision parameter. If a random variable has density (13.2.1), we write $\Theta \sim CN(\mu, \kappa)$. An alternative parameterization is possible which makes it most evident that this corresponds to a two-parameter exponential family; thus

$$f_\Theta(\theta; \alpha, \beta) \propto \exp\left[\alpha \cos\theta + \beta \sin\theta\right] I(0 < \theta < 2\pi), \tag{13.2.2}$$

where $\alpha, \beta \in (-\infty, \infty)$. At the risk of some confusion, we will sometimes, when using this parameterization, write $\Theta \sim CN(\alpha, \beta)$. Since we are dealing with a two-parameter exponential family of densities, if we have a sample of size n from either version of this density, the complete minimal sufficient statistic will be

$$\left(\sum_{i=1}^n \cos\Theta_i, \sum_{i=1}^n \sin\Theta_i\right). \tag{13.2.3}$$

A detailed discussion of the circular normal distribution may be found in [6], using the location and precision parameterization (13.2.1).

Circular normal densities are examples of *symmetric* circular densities. A density $f(\theta)$ on $[0, 2\pi)$ is said to be symmetric if there exists a point $\theta_0 \in [0, 2\pi)$ such that $f((\theta_0 + \tau)\, mod(2\pi)) = f((\theta_0 - \tau)\, mod(2\pi))$ for every $\tau \in [0, 2\pi)$. Typically, the center of symmetry, θ_0, is taken to be the mode of the density, but that is not always the case.

The second symmetric circular model to be considered is what we call a *generalized cardioid* distribution (similar to, but more general than, a model introduced by Jones and Pewsey [7]). It is of the form

$$f_\Theta(\theta; \alpha, \beta, \gamma) \propto [1 + \alpha\cos\theta + \beta\sin\theta]^\gamma\ I(0 < \theta < 2\pi), \tag{13.2.4}$$

where $\alpha, \beta, \gamma \in (-\infty, \infty)$ such that $\alpha^2 + \beta^2 < 1$.

If Θ has (13.2.4) as its density, we will write $\Theta \sim GC(\alpha, \beta, \gamma)$.

A location-precision version of this density is also available; thus

$$f_\Theta(\theta; \mu, \kappa, \gamma) \propto [1 + \kappa \cos(\theta - \mu)]^\gamma \; I(0 < \theta < 2\pi). \tag{13.2.5}$$

As a special case (corresponding to $\gamma = -1$) this generalized cardioid model includes the *wrapped Cauchy* distribution. The name "wrapped Cauchy" arises from the representation of such a random variable in the form

$$\Theta = U \; mod(2\pi), \tag{13.2.6}$$

where $U \sim Cauchy(\mu, \sigma)$ (thus U is wrapped around the circle of radius 1). Other wrapped models can be considered in which U has a distribution that is not of the Cauchy form. The symmetric stable family of distributions (including the normal distribution) is often considered for such "wrapping." Unfortunately, only the Cauchy choice leads to a simple analytic form for the wrapped density (only a series expansion is available for other choices) in this family.

The name "generalized cardioid" distribution has its genesis in the observation that it includes as special case (when $\gamma = 1$) a distribution known as the cardioid distribution which was sometimes considered as a competitor of the von Mises distribution for modeling circular data. Note also that the uniform distribution on the circle is also included as a special case of the generalized cardioid model (corresponding to the case in which $\gamma = 0$). The generalized cardioid family thus includes three popular simpler submodels, cardioid, wrapped Cauchy, and uniform. The circular normal density arises as a limiting case. Thus if $\Theta_n \sim GC(\alpha/n, \beta/n, n)$ for each n, then $\Theta_n \longrightarrow^d \Theta \sim CN(\alpha, \beta)$.

All of the densities thus far discussed are unimodal. Multimodal circular datasets are not uncommon. The usual approach in such situations is to consider mixtures of unimodal densities. Abe et al. [1] discuss a multimodal extension of the cardioid density, originally proposed by Papakonstantinou [11], of the following form:

$$f_\Theta(\theta; \mu.\kappa, \nu) = (1/2\pi)[1 + \kappa \cos\{(\theta - \mu) + \nu \sin(\theta - \mu)\}], \tag{13.2.7}$$

where $0 \le \kappa \le 1, 0 \le \mu < 2\pi$, and $-\infty < \nu < \infty$. These multimodal models are all centered at μ.

A generalized version of this density, as a parallel to the generalized cardioid model, will be of the form:

$$f_\Theta(\theta; \mu.\kappa, \nu) \propto [1 + \kappa \cos\{(\theta - \mu) + \nu \sin(\theta - \mu)\}]^\gamma, \tag{13.2.8}$$

with suitable constraints on the parameters. Only a very preliminary investigation of this general model has been made up till now.

Kato and Jones [8] introduced a four-parameter extension of the wrapped Cauchy distribution that also includes the cardioid model. Suitably parameterized, the parameters admit interpretation as location, concentration, skewness, and kurtosis.

An alternative route to construction of circular distributions may be called the *angular* approach. This involves consideration of a bivariate random variable (X, Y) and the angle that is determined when the point (X, Y) is joined by a line to the origin, i.e., $\Theta = tan^{-1}(Y/X)$. A symmetric distribution for Θ can arise from such a construction for a judicious choice of distribution for (X, Y). More typically, however, the resulting distribution will be asymmetric. An example is the displaced normal model [9] in which for i.i.d. standard normal random variables Z_1, Z_2 and a real constant γ we define $(X, Y) = (\gamma + Z_1, Z_2)$. A symmetric distribution results only if $\gamma = 0$.

Finally, we remark that while a broad spectrum of distributions with support $[0, 2\pi)$ can serve as models for circular data, it is usually reasonable, in the case in which thc distribution is absolutely continuous, to insist that the corresponding density satisfies $\lim_{\theta \to 2\pi} f(\theta) = f(0)$, to reflect the fact that the directions 0 and 2π are indistinguishable.

13.3 Axial Models

Axial random variables take on values in the interval $[0, \pi)$ and correspond to angles determined by undirected vectors.

Any circular density (such as those described in Sect. 13.2) can be used to generate an axial density by the simple device of *doubling*. Thus if $g(\theta)$ is a circular density with support $[0, 2\pi)$, a corresponding axial model will be of the form

$$f_\Theta(\theta) = 2g(2\theta)\, I(0 \leq \theta < \pi). \tag{13.3.1}$$

Note that if the circular density used in this construction is symmetric, then clearly the resulting axial variable will be symmetric.

Axial models can also be constructed by wrapping. Thus for V, a real-valued random variable, we may define an axial variable as

$$\Theta = V\, mod(\pi). \tag{13.3.2}$$

In some circumstances it may be natural to begin in (3.2) with V being a random variable with a circular density. The resulting wrapped density in this case takes the form:

$$f_\Theta(\theta) = [f_V(\theta) + f_V(\theta + \pi)]\, I(0 \leq \theta < \pi). \tag{13.3.3}$$

In this manner, with V being a circular normal random variable, one arrives at an *axial normal* density of the form

$$f_\Theta(\theta) \propto \cosh(\alpha \cos\theta + \beta \sin\theta)\, I(0 \leq \theta < \pi), \tag{13.3.4}$$

where $\alpha, \beta \in (-\infty, \infty)$. This family of axial normal densities, which are symmetric, though arising naturally from circular normal variables, suffers from the fact that it does not correspond to an exponential family. This creates some (not insurmountable) difficulties with regard to estimation and inference.

If an exponential family of symmetric axial densities is desired, one can be constructed by doubling a circular normal random variable and arriving in this way at a density of the form:

$$f_\Theta(\theta; \alpha, \beta) \propto \exp\left[\alpha \cos 2\theta + \beta \sin 2\theta\right] I(0 \leq \theta < \pi). \tag{13.3.5}$$

The angular central Gaussian density can also be considered. To derive it, we begin with a classical bivariate normal density for (X, Y) with mean vector $(0, 0)$ and arbitrary variance-covariance matrix. We then define

$$\Theta = \tan^{-1}\left(\frac{Y}{X}\right) + \frac{\pi}{2}, \tag{13.3.6}$$

where $\tan^{-1}$ has range $(-\pi/2, \pi/2)$. Perhaps surprisingly, this leads to the same density as one obtained by doubling the wrapped Cauchy density, i.e.,

$$f_\Theta(\theta; \alpha, \beta) \propto (1 + \alpha \cos 2\theta + \beta \sin 2\theta)^{-1} I(0 \leq \theta < \pi), \tag{13.3.7}$$

where $\alpha^2 + \beta^2 < 1$.

Note that in (13.3.6) the random variable Y/X has a Cauchy distribution. Other choices for the distribution of Y/X will be discussed in the context of asymmetric distributions.

A natural extension of (13.2.4) is of the form

$$f_\Theta(\theta; \alpha, \beta, \gamma) \propto (1 + \alpha \cos 2\theta + \beta \sin 2\theta)^{\gamma} I(0 \leq \theta < \pi), \tag{13.3.8}$$

where $\gamma \in (-\infty, \infty)$ and $\alpha^2 + \beta^2 < 1$.

In developing the distribution of the classical t distribution, [14, p. 171] introduced an axial distribution that can be called the axial beta distribution. In the context of the classical t distribution the family of axial distributions was indexed by a parameter k corresponding to the degrees of freedom of the model. Typically k assumes positive integer values. However there is no problem considering fractional degrees of freedom in the definition of the t distribution, and consequently we may extend Rao's axial distribution to be parameterized by a positive real-valued parameter which we will denote by α. The resulting family of axial densities is of the form

$$f_\Theta(\theta; \alpha) = [B(1/2, \alpha/2)]^{-1} (\cos(\theta - \pi/2))^{\alpha - 1} I(0 \leq \theta < \pi), \tag{13.3.9}$$

where $\alpha \in (0, \infty)$.

The key feature of axial models is that their support is $[0, \pi)$. Thus, many distributions with this support can serve as axial models. However it may be reasonable, in the case in which the distribution is absolutely continuous, to insist that the corresponding density satisfies $\lim_{\theta \to \pi} f(\theta) = f(0)$, to reflect the fact that the directions 0 and π are indistinguishable.

13.4 Asymmetric Variants

Several authors have suggested ways to modify the popular circular models described in Sect. 13.2 that are symmetric, in such a manner as to obtain asymmetric versions. In completely parallel fashion, asymmetric variants of axial models can be constructed.

Asymmetric densities related to the circular normal model (2.1) can be obtained by introducing higher frequency terms. For example, we might consider a model of the form:

$$f_\Theta(\theta; \alpha, \beta, \delta_1, \delta_2) \propto \exp\left(\alpha \cos\theta + \beta \sin\theta + \delta_1 \cos 2\theta + \delta_2 \sin 3\theta\right) I(0 \le \theta < 2\pi). \tag{13.4.1}$$

Asymmetric versions of the generalized cardioid density (2.2) can be similarly constructed by adding higher frequency terms, such as

$$f_\Theta(\theta; \alpha, \beta, \delta_1, \delta_2, \delta_3) \propto [1 + \alpha \cos\theta + \beta \sin\theta + \delta_1 \cos 2\theta + \delta_2 \sin 3\theta + \delta_3 \cos 4\theta]^\gamma. \tag{13.4.2}$$

The technique of wrapping (i.e., use of (2.3)) will only yield a symmetric circular density if the wrapped variable U itself has a suitably symmetric distribution. Otherwise asymmetry will result. Pewsey [12, 13], for example, obtains an asymmetric circular density by wrapping a skew-normal variable.

Umbach and Jammalamadaka [16] describe a general method of constructing asymmetric circular distributions, motivated by Azzalini's [5] work on construction of skewed distributions on the real line. They begin with a symmetric density on $[0, 2\pi)$ and a possibly different symmetric distribution function G on $[0, 2\pi)$. Then for any odd function w with period 2π and satisfying $0 \le w(\theta) \le \pi$, the function

$$2f(\theta)G(w(\theta)) \tag{13.4.3}$$

is a well-defined asymmetric density on $[0, 2\pi)$.

The angular projection approach, in which (X, Y) has a given bivariate density and a circular variable is defined by the angle in $[0, 2\pi)$ determined by the line joining the origin to the point (X, Y), will typically not lead to a symmetric circular density. It does, as observed in Sect. 13.2, lead to a symmetric model when (X, Y) has bivariate normal density with mean vector $(0, 0)$ and, indeed, it will yield a symmetric model when (X, Y) has an elliptically contoured model centered at $(0, 0)$. But in general, an asymmetric density can be expected. For example, even a change of center for elliptically contoured densities for (X, Y) will destroy symmetry in the resulting circular model. In this approach, asymmetry is the norm rather than the exceptional case.

Finite mixtures of symmetric circular densities will usually be asymmetric as will finite logarithmic mixtures, unless the mixture components share the same center of symmetry. Thus for a given collection $f_1, f_2, \ldots, f_m$ of symmetric densities, not all with the same center of symmetry, we might consider typically asymmetric densities such as

$$f(\theta) \propto \left[\sum_{i=1}^{m} a_i f_i(\theta)\right] I(0 \leq \theta < 2\pi), \tag{13.4.4}$$

or, using a multiplicative scheme,

$$g(\theta) \propto \left[\prod_{i=1}^{m} \{f_i(\theta)\}^{a_i}\right] I(0 \leq \theta < 2\pi). \tag{13.4.5}$$

Yet another approach to constructing asymmetric circular models is to begin with two independent random variables Θ_0 and Z, where Θ_0 has a symmetric circular density and Z is quite arbitrary in nature. Then define

$$\Theta_1 = \Theta_0 + Z\, mod(2\pi). \tag{13.4.6}$$

Only in very special circumstances will such a circular random variable be symmetric.

All of the above techniques for generating asymmetric circular distributions can be used, with obvious modifications, to construct asymmetric axial densities. We will here give details on only one of these constructions. The angular projection method of generating an axial density begins with a given random vector (X, Y) and considers

$$\Theta = \tan^{-1}\left(\frac{Y}{X}\right) + \frac{\pi}{2} \tag{13.4.7}$$

(where $\tan^{-1}$ has range $(-\pi/2, \pi/2)$). In the case in which (X, Y) has a centered bivariate normal density, Y/X has Cauchy density, and consequently the model (13.4.7) can be written in the equivalent form as

$$\Theta = \tan^{-1}(\mu + \sigma Z) + \pi/2, \tag{13.4.8}$$

where Z has a standard $Cauchy(0, 1)$ distribution. One may consider the consequences of replacing the Cauchy random variable Z in (13.4.8) by a random variable W with some other distribution distinct from the Cauchy form. Almost inevitably this yields an asymmetric distribution for Θ. Even if W has a symmetric density, the density of Θ can be surprising in shape (even if W is normal).

Finally we observe that an asymmetric version of the axial beta density is available as

$$f_\Theta(\theta; \alpha, \delta) = e^{-\delta^2/2}(\cos(\theta - \pi/2))^{\alpha-1} \sum_{j=0}^{\infty} B(\alpha/2, (j+1)/2)(2\delta sin(\theta - \pi/2))^j, \tag{13.4.9}$$

where $\alpha \in (0, \infty)$ and $\delta \in (-\infty, \infty)$. Note that if $\delta = 0$, this reduces to yield the axial beta density. In addition, if $\alpha = 1$, the resulting density is uniform on $[0, 2\pi)$.

13.5 In Higher Dimensions

We will focus on bivariate circular and axial distributions, though in most cases, multivariate extensions can be readily visualized. The extensions to higher dimensions are conceptually straightforward but cluttered by a veritable forest of parameters and/or subscripts. We begin by reviewing bivariate circular models. It will be observed that, in the models that have been proposed in bivariate settings, asymmetry is even more pervasive than it was in one dimension. Only in a very few special cases will the bivariate models exhibit features that can be characterized as exhibiting some form of symmetry.

When seeking bivariate extensions of univariate models, there are two general strategies that are frequently used. One is to seek bivariate distributions with marginal densities with specific characteristics or being members of specified parametric families. This can be labeled as a search for distributions with specified marginals. The second approach, somewhat less common, is to seek distributions with specified conditional distributions. Both techniques can be fruitfully applied in the context of directional (or axial) distributions. We first consider marginal specification.

Let F_1 and F_2 be two given circular distributions with corresponding quantile functions F_1^{-1} and F_2^{-1}, respectively. Let (U_1, U_2) be a two-dimensional random variable with *Uniform*$(0, 1)$ marginal distributions. To obtain a bivariate circular random variable with marginal densities given by F_1^{-1} and F_2^{-1}, we define

$$(\Theta_1, \Theta_2) = (F_1^{-1}(U_1), F_2^{-1}(U_2)). \tag{13.5.1}$$

Joint distributions with *Uniform*$(0, 1)$ marginals are often called *copulas*. Different choices of copula for the distribution of (U_1, U_2) will lead to different bivariate circular distributions in (13.5.1). Since many of the popular univariate circular distributions are defined in terms of their densities and do not have readily available expressions for their distribution functions and quantile functions, the approach based on (13.5.1) may be difficult to implement. An exception occurs in the case of the Kato-Jones extended cardioid model for which expressions for the distribution functions are available.

Needless to say, if we instead assume that the quantile functions in (13.5.1) correspond to axial rather than circular distributions, we will generate bivariate axial distributions with given marginals. Next we turn to conditional specification.

Consider distributions for (Θ_1, Θ_2) such that all conditional densities of Θ_1 given $\Theta_2 = \theta_2$ belong to a given parametric family, say $g_1(\theta_1; \underline{\eta}(\theta_2))$, and all conditional densities of Θ_2 given $\Theta_1 = \theta_1$ belong to a second (possibly different) given parametric family, say $g_2(\theta_2; \underline{\tau}(\theta_1))$. Such a joint distribution is said to be conditionally specified. Arnold and Strauss [3] characterized the full class of bivariate exponential family conditionals in $\mathbb{R}^2$. SenGupta [15], exploiting this exponential family result of [3], introduced full conditionally specified families beyond the Euclidean manifold $\mathbb{R}^2$, to more general manifolds, e.g., torus (or bivariate circular), cylinder, sphere, etc. We will illustrate this approach first with circular normal conditionals (an exponential family case—for more such examples the reader is referred to p. 149 of [15], and second with generalized cardioid conditionals (not an exponential family).

First we consider bivariate densities with circular normal conditionals. Thus we wish to identify all bivariate densities for (Θ_1, Θ_2) which have all of their conditional densities in the family (13.2.2). Using the result of [3], the class of such densities is itself an exponential family and is given by

$$f_{\Theta_1,\Theta_2}(\theta_1, \theta_2) = \exp\{(1, \cos\theta_1, \sin\theta_1)A(1, \cos\theta_2, \sin\theta_2)'\}\, I(0 \leq \theta_1, \theta_2 < 2\pi), \tag{13.5.2}$$

where A is a matrix of real parameters, i.e.,

$$A = \begin{pmatrix} a_{00} & a_{01} & a_{02} \\ a_{10} & a_{11} & a_{12} \\ a_{20} & a_{21} & a_{22}. \end{pmatrix}. \tag{13.5.3}$$

Actually there are only eight parameters. a_{00} is a function of the other a_{ij}'s chosen to make the density integrate to 1. It is not difficult to verify that the model (13.5.2) does indeed have circular normal conditionals. The model (13.5.2) includes most of the popular bivariate von Mises distributions in the literature as special cases. Note that it is only in the case of independence that it will have circular normal marginals. The full model (13.5.2) corresponds to the class of all distributions with circular normal conditionals. Two representative densities corresponding to the model (13.5.2) may be found in [15], Figs. 13.1 and 13.2.

If, instead, we seek distributions with generalized cardioid conditionals, i.e., all conditionals in the family (13.2.4), we may consider the family of densities of the following form:

$$f_{\Theta_1,\Theta_2}(\theta_1,\theta_2) \propto [(1,\cos\theta_1,\sin\theta_1)A(1,\cos\theta_2,\sin\theta_2)']^{\gamma}\ I(0 \le \theta_1,\theta_2 < 2\pi). \tag{13.5.4}$$

It has been recently confirmed [4] that the family (13.5.4) includes all densities with conditionals in the family (13.2.4) for a given fixed value of γ. It is a very extensive class of such densities. The entries in the parameter matrix A must be constrained to ensure that the expression (13.5.4) is nonnegative over the region $(0, 2\pi)^2$, but except for that, they may be quite arbitrary. The case $\gamma = -1$ corresponds to wrapped Cauchy conditionals, while $\gamma = 1$ yields cardioid conditionals. Two representative (non-normalized) densities with generalized cardioid conditionals (GCD) (i.e., of the form (13.5.4)) are displayed in Figs. 13.1 and 13.2, where "kpdf" represents the kernel of the pdf with arbitrarily chosen a_{00}. The values of the parameters are $a_{00} = 1, a_{01} = 4, a_{02} = 5, a_{10} = 2, a_11 = 6, a_{12} = 7, a_{20} = 3, a_{21} = 8, a_{22} = 9$. Note that the densities in these figures differ only in the chosen values for the parameter γ, $\gamma = 2$ and 4, in Figs. 13.1 and 13.2, respectively, to illustrate the marked effect of changing that particular parameter value.

The development of conditionally specified axial distributions follows along the same lines to yield for example:

(i) Conditionals in the family (13.3.4):

$$f_{\Theta_1,\Theta_2}(\theta_1,\theta_2) = \cosh\{(\cos\theta_1,\sin\theta_1)B(\cos\theta_2,\sin\theta_2)'\}\ I(0 \le \theta_1,\theta_2 < \pi), \tag{13.5.5}$$

where B is a 2×2 matrix of parameters.

Fig. 13.1 Bivariate GCD kernel for $\gamma = 1$

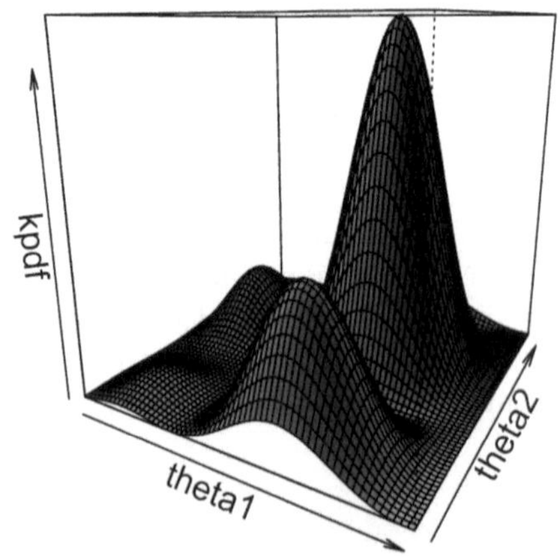

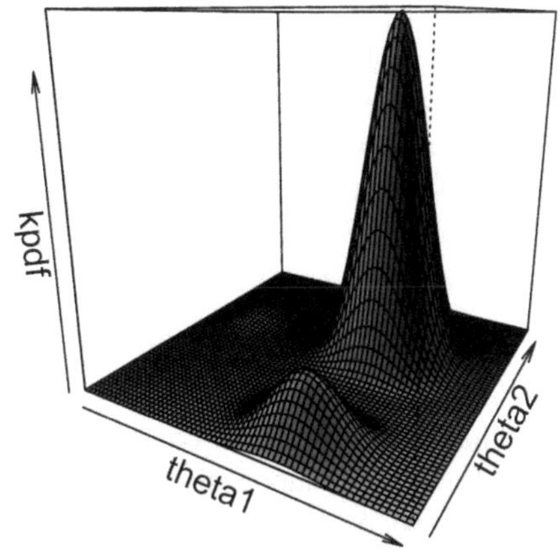

Fig. 13.2 Bivariate GCD kernel for $\gamma = 2$

(ii) Conditionals in the family (13.3.5):

$$f_{\Theta_1,\Theta_2}(\theta_1,\theta_2) = \exp\{(1, \cos 2\theta_1, \sin 2\theta_1)A(1, \cos 2\theta_2, \sin 2\theta_2)'\} \\ I(0 \le \theta_1, \theta_2 < \pi), \quad (13.5.6)$$

where A is a 3×3 matrix of parameters.

(iii) Conditionals in the family (13.3.8):

$$f_{\Theta_1,\Theta_2}(\theta_1,\theta_2) \propto [(1, \cos 2\theta_1, \sin 2\theta_1)A(1, \cos 2\theta_2, \sin 2\theta_2)']^{\gamma} \\ I(0 \le \theta_1, \theta_2 < \pi), \quad (13.5.7)$$

where A is a 3×3 matrix of parameters.

13.5.1 Wrapping

The wrapping technique can be used just as well in two dimensions as it was in one dimension. Begin with an arbitrary two-dimensional random variable (X, Y), and define a two-dimensional circular variable by

$$(\Theta_1, \Theta_2) = (X\ mod(2\pi), Y\ mod(2\pi)). \quad (13.5.8)$$

For example, we might begin with (X, Y) having a bivariate Cauchy distribution. The bivariate wrapped variable (13.5.8) will have wrapped Cauchy marginals (i.e., of the form (13.2.4) with $\gamma = -1$). A variety of candidate bivariate Cauchy models exist and can be used in this construction. Although the resulting marginal densities are recognizable as having a generalized cardioid form, the joint densities are generally not obtainable in a simple form. Of course random variables of the form (biv-wrap) will be easy to simulate provided that realizations of the corresponding random vector (X, Y) can be simulated.

Bivariate axial variables can, of course, be analogously derived by wrapping $mod\ (\pi)$ instead of 2π.

13.5.2 Additive Components

Another easy to simulate construction of bivariate circular variables involves the usc of additive components. For it we begin with three circular random variables Λ_1, Λ_2, and Λ_3 which may be dependent but would usually be independent. We then define a bivariate circular variable by

$$(\Theta_1, \Theta_2) = ((\Lambda_1 + \Lambda_3)\ mod(2\pi), (\Lambda_2 + \Lambda_3)\ mod(2\pi)). \quad (13.5.9)$$

Bivariate axial variable can also be constructed in this fashion.

13.5.3 Bivariate Cardioid Models via Mixing

The circular cardioid model was shown by Ong and SenGupta [10] to be closed under a form of mixing using a beta distribution. They then used this fact to develop a variety of bivariate cardioid models. Their starting point was the consideration of a variable Θ having a conditional distribution given $X = x$ of the following cardioid form:

$$f_{\Theta|X=x}(\theta; \mu, \kappa) = (1/2\pi)[1 + \kappa x \cos(\theta - \mu)]\ I(0 < \theta < 2\pi), \quad (13.5.10)$$

where $\kappa \in (-1, 1)$ and $\mu \in (0, 2\pi)$. They then assumed that X had a four-parameter beta density over the interval $(-1, 1)$ and verified that, unconditionally, Θ had once more a cardioid distribution. Perhaps surprisingly, this closure under mixing result holds much more generally.

We remark at this point that it is clear that analogous discussion could be presented for bivariate axial cardioid models throughout.

Suppose that the concomitant random variable (or mixing variable) X has a quite arbitrary distribution function $G(x)$ on the interval $(-1, 1)$ with finite mean $E_G(X)$; in this case we can identify the corresponding marginal density of Θ as follows:

$$\begin{aligned} f_\Theta(\theta; \kappa, \mu, G) &= (1/2\pi)\int_{-1}^{1} [1 + \kappa x \cos(\theta - \mu) dG(x)\ I(0 < \theta < 2\pi) \\ &= (1/2\pi)[1 + \kappa E_G(X) \cos(\theta - \mu)\ I(0 < \theta < 2\pi). \end{aligned}$$

Thus the mixed density is of the circular cardioid form with a modified precision parameter (κ).

In a completely parallel fashion, the axial cardioid family of densities is closed under mixing of the precision parameter. To see this we consider a variable Θ having a conditional distribution given $X = x$ of the following axial cardioid form:

$$f_{\Theta|X=x}(\theta;\mu,\kappa) = (1/\pi)[1+\kappa x\cos(2(\theta-\mu))]\, I(0<\theta<\pi), \qquad (13.5.11)$$

where $\kappa \in (-1,1)$ and $\mu \in (0,\pi)$. Suppose that the concomitant random variable (or mixing variable) X has a quite arbitrary distribution function $G(x)$ on the interval (-1,1) with finite mean $E_G(X)$; in this case we can identify the corresponding marginal density of Θ as follows:

$$\begin{aligned} f_\Theta(\theta;\kappa,\mu,G) &= (1/\pi)\int_{-1}^{1}[1+\kappa x\cos(2(\theta-\mu))dG(x)\, I(0<\theta<\pi) \\ &= (1/\pi)[1+\kappa E_G(X)\cos(2(\theta-\mu))\, I(0<\theta<\pi), \end{aligned}$$

showing that the mixed density is of the axial cardioid form with a modified precision parameter (κ).

As suggested by Ong and SenGupta [10], the precision mixing provides a fruitful source of bivariate (and potentially, multivariate) circular and axial cardioid models. The first such bivariate model that they considered had Θ_1 and Θ_2 conditionally independent given $X = x$ each having a circular cardioid density. Thus

$$f_{\Theta_1,\Theta_2|X=x} = [(1/2\pi)[1+\kappa_1 x\cos(\theta_1-\mu_1)][(1/2\pi)[1+\kappa_2 x\cos(\theta_2-\mu_2)], \qquad (13.5.12)$$

where $((\theta_1,\theta_2)\in(0,2\pi)^2)$, $(\kappa_1,\kappa_2)\in(-1,1)^2$, and $(\mu_1,\mu_2)\in(0,2\pi)^2$. Instead of assuming that X has a beta distribution, we will assume that X has a quite arbitrary distribution function $G(x)$ on the interval (-1,1) with necessarily finite first and second moments $E_G(X)$ and $E_G(X^2)$. In this case we can identify the corresponding joint density of (Θ_1,Θ_2) as follows:

$$\begin{aligned} &f_{\Theta_1,\Theta_2}(\theta_1,\theta_2;\kappa_1,\kappa_2,\mu_1,\mu_2,G) \\ &= \int_0^1 [(1/2\pi)[1+\kappa_1 x\cos(\theta_1-\mu_1)][(1/2\pi)[1+\kappa_2 x\cos(\theta_2-\mu_2)]dG(x) \\ &\qquad = (1/2\pi)^2[1+E_G(X)\{\kappa_1\cos(\theta_1-\mu_1)+\kappa_2\cos(\theta_2-\mu_2)\} \\ &\qquad\qquad + E_G(X^2)\kappa_1\kappa_2\cos(\theta_1-\mu_1)\cos(\theta_2-\mu_2)]. \end{aligned}$$

It is readily verified that this joint density has marginal densities of the circular cardioid form. The conditional densities are of a more complicated nature.

Following Ong and SenGupta [10] we consider the case in which a two-dimensional mixing variable (X, Y) is utilized, but, whereas they considered various bivariate beta densities for the mixing distribution, it is possible to assume a more general mixing distribution, with certain moment constraints.

The following joint conditional density of (θ_1, θ_2) given $(X, Y) = (x, y)$ is assumed

$$f_{\Theta_1,\Theta_2|X=x,Y=y} = [(1/2\pi)[1 + \kappa_1 x \cos(\theta_1 - \mu_1)][(1/2\pi)[1 + \kappa_2 y \cos(\theta_2 - \mu_2)], \tag{13.5.13}$$

where $(\kappa_1, \kappa_2) \in (-1, 1)^2$ and $(\mu_1, \mu_2) \in (0, 2\pi)^2$. We now assume that (X, Y) has an arbitrary joint distribution $H(x, y)$ on the set $(-1, 1)^2$ with necessarily finite moments $E_H(X)$, $E_H(Y)$, and $E_H(XY)$. The unconditional joint density of (θ_1, θ_2) is then given by

$$f_{\Theta_1,\Theta_2}(\theta_1, \theta_2; \kappa_1, \kappa_2, \mu_1, \mu_2, H)$$

$$= \int_0^1 [(1/2\pi)[1 + \kappa_1 x \cos(\theta_1 - \mu_1)][(1/2\pi)[1 + \kappa_2 y \cos(\theta_2 - \mu_2)]dH(x)$$

$$= (1/2\pi)^2[1 + E_H(X)\kappa_1 \cos(\theta_1 - \mu_1) + E_H(Y)\kappa_2 \cos(\theta_2 - \mu_2)$$

$$+ E_H(XY)\kappa_1\kappa_2 \cos(\theta_1 - \mu_1)\cos(\theta_2 - \mu_2)].$$

It is readily verified that this joint density also has marginal densities of the circular cardioid form, while the conditional densities are of a more complicated nature.

Ong and SenGupta [10] discuss a spectrum of examples in which a variety of bivariate beta models are used for the mixing distribution in the above construction. As has been observed above, it is only certain moments, i.e., $E_H(X)$, $E_H(Y)$, and $E_H(XY)$, that are involved in the resulting models.

A very general bivariate beta model was introduced by Arnold and Ng [2] involving ratios of sums of independent gamma variables with unit scale parameters. It includes, as special cases, many of the bivariate beta models discussed in the literature, such as Dirichlet, Jones-Olkin-Liu, etc.

The construction of the Arnold-Ng model begins with eight independent gamma distributed random variables, $U_1, U_2, \ldots, U_8$ with $U_j \sim \Gamma(\delta_j, 1)$, $j = 1, 2, \ldots, 8$.

Next, define

$$V = \frac{U_1 + U_5 + U_7}{U_3 + U_6 + U_8}, \tag{13.5.14}$$

$$W = \frac{U_2 + U_5 + U_8}{U_4 + U_6 + U_7}. \tag{13.5.15}$$

By construction (V, W) has a bivariate density with marginal densities of the second kind beta form (ratios of independent gamma variables with common scale parameter), and moreover this is the most general such model constructed via sums of ratios of independent gamma variables. We can readily transform this joint density into one with marginals which have beta densities on the interval $(-1, 1)$ by defining

$$X = \frac{V-1}{V+1} = \frac{(U_1 + U_5 + U_7) - (U_3 + U_6 + U_8)}{(U_1 + U_5 + U_7) + (U_3 + U_6 + U_8)} \tag{13.5.16}$$

and

$$Y = \frac{W-1}{W+1} = \frac{(U_2 + U_5 + U_8) - (U_4 + U_6 + U_7)}{(U_2 + U_5 + U_8) + (U_4 + U_6 + U_7)}. \tag{13.5.17}$$

Since (X, Y), defined in this way, has marginals that are beta densities on the interval $(-1, 1)$, it may be used as a mixing distribution for the conditional model (13.5.13). The marginal means of (X, Y) are readily computed, thus

$$E(X) = \frac{(\delta_1 + \delta_5 + \delta_7) - (\delta_3 + \delta_6 + \delta_8)}{(\delta_1 + \delta_5 + \delta_7) + (\delta_3 + \delta_6 + \delta_8)},$$

$$E(Y) = \frac{(\delta_2 + \delta_5 + \delta_8) - (\delta_4 + \delta_6 + \delta_7)}{(\delta_2 + \delta_5 + \delta_8) + (\delta_4 + \delta_6 + \delta_7)}.$$

However, it will only be in simplified submodels (obtained by imposing linear constraints on the δ_j's or by setting some of them equal to 0) that one can obtain relatively simple evaluations of $E(XY)$ which is needed to identify the density resulting from the mixing.

13.5.4 Miscellaneous Techniques

(a) In principle, we could begin with a three-dimensional random vector (X, Y, Z) and consider two angles determined by the line joining the point (X, Y, Z) to the origin $(0, 0, 0)$ to obtain a bivariate circular random vector.
(b) Begin with three independent random variables X_1, X_2, X_3 where $X_i \sim Cauchy(\mu_i, \sigma_i)$, $i = 1, 2, 3...$ Then construct a bivariate axial variable of the form $\Theta_1, \Theta_2)$ where

$$\Theta_1 = \frac{\pi}{2} + \tan^{-1}(X_1 + X_3),$$

$$\Theta_2 = \frac{\pi}{2} + \tan^{-1}(X_2 + X_3).$$

The marginal densities for this model will be of the form (13.3.7).

(c) Let (U_1, U_2) have $Uniform(0, 1)$ marginals (i.e., its distribution will be a copula), and let F_1, F_2, F_3, F_4 be four not necessarily distinct distribution functions with corresponding quantile functions F_i^{-1}, $i = 1, 2, 3, 4$. We may then define a bivariate axial variable (Θ_1, Θ_2), by

$$\Theta_1 = \pi F_1(\mu_1 + \sigma_1 F_2^{-1}(U_1)),$$
$$\Theta_2 = \pi F_3(\mu_2 + \sigma_2 F_4^{-1}(U_2)).$$

13.5.5 Further Comments on Asymmetry

As mentioned earlier, most of the bivariate models exhibit very limited symmetry. The techniques used to introduce asymmetry in univariate models can sometimes be used to further augment available bivariate models by introducing yet more asymmetry. For models (13.5.2) and (13.5.4)–(13.5.7) one can introduce higher frequency components as was described in one dimension, retaining exponential family structure when present but adding flexibility. Most of the other bivariate models described (in equations (13.5.1), (13.5.8), and (13.5.9) and in Sect. 13.5.3) will typically lack symmetry. For even more flexibility and predictably less symmetry, mixtures of the various models can be considered.

13.6 Closing Comments

In this whirlwind tour of circular and axial models in one or more dimensions, essentially no attention has been given to inferential issues that might arise in application of the models. A detailed discussion of such issues would render the presentation unacceptably long. There are indeed many questions regarding parameter estimation, model fitting, and model comparison that require careful consideration. The use of maximum likelihood, conditional likelihood, and variations on the method of moments theme will undoubtedly be involved in first addressing these questions. Before this is attempted, it will be, in many cases, necessary to explicitly identify the acceptable ranges of all parameters appearing in each model in order that they will represent valid univariate or bivariate densities. There is much work to be done.

References

1. Abe, T., Pewsey, A., Shimizu, K.: On Papakonstantinou's extension of the cardioid distribution. Statist. Probab. Lett. **79**, 2138–2147 (2009)
2. Arnold, B.C., Ng, H.K.T.: Flexible bivariate beta distributions. J. Multivariate Anal. **102**, 1194–1202 (2011)
3. Arnold, B.C., Strauss, D.J.: Bivariate distributions with conditionals in prescribed exponential families. J. Roy. Statist. Soc. Ser. B **53**, 365–375 (1991)
4. Arnold, B.C., SenGupta, A.: On generalized cardioid distributions. Submitted (2024)
5. Azzalini, A.: The skew-normal distribution and related multivariate families. Scand. J. Statist **32**, 159–200 (2005)
6. Jammalamadaka, S.R., SenGupta, A.: Topics in Circular Statistics. World Scientific, Singapore (2001)
7. Jones, M.C., Pewsey, A.: A family of symmetric distributions on the circle. J. Am. Stat. Assoc. **100**, 1422–1428 (2005)
8. Kato, S., Jones, M.C.: A tractable and interpretable four-parameter family of unimodal distributions on the circle. Biometrika **102**, 181–190 (2015)
9. Kendall, D.G.: Pole-seeking Brownian motion and bird navigation. J. Roy. Statist. Soc. Ser. B **36**, 365–417 (1974)
10. Ong, S.-H., SenGupta, A.: Bivariate cardioid distributions. In: Forum Interdisciplinary Mathematics Springer, Singapore, pp. 245–262 (2022)
11. Papakonstantinou, G.: Cascade transformation. IEEE Trans. Comput. **C-25**, 93–95 (1976)
12. Pewsey, A.: Modelling asymmetrically distributed circular data using the wrapped skew-normal distribution. Environ. Ecol. Stat. **13**, 257–269 (2006)
13. Pewsey, A.: Some observations on a simple means of generating skew distributions. In: Statistics for Industry and Technology. Birkhäuser Boston, Inc., Boston, MA., pp. 75–84 (2006)
14. Rao, C.R.: Linear statistical inference and its applications, 2nd edn. In: Wiley Series in Probability and Mathematical Statistics. Wiley, New York (1973)
15. SenGupta, A.: On the construction of probability distributions for directional data. Bull. Cal. Math. Soc. **96**, 139–154 (2004)
16. Umbach, D., Jammalamadaka, S.R.: Building asymmetry into circular distributions. Statist. Probab. Lett. **79**, 659–663 (2009)

Chapter 14
Classification and Regression Error Bounds for Inhomogeneous Data with Applications to Wireless Networks

Ghurumuruhan Ganesan

Abstract In this chapter, we study classification and regression error bounds for inhomogeneous data that are independent but not necessarily identically distributed. First, we consider classification of data in the presence of nonstationary noise and establish ergodic type sufficient conditions that guarantee the achievability of the Bayes error bound, using universal rules. We then perform a similar analysis for k-nearest neighbour regression and obtain optimal error bounds for the same. Finally, we illustrate applications of our results in the context of wireless networks.

Keywords Classification · Regression · Error bounds · Inhomogeneous data · Wireless networks

14.1 Introduction

Classification and regression are two important problems from both application and theoretical perspectives and have extensive applications in machine learning. It is well known [5] that the minimum possible error is obtained by the Bayes classifier that assumes complete knowledge of the data distribution. As mentioned before, since this information is usually not available, it is of interest to compare the performance of universal rules against the Bayes classifier. When the data are independent and identically distributed (i.i.d.), it is well known [5] that the k-nearest neighbour rule is universally consistent in the sense that it achieves the Bayes bound asymptotically.

Recently, there has been interest in studying the *rate of convergence* of the error probability toward the Bayes bound, and various results are known under varying assumptions on the conditional probabilities (see, e.g., [2, 6], and the references therein). We also remark that inhomogeneous regression has been studied from other application perspectives, and in this regard, we refer to [9, 10].

G. Ganesan (✉)
University of Bristol, Bristol, England

A. Adhikari et al. (eds.), *Quantitative Frontiers*, Industrial and Applied Mathematics, https://doi.org/10.1007/978-981-96-9435-8_14

In this chapter, we study classification and regression of data that are not necessarily i.i.d. and obtain ergodic type sufficient conditions for minimizing the error probability. At the end, we discuss detailed applications of our results in the context of wireless networks.

The chapter is organized as follows: In Sect. 14.2, we obtain error bounds for regression of inhomogeneous data under ergodic type conditions. Next, in Sect. 14.3, we derive bounds for the minimum classification error probability of inhomogeneous data when the noise statistics satisfy an ergodic type condition. Finally, in Sect. 14.4 we briefly illustrate how our results are applicable in wireless networks and conclude with potential future directions.

14.2 Regression of Inhomogeneous Data

For $j \geq 1$ let $X_j \in \mathbb{R}^d$ be a random vector with density f_j, and let $Y_j = h(X_j)+N_j$, where h is a deterministic but unknown function and N_j is the noise random variable that is bounded (i.e., $|N_j| \leq C$ for some constant $C > 0$) with zero mean and variance σ_N^2. We discuss the application of this setup in Sect. 14.4, in the context of estimating transmission power levels in wireless networks.

A *regressor* is a sequence of maps $g_n : (\mathbb{R}^d \times \mathbb{R})^n \times \mathbb{R}^d \rightarrow \mathbb{R}$, which takes $(X_j, Y_j), 1 \leq j \leq n$ and X_{n+1} as input and outputs an estimate Z_{n+1} of Y_{n+1}. We say that $\{g_n\}$ is *universal* if no g_n depends on the distributions of $(X_j, Y_j), j \geq 1$. To reduce the effect of local fluctuations in the data, we use *average* variance as a comparison metric and say that $\{g_n\}$ is *consistent* if the average variance

$$\frac{1}{n}\sum_{i=1}^{n}\mathbb{E}(Y_i - Z_i)^2 \longrightarrow \sigma_N^2 \tag{14.2.1}$$

as $n \rightarrow \infty$.

When the data is i.i.d., there are many well-known universally consistent regressors like classification and regression trees, k-nearest neighbour rules [3], etc. Under an ergodic type condition on the data densities, we have the following result for nonstationary data.

Theorem 1 *If h is uniformly continuous and*

$$\sup_x \left|\frac{1}{n}\sum_{i=1}^{n} f_i(x) - f(x)\right| \longrightarrow 0 \tag{14.2.2}$$

as $n \rightarrow \infty$, where f is a uniformly continuous density, then there is a universally consistent regressor that achieves (14.2.1).

Proof of Theorem 1 Let $\mathcal{S}(f) := \{x : f(x) > 0\}$ be the support of the density f, and assume there are constants $0 < c_1 \leq c_2 < \infty$ such that $c_1 \leq f(x) \leq c_2$ for all $x \in \mathcal{S}(f)$. Also assume that $0 < \eta_1 \leq h(x) \leq \eta_2 < \infty$ for all x. We relax these conditions later.

The regression problem states that given $(X_j, Y_j), 1 \leq j \leq n$, and X_{n+1}, we are to estimate the value of Y_{n+1}, using the k-nearest neighbours. We begin with some preliminary computations.

Let $k = k(n)$ be such that $\frac{k}{n} \longrightarrow 0$ as $n \to \infty$. Let $x \in \mathbb{R}^d$ be such that $f(x) > 0$, and let $B_r(x)$ be the d-dimensional ball of radius r centred at x, where $r = r(n, x) := \left(\frac{k}{\pi_d n f(x)}\right)^{1/d}$ and π_d is the volume of the d-dimensional ball of unit radius. Since $f(x) \geq c_1$, we get that $\sup_x r(n, x) \longrightarrow 0$ as $n \to \infty$. Now if $N_X(x) := \sum_{i=1}^n \mathbb{1}(X_i \in B_r(x))$ is the number of data points in $B_r(x)$, then $\mathbb{E}N_X(x) = \sum_{i=1}^n \int_{B_r(x)} f_i(y)dy$ and so using (14.2.2), we get that

$$\sup_x \left| \frac{\mathbb{E}N_X(x)}{n\pi_d r^d} - \int_{B_r(x)} f(y)dy \right| \longrightarrow 0 \tag{14.2.3}$$

as $n \to \infty$. By the uniform continuity of the density f, we know that

$$\sup_x \left| \frac{1}{\pi_d r^d} \int_{B_r(x)} f(y)dy - f(x) \right| \longrightarrow 0$$

as $n \to \infty$, and by definition $nf(x)\pi_d r^d = k$. Combining these observations with (14.2.3), we get

$$\sup_x \left| \frac{\mathbb{E}N_X(x)}{k} - 1 \right| \longrightarrow 0 \tag{14.2.4}$$

as $n \to \infty$.

From the deviation estimate (14.3.11), we also know that

$$\mathbb{P}\left(|N_X(x) - \mathbb{E}N_X(x)| \geq \epsilon \mathbb{E}N_X(x)\right) \leq \exp\left(-\frac{\epsilon^2}{4}\mathbb{E}N_X(x)\right)$$

and so from (14.2.4), we get that

$$\sup_x \mathbb{P}\left(|N_X(x) - k| \geq 2\epsilon k\right) \leq \exp\left(-\frac{k\epsilon^2}{8}\right) \tag{14.2.5}$$

for all n large.

Letting $Z_{est} := \sum_{i=1}^n Y_i \mathbb{1}(X_i \in B_r(x))$ be the sum of the observations whose corresponding data points lie within $B_r(x)$, we have that

$$\mathbb{E}Z_{est} = \sum_{i=1}^n \mathbb{E}Y_i \mathbb{1}(X_i \in B_r(x)) = \sum_{i=1}^n \int_{B_r(x)} h(y) f_i(y)dy$$

since the noise N_i has zero mean and is independent of X_i, for each i. Again using (14.2.2) and the uniform continuity of $h(x)$, we then get that

$$\sup_x \left| \frac{1}{n\pi_d r^d} \sum_{i=1}^{n} \int_{B_r(x)} h(y) f_i(y) dy - h(x) f(x) \right| \longrightarrow 0$$

as $n \to \infty$ and so an analogous argument as in (14.2.4) gives us

$$\sup_x \left| \frac{\mathbb{E} Z_{est}}{k} - h(x) \right| \longrightarrow 0 \tag{14.2.6}$$

as $n \to \infty$. The function $h(.)$ and the noise N_i are both bounded, and therefore each $|Y_i| \leq B$ for some constant $B > 0$. This allows us to use the Azuma-Hoeffding inequality [1] to obtain that

$$\begin{aligned} \sup_x \mathbb{P}\left(|Z_{est} - kh(x)| \geq 2\epsilon kh(x)\right) &\leq \sup_x \exp\left(-C_1 \epsilon^2 kh(x)\right) \\ &\leq \exp\left(-C_2 \epsilon^2 k\right), \end{aligned} \tag{14.2.7}$$

for some constants $C_i = C_i(\eta_1, \eta_2) > 0$, where the final relation in (14.2.7) is true since $h(x) \geq \eta_1 > 0$.

Defining

$$E_{tot}(x) := \{|N_X(x) - k| \leq 2\epsilon k\} \bigcap \{|Z_{est} - kh(x)| \leq 2\epsilon kh(x)\},$$

we get from (14.2.5), (14.2.7), and the union bound that

$$\inf_x \mathbb{P}(E_{tot}(x)) \geq 1 - \exp\left(-\frac{k\epsilon^2}{8}\right) - \exp\left(-C_2 \epsilon^2 k\right) \geq 1 - e^{-C_3 \epsilon^2 k} \tag{14.2.8}$$

for some constant $C_3 > 0$. We assume henceforth that $E_{tot}(x)$ occurs, let $\mathcal{N}_k(x)$ be the k closest data points to x in the set $\{X_i\}_{1 \leq i \leq n}$, and define

$$Y_{kNN} := \frac{1}{k} \sum_{i=1}^{n} Y_i \mathbb{1}(X_i \in \mathcal{N}_k(x)) \tag{14.2.9}$$

to be the corresponding sum of the observations. Because $E_{tot}(x)$ occurs, the number of data points within $B_r(x)$ lies between $k - k\epsilon$ and $k + k\epsilon$ and so using $|Y_i| \leq B$ for each i, we see that

$$\frac{1}{k} Z_{est} - B\epsilon \leq Y_{kNN} \leq \frac{1}{k} Z_{est} + B\epsilon$$

and therefore that

$$h(x) - 3B\epsilon \le Y_{kNN} \le h(x) + 3B\epsilon.$$

Consequently, we get from (14.2.7) that

$$\sup_x \mathbb{P}\left(|Y_{kNN} - h(x)| \ge 3B\epsilon\right) \le e^{-C_3\epsilon^2 k} \tag{14.2.10}$$

and therefore that

$$\sup_x \mathbb{E}\,|Y_{kNN} - h(x)|^2 \le 9B^2\epsilon^2 + e^{-C_3\epsilon^2 k}. \tag{14.2.11}$$

Setting $\epsilon = \frac{(\log k)}{\sqrt{k}}$, the right side of (14.2.11) is at most

$$\frac{9B^2(\log k)^2}{k} + \exp\left(-C_3(\log k)^2\right) \le \frac{C_4(\log k)^2}{k} =: b_n(k) = b_n$$

for all n large and some constant $C_4 > 0$. Therefore using the mutual independence of Y_{kNN} and N_{n+1} and the fact that N_{n+1} has zero mean, we get that

$$\mathbb{E}(Y_{kNN} - h(x) - N_{n+1})^2 = \mathbb{E}(Y_{kNN} - h(x))^2 + \mathbb{E}N_{n+1}^2 \le b_n + \sigma_N^2.$$

Finally, using Fubini's theorem, we see that the average variance as defined in (14.2.1) is at most $\frac{1}{n}\sum_{i=1}^n b_i + \sigma_N^2 \longrightarrow \sigma_N^2$ as $n \to \infty$, since $b_n \longrightarrow 0$ as $n \to \infty$.

Next we relax the condition on the lower and upper bounds for the density f. For $\zeta > 0$ small, we define the open set

$$A_{good} = \left\{\zeta < f < \zeta^{-1}\right\} \bigcap \left\{\zeta < h < \zeta^{-1}\right\}$$

and perform an analogous analysis as above for $x \in A_{good}$. Using (14.2.11) and arguing as above, we then get that

$$\mathbb{E}|Y_{kNN} - Y_{n+1}|^2 \le \sigma_N^2 + b_n + \int_{A_{good}^c} f_{n+1}(x)dx$$

and so the average variance is at most

$$\sigma_N^2 + \frac{1}{n}\sum_{i=1}^n b_i + \int_{A_{good}^c} \frac{1}{n}\sum_{i=1}^n f_i(x)dx.$$

Again using (14.2.2) and the fact that $b_n \longrightarrow 0$, we get that the average variance is at most

$$\sigma_N^2 + \int_{A_{good}^c} f(x)dx \longrightarrow \sigma_N^2$$

as $\zeta \to 0$. □

14.3 Classification of Inhomogeneous Data

Let μ be a probability measure in $\mathbb{R}^d$, and let X be a random variable with measure μ. Let $X_i, 1 \leq i \leq n$, be independent and identically distributed (i.i.d.) with the same distribution as X. For $1 \leq i \leq n$, let $Y_i = h_i(X_i) \in \{0, 1\}$ be a binary random variable depending only on X_i and independent of all $(X_j, Y_j), j \neq i$. We discuss the application of this setup in Sect. 14.4 in the context of primary user detection in cognitive radio networks.

The random variables $\{Y_i\}$ need *not* be identically distributed, and we refer to $(X_i, Y_i), 1 \leq i \leq n$, henceforth as the *data*. From Fubini's theorem, we have that

$$\mathbb{P}(Y_i = 1, X_i \in A) = \int_A h_i(x)\mu(dx) \tag{14.3.1}$$

for all measurable sets A.

A classification rule is a sequence of deterministic measurable functions $g_n, n \geq 1$, such that $g_n : \mathbb{R}^{2nd+1} \to \{0, 1\}$ takes $(X_i, Y_i), 1 \leq i \leq n$, and X_{n+1} as input and outputs its prediction of Y_{n+1}. We say that the rule $\{g_n\}$ is *universal* if no g_n depends on either μ or any of the h_i. For example, the k-nearest neighbours rule and the histogram rule [5] are examples of universal classifiers. Formally, the k-NN rule is described as follows. For each point x, let N_1 be the number of 1's amongst the k-nearest neighbours of x, in terms of Euclidean distance. We then define the k-NN rule as

$$w_n(x) := \begin{cases} 0 & \text{if } N_1 \leq \frac{k}{2} \\ 1 & \text{otherwise.} \end{cases} \tag{14.3.2}$$

For any classification rule $\{g_n\}$, we define

$$\varepsilon_n := \mathbb{P}\left(Y_{n+1} \neq g_n(X_{n+1})\right)$$

to be the error probability at the $(n + 1)$th data point, where $g_n(X_{n+1}) = g_n\left(X_{n+1}; X_1, Y_1, \ldots, X_n, Y_n\right)$ possibly depends on all or some of the previous data points. In general, it is of interest to minimize ε_n and if $\{Y_i\}$ are i.i.d., i.e., $Y_i = h(X_i)$

for some deterministic function h, then it is well known (Theorem 2.1, p. 10 [5]) that the desired minimization is achieved using the Bayes classification rule, i.e., $g_n = \mathbb{1}\left(h > \frac{1}{2}\right)$, where $\mathbb{1}(.)$ refers to the indicator function. In this case

$$\varepsilon_n = \int \min(h(x), 1 - h(x))\mu(dx) =: L^*, \tag{14.3.3}$$

and the minimum error remains invariant, i.e., it does not change with the index n.

If $\{Y_i\}$ are not identically distributed, then the minimum error might fluctuate, i.e., vary with n. To counter the effect described above, in this chapter, we define and study the *average* error probability

$$T_n := \frac{1}{n}\sum_{i=1}^{n} \varepsilon_i. \tag{14.3.4}$$

Also, because it may not be feasible to estimate η_i for each i, we define our error probabilities of interest as

$$E^* := \inf_{\{g_n\}} \liminf_n T_n \text{ and } U^* = \inf_{\{u_n\}} \limsup_n T_n, \tag{14.3.5}$$

where the infimum for E^* is over all possible classification rules and the infimum for U^* is over all possible *universal* classification rules. We say that a universal classification rule $\{u_n\}$ is *universally consistent* [5] if it achieves U^*, i.e., the error probability $\limsup_n T_n = U^*$.

Our main result below obtains the minimum possible average error probability under an ergodic type condition on the noise statistics $\{\eta_i\}$.

Theorem 2 *If*

$$\sup_x \frac{1}{n}\sum_{i=1}^{n} |h_i(x) - h(x)| \longrightarrow 0, \tag{14.3.6}$$

then $E^* = U^* = L^*$, *where* L^* *is the Bayes classification error as defined in (14.3.3). If in addition*

$$\mathbb{P}\left(h(X) = \frac{1}{2}\right) = 0, \tag{14.3.7}$$

then there is a universally consistent rule $\{u_n\}$ *that achieves* L^*.

Condition (14.3.6) is an ergodic type condition for the existence of universal classifiers performing arbitrarily close to optimality as possible. The stronger condition (14.3.7) states that as long as the output Y is never “truly” random, there are universal classifiers that in fact achieve the Bayes bound L^*.

In our proof of Theorem 2 below, we use a majority rule to estimate the error probabilities and obtain the bound $U^* \leq L^*$. This is described in the following auxiliary result of independent interest.

Lemma 1 *If*

$$\sup_x \left| \frac{1}{n} \sum_{i=1}^{n} h_i(x) - h(x) \right| \longrightarrow 0, \tag{14.3.8}$$

then for every $\zeta > 0$ *there is a sequence of universal classifiers* $\{g_n\}$ *whose error probability satisfies* $\limsup_n T_n \leq L^* + b(\zeta)$, *where* $b(\zeta) \longrightarrow 0$ *as* $\zeta \to 0$.

Proof of Lemma 1 We use a classifier based on the following majority rule: Let $r = r_n > 0$ be such that

$$r_n \to 0 \text{ and } \frac{nr_n^d}{\log n} \longrightarrow \infty, \tag{14.3.9}$$

as $n \to \infty$, and for $x \in \mathbb{R}^d$, let $S_r(x)$ be the Euclidean ball of radius r centred at x. Let N_{tot} be the total number of data points of $\{X_i\}_{1\leq i\leq n}$, present within $S_r(x)$, and for $k \in \{0, 1\}$ let N_k be the number of k's amongst the data points $\{Y_i\}_{1\leq i\leq n}$, within $S_r(x)$. For $0 < \zeta < \frac{1}{4}$, we define our classification rule as

$$g_n(x) := \begin{cases} 1, & \text{if } N_1 - N_0 > \zeta N_{tot} \\ Z_n, & \text{if } |N_1 - N_0| \leq 2\zeta^2 N_{tot} \\ 0, & \text{otherwise}, \end{cases} \tag{14.3.10}$$

where $\{Z_j\}$ are i.i.d. random binary random variables also independent of the data $\{(X_i, Y_i)\}$, satisfying

$$\mathbb{P}(Z_j = 1) = \frac{1}{2} = \mathbb{P}(Z_j = 0).$$

In words, as long as the number of $k's$ is reasonably larger than the number of $(1-k)$'s, our classifier outputs $k \in \{0, 1\}$ as the predicted value. Otherwise, our classifier plays it safe and outputs the result of a random coin toss in the form of Z_n.

Throughout, we use the following standard deviation estimate which we state separately for convenience: Let $\{Z_j\}_{1\leq j\leq r}$ be independent Bernoulli random variables with $\mathbb{P}(Z_j = 1) = 1 - \mathbb{P}(Z_j = 0) > 0$. If $U_r := \sum_{j=1}^{r} Z_j, \theta_r := \mathbb{E}U_r$, and $0 < \gamma \leq \frac{1}{2}$, then

$$\mathbb{P}\left(|U_r - \theta_r| \geq \theta_r \gamma\right) \leq 2\exp\left(-\frac{\gamma^2}{4}\theta_r\right) \tag{14.3.11}$$

for all $r \geq 1$. For a proof of (14.3.11), we refer to Corollary A.1.14, pp. 312 of [1].

We prove the desired result in three steps. In the first step, we prove certain convergence results used in the final evaluation of the error probability, and in the second step, we derive analytically tractable expressions for the error probability. In our third step, we combine the results of Steps 1 and 2 to obtain the desired bound on U^*.

Step 1: For μ-a.e. x, we show that

$$\frac{N_1}{N_{tot}} \longrightarrow \eta(x) \text{ a.s.} \tag{14.3.12}$$

as $n \to \infty$. To prove (14.3.12), we use the following estimates that hold μ-a.e. x:

$$\lim_{r\to 0} \frac{1}{\mu(S_r(x))} \int_{S_r(x)} h d\mu = h(x) \text{ and } \lim_{r\to 0} \frac{r^d}{\mu(S_r(x))} < \infty. \tag{14.3.13}$$

The first result in (14.3.13) is the generalization of Lebesgue differentiation theorem for Radon measures (see Theorem 11.1, p. 192 in [4]), and the second relation in (14.3.13) is a consequence of the almost sure finiteness of the derivatives of Radon measures (set ν to be the Lebesgue measure in Proposition 9.1, p. 186 and Equation 9.1, p. 187 of [4]).

Using (14.3.13), we obtain (14.3.12) as follows. By definition

$$\mathbb{E}N_{tot} = n\mu(S_r(x)) =: n\mu_r$$

and so from the deviation estimate (14.3.11), we get

$$\begin{aligned}\mathbb{P}((1-\epsilon)n\mu_r \le N_{tot} \le (1+\epsilon)n\mu_r) &\ge 1 - \exp(-C_1 n\mu_r)\\ &\ge 1 - \exp\left(-C_2 n r^d\right)\end{aligned} \tag{14.3.14}$$

for some constants $C_1, C_2 > 0$, by the second relation in (14.3.13). From the second relation in (14.3.9), we get that $C_2 n r^d \ge 2\log n$ for all n large and so the Borel-Cantelli lemma implies that

$$\frac{N_{tot}}{\mathbb{E}N_{tot}} = \frac{N_{tot}}{n\mu_r} \longrightarrow 1 \text{ a.s.} \tag{14.3.15}$$

as $n \to \infty$.

Next we have that $\mathbb{E}N_1 = \int_{S_r(x)} \sum_{i=1}^n h_i d\mu$ and so using the uniform convergence condition (14.3.8), we have for $\epsilon > 0$ that

$$\begin{aligned}\left|\frac{1}{n\mu_r}\mathbb{E}N_1 - \frac{1}{\mu_r}\int_{S_r(x)} h d\mu\right| &= \left|\frac{1}{\mu_r}\int_{S_r(x)} \left(\frac{1}{n}\sum_{i=1}^n h_i - h\right) d\mu\right| \\ &\leq \frac{1}{\mu_r}\int_{S_r(x)} \left|\frac{1}{n}\sum_{i=1}^n h_i - h\right| d\mu \\ &\leq \epsilon \end{aligned} \tag{14.3.16}$$

for all n large. Combining this with the first relation in (14.3.13), we then get that

$$\frac{\mathbb{E}N_1}{n\mu_r} \longrightarrow h(x) \tag{14.3.17}$$

as $n \to \infty$.

If $h(x) > 0$, then

$$\mathbb{E}N_1 \geq D_1 n\mu_r \geq D_2 n r^d \tag{14.3.18}$$

for all n large and some constants $D_1, D_2 > 0$, where the second inequality in (14.3.18) is true by the second condition in (14.3.13). Consequently, the deviation estimate (14.3.11) gives us that

$$\mathbb{P}\left(\mathbb{E}N_1(1-\epsilon) \leq N_1 \leq \mathbb{E}N_1(1+\epsilon)\right) \geq 1 - \exp\left(-D_3 n r^d\right) \geq 1 - \frac{1}{n^2} \tag{14.3.19}$$

for some constant $D_3 > 0$, where the final estimate in (14.3.19) is true by the second condition in (14.3.9). Since $\epsilon > 0$ is arbitrary, the Borel-Cantelli lemma then implies that $\frac{N_1}{\mathbb{E}N_1} \longrightarrow 1$ a.s. as $n \to \infty$, and combining this with (14.3.17) and (14.3.15), we get (14.3.12).

If $h(x) = 0$, then we use Chernoff bound to obtain the almost sure convergence. Indeed, letting $p_i := \mathbb{P}(Y_i = 1, X_i \in S_r(x))$, we have that

$$\mathbb{E}e^{N_1} = \prod_{i=1}^n (1 + (e-1)p_i) \leq \exp\left((e-1)\sum_i p_i\right) = \exp\left((e-1)\mathbb{E}N_1\right).$$

For any $\epsilon > 0$, we use (14.3.17) and the second relation in (14.3.13) to get that $\mathbb{E}N_1 \leq \epsilon n r^d$ for all n large. Consequently, the Chernoff bound gives us that

$$\mathbb{P}(N_1 \geq 2\epsilon n r^d) \leq e^{-2\epsilon n r^d}\mathbb{E}e^{N_1} \leq e^{-\epsilon n r^d}$$

and so again using the second conditions in (14.3.13) and (14.3.9), we get that

$$\mathbb{P}(N_1 \geq C\epsilon n\mu_r) \leq \frac{1}{n^2}$$

for some constant $C > 0$. As before, the Borel-Cantelli lemma and the fact that $\epsilon > 0$ is arbitrary imply that $\frac{N_1}{n\mu_r} = \frac{N_1}{\mathbb{E}N_{tot}} \longrightarrow 0$ a.s. and so from (14.3.15), we get (14.3.12).

Step 2: In this step, we obtain an expression for T_n in terms of the classifier error probability. We first rewrite ε_n in terms of the classifier outputs $g_n(.)$ as

$$\varepsilon_n = \mathbb{P}(Y_{n+1} \neq g_n(X_{n+1})) = \mathbb{P}(Y_{n+1} = 1, g_n = 0) + \mathbb{P}(Y_{n+1} = 0, g_n = 1). \tag{14.3.20}$$

By Fubini's theorem we have

$$\mathbb{P}(Y_{n+1} = 1, g_n = 0) = \int h_{n+1}(x)\mathbb{P}(g_n(x) = 0)\mu(dx)$$

and

$$\mathbb{P}(Y_{n+1} = 0, g_n = 1) = \int (1 - h_{n+1}(x))\mathbb{P}(g_n(x) = 1)\mu(dx).$$

Plugging these into (14.3.20), we get that

$$\varepsilon_n = \int \theta_n(x)\mu(dx), \tag{14.3.21}$$

where

$$\theta_n(x) := h_{n+1}(x)\mathbb{P}(g_n(x) = 0) + (1 - h_{n+1}(x))\mathbb{P}(g_n(x) = 1). \tag{14.3.22}$$

Consequently,

$$T_n = \int \frac{1}{n}\sum_{i=1}^{n} \theta_i(x)\mu(dx). \tag{14.3.23}$$

Step 3: We now combine the results of Steps 1 and 2 to complete the proof of the lemma. From (14.3.12), we see that $\frac{N_1 - N_0}{N_{tot}} \longrightarrow 2h(x) - 1$ a.s. as $n \to \infty$. Therefore if $h(x) > \frac{1+2\zeta}{2}$, then

$$N_1 - N_0 > (2h(x) - 1 - \zeta^2)N_{tot} > (2\zeta - \zeta^2)N_{tot} > \zeta N_{tot}$$

for all n large a.s. since $0 < \zeta < \frac{1}{4}$. Therefore we get that

$$\mathbb{P}(g_n(x) = 1) \longrightarrow 1. \tag{14.3.24}$$

Similarly if $h(x) < \frac{1-2\zeta}{2}$, then

$$N_0 - N_1 > (1 - 2h(x) - \zeta^2)N_{tot} > 2\zeta - \zeta^2 N_{tot} > 2\zeta^2 N_{tot}$$

for all n large a.s. and so

$$\mathbb{P}(g_n(x) = 0) \longrightarrow 1 \tag{14.3.25}$$

as $n \to \infty$. Finally, if $|2h(x) - 1| \leq \zeta^2$, then again using (14.3.12) we get that $|N_1 - N_0| \leq 2\zeta^2 N_{tot}$ for all n large a.s. Thus $\mathbb{P}(g_n = Z_n \text{ eventually}) \longrightarrow 1$ as $n \to \infty$ and so $\mathbb{P}(g_n = Z_n) \longrightarrow 1$ as $n \to \infty$. Consequently

$$\mathbb{P}(g_n = 0) = \mathbb{P}(Z_n = 0) + \mathbb{P}(g_n = 0, g_n \neq Z_n) - \mathbb{P}(Z_n = 0, g_n \neq Z_n) \longrightarrow \frac{1}{2}$$

as $n \to \infty$ and hence

$$\frac{1}{n}\sum_{i=1}^{n} \theta_i(x) \longrightarrow \frac{1}{2} \tag{14.3.26}$$

as $n \to \infty$.

Now if $\{a_n\}$ and $\{b_n\}$ are bounded sequences satisfying

$$a_n \to a \text{ and } \frac{1}{n}\sum_{i=1}^{n} b_i \longrightarrow b,$$

then

$$\frac{1}{n}\sum_{i=1}^{n} a_i b_i \longrightarrow ab. \tag{14.3.27}$$

Therefore plugging (14.3.24), (14.3.25), and (14.3.26) into (14.3.22) and using the convergence condition (14.3.8), we get for μ-a.e. x that

$$\lim_n \frac{1}{n}\sum_{i=1}^{n} \theta_i(x) = \begin{cases} 1 - h(x), & \text{if } h(x) > \frac{1}{2} + \zeta \\ h(x), & \text{if } h(x) < \frac{1}{2} - \zeta \\ \frac{1}{2}, & \text{if } \frac{1}{2} - \zeta^2 < h(x) < \frac{1}{2} + \zeta^2. \end{cases}$$

From (14.3.23) and Fatou's lemma, we then obtain $\limsup_n T_n \le L^* + b(\zeta)$, where

$$b(\zeta) := \mathbb{P}\left(\eta(X) \in \left[\frac{1}{2} - \zeta, \frac{1}{2}\right) \cup \left(\frac{1}{2}, \frac{1}{2} + \zeta\right]\right) \longrightarrow 0$$

as $\zeta \to 0$. □

Proof of Theorem 2 We first show that $\liminf_n T_n \ge L^*$. From (14.3.3), we immediately see that

$$\liminf_n T_n \ge M^* := \liminf_n \frac{1}{n} \sum_{i=1}^{n} \mathbb{E} \min(h_i, 1 - h_i).$$

Using $\min(a, b) = \frac{a+b-|a-b|}{2}$, we get that

$$\begin{aligned} M^* &= \liminf_n \left(\frac{1}{2} - \frac{1}{2n} \sum_{i=1}^{n} \mathbb{E}|2h_i - 1| \right) \\ &= \frac{1}{2} - \limsup_n \frac{1}{n} \sum_{i=1}^{n} \mathbb{E} \left| h_i - \frac{1}{2} \right|. \end{aligned} \tag{14.3.28}$$

Similarly writing $L^* = \frac{1}{2} - \mathbb{E}\left|h - \frac{1}{2}\right|$, it suffices to show that

$$\frac{1}{n} \sum_{i=1}^{n} \left| h_i - \frac{1}{2} \right| \longrightarrow \left| h - \frac{1}{2} \right|$$

as $n \to \infty$. Indeed, using $||a| - |b|| \le |a - b|$, we get that

$$\left| \left| h_i - \frac{1}{2} \right| - \left| h - \frac{1}{2} \right| \right| \le |h_i - h|$$

and so using the convergence condition (14.3.6) and the bounded convergence theorem, we see that $M^* = L^*$.

If (14.3.7) holds, then we define our universal classification rule as

$$g_n(x) := \begin{cases} 0, & \text{if } N_1 \le \frac{N_{tot}}{2} \\ 1, & \text{otherwise.} \end{cases} \tag{14.3.29}$$

From (14.3.12) in the proof of Lemma 1, we see that if $\eta(x) \ne \frac{1}{2}$, then

$$\mathbb{P}(g_n(x) = 0) \longrightarrow \mathbb{1}\left(h(x) < \frac{1}{2}\right) \text{ and } \mathbb{P}(g_n(x) = 1) \longrightarrow \mathbb{1}\left(h(x) > \frac{1}{2}\right). \tag{14.3.30}$$

Therefore from the convergence condition (14.3.8), (14.3.30), (14.3.22), and (14.3.27), we see that if $h(x) \neq \frac{1}{2}$, then

$$\frac{1}{n}\sum_{i=1}^{n}\theta_i(x) \longrightarrow h(x)\mathbb{1}\left(h(x) < \frac{1}{2}\right) + (1 - h(x))\mathbb{1}\left(h(x) > \frac{1}{2}\right). \tag{14.3.31}$$

Plugging (14.3.31) into (14.3.23) and using Fatou's lemma and (14.3.7), we get the desired bound

$\limsup_n T_n \leq L^*$. □

14.4 Applications and Future Directions

In this section, we briefly describe applications of our results in the context of wireless networks.

14.4.1 Regression Application

Consider a typical wireless network scenario, where mobile nodes are distributed randomly in a geographical region and each node is transmitting data to the base station at a certain power level that depends on the location of the node. The farther the node from the base station, the larger is the transmitting power required to maintain a minimum quality of service (QoS).

In the above context, the power level Y_j of the jth node is a function $h(X_j)$ of the node location X_j, and in practical situations where measurement of power level has small error, we can in fact assume that there is no noise, i.e., $N_j = 0$.

Now if there were no shadowing and fading, then we can set $h(x) = \frac{1}{x^\delta}$ for some $\delta > 2$ (see [8], Chapter 2 for more details). However, in practice channel fading and shadowing affect the received power level significantly (Chapter 3 [8]) and so usually the received power level is a deterministic function h that solely depends on the environment.

Suppose now a new node enters the network and would like to interact and transmit data via the base station. Because, it has no a priori knowledge of the environment, it could gather information about the transmitting power levels of its k-nearest neighbours for some suitably chosen k and estimate its own required transmission power using regression. In other words, the new node "discovers" its surrounding environment using regression.

14.4.2 Classification Application

We demonstrate how our classification error bounds have application in cognitive radio (CR) networks consisting of unlicensed users that are allowed to use licensed spectrum with the understanding that they vacate upon arrival of the licensed user [7].

Unlicensed users "far" from the licensed user could continue to use the spectrum since they would not cause any interference. However, unlicensed users close to the licensed user must vacate. Therefore in this context, we let $h_i(.)$ denote the indicator function that is one if ith user is a disturbance to the licensed user and is zero otherwise. Again $h_i(.)$ depends on the geography and also the behaviour of the ith unlicensed user (like if it is transmitting at high power levels, etc.).

As before, a new unlicensed user could use neighbourhood information to determine whether or not it would be a disturbance to the licensed user and proceed to either transmit or not, based on the decision outcome.

14.4.3 Future Directions

In this chapter, we have considered regression and classification for inhomogeneous data with applications to wireless networks. We have also discussed the applications above. It would be interesting to relax the assumptions in our main theorems and see how performance degrades as a result. Moreover, it would be even interesting to obtain partial or full converses that would essentially demonstrate the optimality of our results.

Acknowledgments I thank Professors Rahul Roy, Thomas Mountford, Federico Camia, Alberto Gandolfi, and C. R. Subramanian for crucial comments and also thank IMSc and IISER Bhopal for my fellowships.

References

1. Alon, N., Spencer, J.: The Probabilistic Method, 3rd edn. Wiley Interscience, New York (2008)
2. Chaudhuri, K., Dasgupta, S.: Rates of convergence for nearest neighbor classification. In: Ghahramani, Z., Welling, M. et al. (eds.) Neural Information Processing Systems, vol. 27, pp. 3437–3445 (2016). Preprint at: https://arxiv.org/abs/1407.0067v2
3. Chen, G.H., Shah, D.: Explaining the success of nearest neighbour methods in prediction. Found. Trends Mach. Learn. **10**, 337–588 (2018)
4. DiBenedetto, E.: Real analysis, 1st edn. Birkhäuser Advanced Texts, Springer, Berlin (2002)
5. Devroye, L., Györfi, L, Lugosi, G.: A probabilistic theory of pattern recognition. In: Applications of Mathematics, 1st edn. Springer, Berlin (1996)
6. Döring, M., Györfi, L., Walk, H.: Rate of convergence of k-nearest-neighbor classification rule. J. Mach. Learn. Res. **18**, 1–16 (2018)

7. Ganesan, G., Li, Y.G.: Cooperative spectrum sensing in cognitive radio, Parts I and II: two user networks. IEEE Trans. Wirel. Commun. **6**, 2204–2222 (2007)
8. Goldsmith, A.: Wireless Communications, 2nd edn. Cambridge University Press, Cambridge (2005)
9. Lerch, S., Thorarinsdottir, T.L.: Comparison of non-homogeneous regression models for probabilistic wind speed forecasting. Tellus A: Dyn. Meteorol. Oceanogr. **65**, 1–13 (2013)
10. Messner, J.W., Mayr, G.J., Zeileis, A.: Homogeneous boosting for predictor selection in ensemble post-processing. In: Working Papers in Economics and Statistics, University of Innsbruck, Research Platform Empirical and Experimental Economics (EEECON), Innsbruck (2016)

Chapter 15
Probabilistic Bounds for Data Storage with Feature Selection and Undersampling

Ghurumuruhan Ganesan

Abstract In this chapter we consider data storage from a probabilistic point of view and obtain bounds for efficient storage in the presence of feature selection and undersampling, both of which are important from the data science perspective. First, we consider encoding of correlated sources for nonstationary data and obtain a Slepian-Wolf type result for the probability of error. We then reinterpret our result by allowing one source to be the set of features to be discarded and other source to be remaining data to be encoded. Next, we consider neighbourhood domination in random graphs where we impose the condition that a fraction of neighbourhood must be present for each vertex and obtain optimal bounds on the minimum size of such a set. We show how such sets are useful for data undersampling in the presence of imbalanced datasets and briefly illustrate our result using k-nearest neighbours type classification rules as an example.

Keywords Data storage bounds · Slepian-Wolf coding · Feature selection · Graph-based undersampling

15.1 Introduction

Data feature selection and undersampling are important topics from both theoretical and practical perspectives, and in this chapter, we use a probabilistic approach to obtain theoretical bounds for data storage after feature selection and/or undersampling. Feature selection broadly falls into two categories: supervised and unsupervised. In supervised feature selection, the target variable is used as a guide in deciding which features of the data to retain and which features to discard. On the other hand, as the name suggests, unsupervised feature selection is done without using the target variable. There are many statistical methods available for both

G. Ganesan (✉)
University of Bristol, Bristol, England

A. Adhikari et al. (eds.), *Quantitative Frontiers*, Industrial and Applied Mathematics, https://doi.org/10.1007/978-981-96-9435-8_15

supervised and unsupervised feature selection, and for more details, we refer to Chapter 19 of [9].

We are interested in the problem of efficient data storage *after* feature selection using Slepian-Wolf coding. Apart from traditional source coding, Slepian-Wolf coding has also been extensively used in wireless networks. For example, Pradhan et al. [12] described distributed compression techniques for dense microsensor networks using syndromes, and later Hu and Li [8] proposed Slepian-Wolf cooperation to improve inter-user outage performance. In a related work the authors [17] used Slepian-Wolf encoding for data aggregation in cluster-based wireless networks, and Timo and Wigger [14] studied the performance of broadcasting with base-station cooperation and Slepian-Wolf coding.

In this chapter, we obtain a Slepian-Wolf type encoding result for nonstationary data and illustrate how information from the discarded features could be used to store the remaining data in a more efficient manner.

In the next part of our chapter, we study neighbourhood domination in random graphs with applications to data undersampling. Graph domination is important from both theoretical and application perspectives, and many variants of graph domination have been studied in different contexts. For domination in random graphs, the work [16] obtains two-point concentration for the domination number of G when p is essentially a constant and then extended for a wider range of p in [6]. Since then many other variants of domination have also been studied (see, e.g., [2] and [15]), and recently, the author [5] studies robustness of dominating sets when a small deterministic set of edges is removed from the parent complete graph.

In this chapter, we study neighbourhood domination in random graphs where we impose the condition that a fraction of the neighbourhood of each vertex is present in the resulting dominating set. We use the probabilistic method to obtain sufficient conditions for the existence of neighbourhood dominating sets of minimum possible size and demonstrate optimality of our estimates by obtaining a lower bound of the same order, for homogenous random graphs with common edge probability p. We also briefly illustrate a data undersampling methodology using neighbourhood domination and explain its applications to imbalanced learning.

The chapter is organized as follows: In Sect. 15.2, we state and prove our first result regarding a Slepian-Wolf type bound for nonstationary data. As a consequence, we also demonstrate how savings in data storage could be achieved by using information from the discarded features. Next in Sect. 15.3, we describe our second result involving the minimum size of neighbourhood dominating sets in random graphs. We also illustrate an application of our methodology in constrained undersampling of imbalanced data. Finally in Sect. 15.4, we state our conclusion and propose potential future directions.

15.2 Dependent Encoding Based on Feature Selection

Let $\mathcal{X}$ and $\mathcal{Y}$ be arbitrary finite sets, and for $1 \leq i \leq n$, let $(X_i, Y_i) \in \mathcal{X} \times \mathcal{Y}$ be a random element with distribution $\mathbb{P}(X_i = a, Y_i = b) = p_i(a, b)$ and corresponding marginals

$$p_{i,X}(.) := \sum_{b \in \mathcal{Y}} p_i(., b) \text{ and } p_{i,Y}(.) := \sum_{a \in \mathcal{X}} p(x, .).$$

The tuples $(X_i, Y_i), 1 \leq i \leq n$ are independent but not necessarily identically distributed.

An n-length binary code of rate R is a deterministic set $\mathcal{C} \subset \{0, 1\}^n$ of size 2^{nR}.

Definition 1 An n-length, Y-dependent encoder based on $\mathcal{C}$ is a set of one-to-one maps $f := \{f_y(.)\}_{y \in \mathcal{Y}^n}$ such that $f_y : \mathcal{C} \to \mathcal{X}^n$ for each y. We define the probability of error for the encoding scheme $(f, \mathcal{C})$ as

$$q(f, \mathcal{C}) := \mathbb{P}\left(U \notin f_V(\mathcal{C})\right), \tag{15.2.1}$$

where $U := (X_1, \ldots, X_n)$ and $V := (Y_1, \ldots, Y_n)$.

In the context of data analysis, (X_i, Y_i) represents the ith data point, where X_i represents the part of the data to be encoded and Y_i is the part of the data that is to be discarded based on feature selection. The goal is to use the information from $(Y_1, \ldots, Y_n)$ to encode $(X_1, \ldots, X_n)$ as efficiently as possible, and we explain this in more detail at the end of this section.

We also remark that we do not assume that (X_i, Y_i) are identically distributed to allow for the possibility of measurement errors in data due to inherent statistical noise and/or human errors.

Defining

$$H(X_i, Y_i) := - \sum_{x \in \mathcal{X}, y \in \mathcal{Y}} p_i(x, y) \log p_i(x, y) \text{ and}$$

$$H(Y_i) := - \sum_{y \in \mathcal{Y}} p_{i,Y}(y) \log p_{i,Y}(y) \tag{15.2.2}$$

to be the respective entropies of (X_i, Y_i) and Y_i (see Chapter 2 [3]), we have the following result. Throughout, logarithms are to base 2, and constants do not depend on n.

Theorem 1 *Suppose*

$$\epsilon_1 \leq \min_{i,x,y} p_i(x, y) \leq \max_{i,x,y} p_i(x, y) \leq \epsilon_2, \quad \epsilon_1 \leq \min_{i,y} p_{i,Y}(y) \leq \max_{i,y} p_{i,Y}(y) \leq \epsilon_2 \tag{15.2.3}$$

and

$$\frac{1}{n}\sum_{i=1}^{n} H(X_i, Y_i) \longrightarrow H_{XY} \text{ and } \frac{1}{n}\sum_{i=1}^{n} H(Y_i) \longrightarrow H_Y \tag{15.2.4}$$

for some positive finite constants $\epsilon_1, \epsilon_2, H_{XY}$*, and* H_Y*. Let* $\epsilon > 0$ *be an arbitrary constant, and let* $\mathcal{C}$ *be any deterministic n-length binary code of size* 2^{nR} *for some constant* $R > 0$.

(a) If $R > H_{XY} - H_Y + 5\epsilon$*, then there exists an encoding scheme* $(f_0, \mathcal{C})$ *such that the error probability*

$$q(f_0, \mathcal{C}) \leq 4\epsilon. \tag{15.2.5}$$

(b) If $R < H_{XY} - H_Y - 3\epsilon$*, then for any encoding scheme* $(f, \mathcal{C})$ *we have*

$$q(f, \mathcal{C}) \geq 1 - 4\epsilon. \tag{15.2.6}$$

If the (X_i, Y_i) are independent and identically distributed (i.i.d.), then $H_{XY} = H(X, Y)$ and $H_Y = H(Y)$ are entropies as in (15.2.2) and so $H_{XY} - X_Y = H(X \mid Y)$ is the conditional entropy of X given Y. This is the usual Slepian-Wolf encoding bound for correlated sources, and for more details in this setting, we refer to Chapter 15, p. 549 of [3].

In general, for non-i.i.d. distributions, Theorem 1 obtains a sharp threshold for the probability of encoding error for rates above and below the threshold value of $H_{XY} - H_Y$.

Proof of Theorem 1 We begin with some preliminary computations regarding the size of certain typical sets used in our encoding process. Let $U = (X_1, \ldots, X_n)$, $V = (Y_1, \ldots, Y_n)$, where $\{(X_i, Y_i)\}$ are as described prior to (15.2.1), and let $p_{UV}(.,.)$ be the distribution of (X, Y) with corresponding marginals

$$p_U(.) := \sum_{y \in \mathcal{Y}^n} p_{UV}(, y) \text{ and } p_V(.) := \sum_{x \in \mathcal{X}^n} p_{UV}(x, .).$$

For $\epsilon > 0$ arbitrary, we define the typical set

$$E_Y := \{y \in \mathcal{Y}^n : 2^{-n(H_Y + \epsilon)} \leq p_V(y) \leq 2^{-n(H_Y - \epsilon)}\}, \tag{15.2.7}$$

where H_Y is as in (15.2.4). To estimate the probability of the event E_Y, we first use the fact that $\{Y_i\}$ is independent to get that

$$\mathbb{E}\left(-\frac{1}{n}\sum_{i=1}^{n}(\log p_{i,Y}(Y_i) - H(Y_i))\right)^2 = \frac{1}{n^2}\sum_{i=1}^{n}\mathbb{E}(\log p_{i,Y}(Y_i) - H(Y_i))^2. \tag{15.2.8}$$

Now from (15.2.3), we know that $\mathbb{E}(\log p_{i,Y}(Y_i) - H(Y_i))^2 \leq C$ for some constant $C > 0$ and so

$$\mathbb{E}\left(-\frac{1}{n}\sum_{i=1}^{n}(\log p_{i,Y}(Y_i) - H(Y_i))\right)^2 \leq \frac{C}{n} \longrightarrow 0 \tag{15.2.9}$$

as $n \to \infty$. Combining (15.2.9) with (15.2.4), we then get that

$$\mathbb{E}\left(-\frac{1}{n}\log p_V(V) - H_Y\right)^2 \longrightarrow 0$$

as $n \to \infty$. Thus $-\frac{1}{n}\log p_V(V) \longrightarrow H_Y$ in probability and so

$$\mathbb{P}(E_Y) \geq 1 - \epsilon \tag{15.2.10}$$

for all n large.

Similarly defining

$$E_{XY} := \{(x, y) \in \mathcal{X}^n \times \mathcal{Y}^n : 2^{-n(H_{XY}+\epsilon)} \leq p_{UV}(x, y) \leq 2^{-n(H_{XY}-\epsilon)}\} \tag{15.2.11}$$

and performing an analogous analysis as above, we have for all n large that$\mathbb{P}(E_{XY}) \geq 1 - \epsilon$. From (15.2.10) and the union bound, we therefore get that

$$\mathbb{P}(E_{XY} \cap E_Y) \geq 1 - 2\epsilon. \tag{15.2.12}$$

If $\#E_Y$ denotes the cardinality of E_Y, then by definition

$$1 \geq \mathbb{P}(E_Y) = \sum_{y \in E_Y} p_V(y) \geq \#E_Y \cdot \frac{1}{2^{n(H_Y+\epsilon)}}$$

and so $\#E_Y \leq 2^{n(H_Y+\epsilon)}$. Similarly using (15.2.10) we also have that

$$1 - \epsilon \leq \mathbb{P}(E_Y) = \sum_{y \in E_Y} p_V(y) \leq \#E_Y \cdot \frac{1}{2^{n(H_Y-\epsilon)}}$$

and so $\#E_Y \geq (1 - \epsilon)2^{n(H_Y-\epsilon)}$. Combining we get

$$(1 - \epsilon)2^{n(H_Y-\epsilon)} \leq \#E_Y \leq 2^{n(H_Y+\epsilon)}, \tag{15.2.13}$$

and arguing similarly we also get that

$$(1-\epsilon)2^{n(H_{XY}-\epsilon)} \leq \#E_{XY} \leq 2^{n(H_{XY}+\epsilon)}. \tag{15.2.14}$$

This completes the preliminary computation part of our proof.

For obtaining Theorem 1(a), we argue as follows. For $y \in E_Y$ define

$$A_y := \{x \in \mathcal{X}^n : (x, y) \in E_{XY}\} \tag{15.2.15}$$

to be the slice of E_{XY} for a given y. If

$$B_Y := \{y \in E_Y : \#A_y \geq 2^{n(H_{XY}-H_Y+5\epsilon)}\},$$

then necessarily $\#B_Y \leq 2^{n(H_Y-3\epsilon)}$; for otherwise, we would get

$$\#E_{XY} \geq \#B_Y \cdot 2^{n(H_{XY}-H_Y+5\epsilon)} \geq 2^{n(H_{XY}+2\epsilon)},$$

and this contradicts (15.2.14). Thus $\#B_Y \leq 2^{n(H_Y-3\epsilon)}$, and so from the definition of E_Y in (15.2.7) we get that

$$\mathbb{P}(B_Y) \leq 2^{n(H_Y-3\epsilon)} \cdot 2^{-n(H_Y-\epsilon)} = \frac{1}{2^{2n\epsilon}}.$$

Since $B_Y \subset E_Y$, we get from (15.2.10) that

$$\mathbb{P}(E_Y \setminus B_Y) \geq 1 - \epsilon - \frac{1}{2^{2n\epsilon}}. \tag{15.2.16}$$

For each $y \in E_Y \setminus B_Y$, we have that $\#A_y \leq 2^{n(H_{XY}-H_Y+5\epsilon)}$, and so letting $\mathcal{C}$ be any n-length binary code of size $2^{n(H_{XY}-H_Y+5\epsilon)}$ we define a one-to-one map $f_y : \mathcal{C} \to A_y$, for each $y \in E_Y \setminus B_Y$. If (U, V) is the random tuple as described in the first paragraph of this proof, then from (15.2.12), (15.2.16), and the union bound, we see that the probability of error $q(f, \mathcal{C})$ is at most

$$\mathbb{P}((U, V) \notin E_{XY}) + \mathbb{P}(V \notin E_Y \setminus B_Y) \leq 3\epsilon + \frac{1}{2^{2n\epsilon}} \leq 4\epsilon$$

for all n large. This obtains the upper bound (15.2.5) and therefore completes the proof of part (a).

For the lower bound (15.2.6) in part (b), we argue as follows. Again let (U, V) be the random tuple as described in the first paragraph of this proof, and for $y \in \mathcal{Y}^n$, let A_y be the "slice" of the set E_{XY} as defined in (15.2.15). We have that

$$\mathbb{P}(A_V \cap E_Y) = \mathbb{P}(\{(U, V) \in E_{XY}\} \cap \{V \in E_Y\}) \geq 1 - 2\epsilon, \tag{15.2.17}$$

as seen from (15.2.12). Expanding in terms of the elements of E_Y, we also get

$$\mathbb{P}(A_V \cap E_Y) = \sum_{y \in E_Y} \mathbb{P}_y(A_y) p_V(y),$$

where

$$\mathbb{P}_y(A_y) := \mathbb{P}(A_V \mid V = y) = \sum_{x \in A_y} p_{U|V}(x \mid y) \tag{15.2.18}$$

and $p_{U|V}(x \mid y) := \frac{p_{UV}(x,y)}{p_V(y)}$. Therefore defining

$$C_Y := \{y \in E_Y : \mathbb{P}_y(A_y) \geq \epsilon\},$$

we also get the upper bound

$$\begin{aligned}\mathbb{P}(A_V \cap E_Y) &\leq \sum_{y \in C_Y} p_V(y) + \epsilon \sum_{y \in E_Y \setminus C_Y} p_V(y) \\ &\leq \mathbb{P}(C_Y) + \epsilon. \end{aligned} \tag{15.2.19}$$

Combining (15.2.17) and (15.2.19), we get that

$$\mathbb{P}(C_Y) \geq 1 - 3\epsilon. \tag{15.2.20}$$

For any $y \in C_Y \subset E_Y$ and $x \in A_y$, we have by definition of E_Y and E_{XY} (see (15.2.7) and (15.2.11)) that

$$\begin{aligned}p_{U|V}(x \mid y) &= \frac{p_{UV}(x, y)}{p_V(y)} \\ &\leq \frac{2^{-n(H_{XY} - \epsilon)}}{2^{-n(H_Y + \epsilon)}} \\ &= 2^{-n(H_{XY} - H_Y - 2\epsilon)}. \end{aligned} \tag{15.2.21}$$

Thus

$$\epsilon \leq \mathbb{P}_y(A_y) = \sum_{x \in A_y} p_{U|V}(x \mid y) \leq \#A_y \cdot 2^{-n(H_{XY} - H_Y - 2\epsilon)}$$

and so we get that $\#A_y \geq \epsilon \cdot 2^{n(H_{XY} - H_Y - 2\epsilon)}$ for each $y \in C_Y$.

Suppose now that $\mathcal{C}$ is any deterministic n-length code of size at most $2^{n(H_{XY}-H_Y-3\epsilon)}$, and let f be any Y-dependent encoding scheme based on $\mathcal{C}$ as in Definition 1. As in (15.2.18), we have for each $y \in C_Y$ that

$$\begin{aligned}\mathbb{P}_y\left(f_y(\mathcal{C})\right) &= \sum_{x \in f_y(\mathcal{C})} p_{U|V}\, p(x \mid y) \\ &\le 2^{n(H_{XY}-H_Y-3\epsilon)} \cdot 2^{-n(H_{XY}-H_Y-2\epsilon)} \\ &= \frac{1}{2^{n\epsilon}}, \qquad (15.2.22)\end{aligned}$$

where the inequality in (15.2.22) follows from (15.2.21). Therefore

$$\begin{aligned}\mathbb{P}\left(U \in f_V(\mathcal{C})\right) &\le \mathbb{P}(V \notin C_Y) + \sum_{y \in C_Y} p(y)\frac{1}{2^{n\epsilon}} \\ &\le 3\epsilon + \frac{1}{2^{n\epsilon}} \\ &\le 4\epsilon \qquad (15.2.23)\end{aligned}$$

for all n large. Thus the error probability $q(f, \mathcal{C})$ for any encoding scheme $(f, \mathcal{C})$ is at least $1 - 4\epsilon$ for all n large. This obtains the lower bound in (15.2.6) and therefore completes the proof of the theorem. □

15.2.1 Data Storage with Feature Selection

Let $Z = (Z(1), \ldots, Z(m))$ be a random element chosen from some set $\mathcal{Z}$. We refer to the indices $i = 1, 2, \ldots, m$ as features and use statistical methods [9] to determine whether $Z(i)$ and $Z(j)$ are correlated and obtain the dependency graph G_{dep}. The vertex set of G_{dep} is $\{1, 2, \ldots, m\}$, and an edge with endvertices i and j belongs to G_{dep} if and only if $Z(i)$ and $Z(j)$ are determined to be *correlated.*

The rough idea behind the dependency graph is that if i and j are neighbours in G_{dep}, then $Z(i)$ and $Z(j)$ are highly correlated, and we should lose little information if we were to discard either feature i or feature j. In effect, given G_{dep} and an integer $1 \le d \le m$, we would like to obtain a subset S containing at most d features that satisfy the following properties:

(i) The features present in S are nearly uncorrelated.
(ii) Each feature not present in S is highly correlated with some feature present in S.

For simplicity assume that using standard statistical methods (like, e.g., filters or wrappers, see Chapter 19 [9]), we have determined $S = \{1, 2, \ldots, d\}$ to be "best" feature set and set $X := (Z(1), \ldots, Z(d))$ and $Y = (Z(d+1), \ldots, Z(m))$. Using information from both X and Y, we would now like to find an encoding scheme that allows us to store X using as few bits as possible. Indeed, suppose we have n data points (X_i, Y_i), $1 \leq i \leq n$, and (15.2.4) holds together with $\frac{1}{n}\sum_{i=1}^{n} H(X_i) \longrightarrow H_X$ for some $H_X > 0$. Applying Theorem 1 with $\mathcal{Y} = \emptyset$, we see that the relevant information $(X_1, \ldots, X_n)$ can be stored using a binary code of length roughly nH_X, with very small encoding error and without using any information from $(Y_1, \ldots, Y_n)$.

On the other hand, if we use the "side" information $(Y_1, \ldots, Y_n)$, then we can store $(X_1, \ldots, X_n)$ using a binary code of length roughly $n(H_{XY} - H_Y)$. Using the fact that $H(X_i, Y_i) \leq H(X_i) + H(Y_i)$, we see that $H_{XY} \leq H_X + H_Y$ and so $1 - \frac{H_{XY} - H_Y}{H_X}$ could be interpreted as the "savings" obtained via dependent encoding resulting in "smart" storage of data.

15.3 Neighbourhood Domination-Based Undersampling

We begin with the definition and properties of neighbourhood domination in graphs and at the end of this section explain its applications to data undersampling.

Let K_n be the complete graph with vertex set $\{1, 2, \ldots, n\}$, and let H be any deterministic subgraph of K_n. We say that $\mathcal{T} \subset \{1, 2, \ldots, n\}$ is a *dominating* set of H if each vertex v is either present in $\mathcal{T}$ or is adjacent in H to some vertex $u \in \mathcal{T}$. Let $N(v)$ be the set of neighbours of v in the graph H, and let $d(v) = \#N(v)$. For $0 < \theta < 1$, we say that $\mathcal{T}$ is a *θ-neighbourhood dominating* set of H if for each vertex v, there are at least $\theta d(v)$ vertices of $N(v)$ present in $\mathcal{T}$.

Let $M_n = M_n(\theta)$ be the minimum size of a θ-neighbourhood dominating set of H. In the following result below, we obtain an upper bound for M_n with conditions on the minimum and maximum vertex degree of H and then show that the bound is essentially optimal, using random graphs. Formally, let $Y(f)$, $f \in K_n$ be independent random variables indexed by the edge set of K_n and having distribution

$$\mathbb{P}(Y(f) = 1) = p = 1 - \mathbb{P}(Y(f) = 0),$$

where $0 < p = p(n) < 1$. Let G be the homogenous random subgraph of K_n formed by the set of all edges e satisfying $Y(f) = 1$.

We have the following bounds for M_n in terms of the degree parameters of the deterministic graph H and the random graph realization G.

Theorem 2 *We have:*

(a) Let Δ and δ, respectively, denote the maximum and minimum vertex degree in H, and for $\eta > 0$ let $z := \frac{\eta^2(\theta+\eta)}{4}$. Any θ-neighbourhood dominating set of H has size at least $\frac{\theta\delta n}{\Delta^2}$, and if

$$4\Delta^2 e^{-z\delta} \leq 1 \textit{ and } e^{-z} + 2e^{-z\delta} < 1 \tag{15.3.1}$$

strictly, then there is a θ-neighbourhood dominating set of size at most $(\theta + 2\eta)n$.

(b) *For every $\epsilon > 0$ there are constants $M, D > 0$ such that if $p \geq \frac{M \log n}{n}$, then*

$$\mathbb{P}\left((\theta - \epsilon)n \leq M_n \leq (\theta + \epsilon)n\right) \geq 1 - e^{-Dnp}. \tag{15.3.2}$$

In other words, with high probability, i.e., with probability converging to one as $n \to \infty$, we see that the size M_n of a θ-neighbourhood dominating set is roughly of the order of θn.

In the context of data analysis, the vertices of K_n represent data points, and for practical reasons explained at the end of this section, it is often important to undersample or reduce the dataset size to a fraction θn while allowing for enough neighbours per data point. Theorem 2 ensures that there exists such a set under the sufficient condition (15.3.1) which usually holds for most cases of interest as demonstrated at the section end.

In our proof of Theorem 2 below, we use the following results regarding the deviation estimates of sums of independent Bernoulli random variables and the Lovász Local Lemma, which we state together as a lemma for convenience.

Lemma 1

(a) *Let $\{W_j\}_{1 \leq j \leq r}$ be independent Bernoulli random variables satisfying $\mathbb{P}(W_j = 1) = 1 - \mathbb{P}(W_j = 0) > 0$. If $S_r := \sum_{j=1}^{r} W_j, \theta_r := \mathbb{E}S_r$, and $0 < \gamma \leq \frac{1}{2}$, then*

$$\mathbb{P}\left(|S_r - \theta_r| \geq \theta_r \gamma\right) \leq 2\exp\left(-\frac{\gamma^2}{4}\theta_r\right) \tag{15.3.3}$$

for all $r \geq 1$.

(b) *Let $A_1, \ldots, A_t$ be events in an arbitrary probability space. Let Γ be the dependency graph for the events $\{A_i\}$, with vertex set $\{1, 2, \ldots, t\}$ and edge set $\mathcal{E}$; that is, assume that each A_i is independent of the family of events A_j, $(i, j) \notin \mathcal{E}$. If there are reals $0 \leq y(i) < 1$ such that $\mathbb{P}(A_i) \leq y(i) \prod_{(i,j) \in \mathcal{E}} (1 - y(j))$, for each i, then*

$$\mathbb{P}\left(\bigcap_i A_i^c\right) \geq \prod_{1 \leq i \leq n} (1 - y(i)) > 0.$$

For proofs of Lemma 1(a) and (b), we refer, respectively, to Corollary A.1.14, p. 312 and Lemma 5.1.1, p. 64 of [1].

Proof of Theorem 2(a) Say that a set V of vertices is 3-far if the graph distance (i.e., the number of edges in the shortest path) between any two vertices is at least 3. For any vertex v, there are at most Δ^2 vertices at a distance 2 from v where Δ is the maximum vertex degree in H. Therefore from Theorem 3.2.1, p. 27 of [1], we know that the graph H contains a 3-far set $\mathcal{T}$ of size at least $\frac{n}{2\Delta^2}$. For any two vertices $u, v \in \mathcal{T}$, the corresponding neighbourhoods are disjoint; that is, $N(u) \cap N(v) = \emptyset$ and so

$$M_n \geq \sum_{v \in \mathcal{T}} \theta d(v) \geq \frac{\theta \delta n}{2\Delta^2}.$$

For the upper bound on M_n, we use the probabilistic method. Select each vertex of H with probability x, independent of other vertices as follows. Let Z_j, $1 \leq j \leq n$ be independent and identically distributed (i.i.d.) Bernoulli random variables, with

$$\mathbb{P}_Z(Z_j = 1) = x = 1 - \mathbb{P}_Z(Z_j = 0),$$

and let $S = \{v : Z_v = 1\}$ be the random set of chosen vertices. From the standard deviation estimate (15.3.3), we get that

$$\mathbb{P}_Z\left(\#S \leq (x + \eta)n\right) \geq 1 - \exp\left(-\frac{\eta^2}{4} xn\right), \tag{15.3.4}$$

where $\#S$ is the cardinality of S.

Our goal is to show that if x and η are chosen appropriately, then S is a θ-neighbourhood dominating set with positive $\mathbb{P}_Z$-probability. Let $N_Z(v) = \{u \sim v : Z_u = 1\}$ be the set of all vertices adjacent to v in G, which are also present in S, and let A_v be the event that $\#N_Z(v) \geq (x - \eta)d(v)$. From the standard deviation estimate (15.3.3), we get that

$$\mathbb{P}_Z(A_v) \geq 1 - \exp\left(-\frac{\eta^2 x d(v)}{4}\right) \geq 1 - \alpha, \tag{15.3.5}$$

where $\alpha := \exp\left(-\frac{\eta^2 x}{4}\delta\right)$ and δ is the minimum vertex degree in G.

The events A_v and A_u are independent if the graph distance between u and v satisfies $d_G(u, v) \geq 3$ and so if Δ denotes the maximum vertex degree in H, then each A_u depends on at most Δ^2 of the events in $\{A_v\}$, which we denote as $\mathcal{E}(v)$. This allows us to use Lovász Local Lemma in Lemma $1(b)$ under the assumption that

$$4\Delta^2 \alpha \leq 1. \tag{15.3.6}$$

Indeed, setting $y(v) = 2\alpha$, we see that

$$y(v) \prod_{u \in \mathcal{E}(v)} (1 - y(u)) = 2\alpha(1 - 2\alpha)^{\Delta^2} \geq 2\alpha(1 - 2\alpha\Delta^2) \geq \alpha \geq \mathbb{P}_Z(A_v^c), \tag{15.3.7}$$

where the third relation in (15.3.7) is true by (15.3.6) and the final relation in (15.2.8) follows from (15.3.5).

From Lemma 1(b), we therefore see that $\mathbb{P}_Z\left(\bigcap_v A_v\right) \geq (1-2\alpha)^n$. Choosing $x - \eta = \theta$ and combining with (15.3.4), we then obtain that S is a θ-neighbourood dominating set of size at most $(\theta + 2\eta)n$, with $\mathbb{P}_Z$-probability at least

$$(1-2\alpha)^n - \exp\left(-\frac{\eta^2}{4}xn\right) > 0$$

if

$$\exp\left(-\frac{\eta^2}{4}x\right) \leq 1 - 2\alpha - y \tag{15.3.8}$$

for some small constant $y > 0$. The conditions (15.3.6) and (15.3.8) complete the proof of part (a) of the theorem. □

Proof of Theorem 2 (b) The neighbourhood $N(v)$ of a vertex v is the set of all vertices adjacent to v in G, and the degree of v is defined as $d(v) = \#N(v)$, the cardinality of $N(v)$.

For the upper bound, we use the conditions in the theorem statement to show that G satisfies both the conditions (15.3.6) and (15.3.8) with high probability, i.e., with probability converging to one as $n \to \infty$. The expected degree of any vertex v is $(n-1)p$ and so using the deviation estimate (15.3.3), we see that

$$\mathbb{P}\left(d(v) \geq \frac{np}{2}\right) \geq 1 - e^{-Cnp}$$

for some constant $C > 0$. Setting $E_{deg} := \bigcap_{v=1}^{n}\left\{d(v) \geq \frac{np}{2}\right\}$ and using the union bound, we then get that

$$\mathbb{P}\left(E_{deg}\right) \geq 1 - ne^{-Cnp} \geq 1 - ne^{-CM\log n} \geq 1 - \frac{1}{n}, \tag{15.3.9}$$

provided M is chosen sufficiently large. We fix such an M henceforth and see that if E_{deg} occurs, then the minimum vertex degree in G is at least $\frac{np}{2} \geq \frac{M\log n}{2}$. For all n large, this implies that (15.3.8) holds, provided we fix y small enough. Also, since the maximum vertex degree $\Delta \leq n$, we can choose M larger if necessary and ensure that (15.3.6) also holds. In effect, we see that if E_{deg} holds, then there is a θ-neighbourhood dominating set of G containing at most $(\theta + 2\eta)n$ vertices, and this obtains the upper bound in (15.3.2).

We prove the lower bound as follows. For any vertex v the expected degree $\mathbb{E}d(v) = (n-1)p$ and so using the standard deviation estimate (15.3.3), we get for $\epsilon > 0$ that

$$\mathbb{P}\left(d(v) \geq np(1-\epsilon)\right) \geq 1 - e^{-Cnp}$$

for some constant $C > 0$. Letting $E_{deg} := \bigcap_{v=1}^{n} \{d(v) \geq np(1-\epsilon)\}$, we then get from the union bound that

$$\mathbb{P}(E_{deg}) \geq 1 - ne^{-C_1 np} \tag{15.3.10}$$

for some constant $C_1 > 0$.

Next, let S be any set containing ζn vertices with $\zeta < \theta$ strictly, and let $m(S)$ be the number of edges of G having both endvertices in S. We know that

$$\frac{\zeta^2 n^2 p}{2} \geq \mathbb{E}m(S) = \binom{\zeta n}{2} p \geq \frac{\zeta^2 n^2 p}{4}$$

and so by the deviation estimate (15.3.3), we get that

$$\mathbb{P}\left(m(S) \leq \frac{\zeta^2 n^2 p}{2}\right) \geq 1 - e^{-C_2 n^2 p} \tag{15.3.11}$$

for some constant $C_2 > 0$. The number of choices for S is at most 2^n and so letting E_{edge} be the event that $m(S) \leq \frac{\zeta^2 n^2 p}{2}$ for each S containing ζn vertices, we get from the union bound and (15.3.11) that

$$\mathbb{P}(E_{edge}) \geq 1 - 2^n e^{-C_2 n^2 p}. \tag{15.3.12}$$

We assume henceforth that $E_{deg} \cap E_{edge}$ occurs, which by the union bound (15.3.10) and (15.3.12) happens with probability

$$\mathbb{P}(E_{deg} \cap E_{edge}) \geq 1 - ne^{C_1 np} - 2^n e^{-C_2 n^2 p} \geq 1 - e^{-C_3 np} \tag{15.3.13}$$

for some constant $C_3 > 0$ provided $np > C_4 \log n$ for some sufficiently large constant $C_4 > 0$. Let S be any set of ζn vertices. Because $m(S) \leq \frac{\zeta^2 n^2 p}{2}$ and the sum of the vertex degrees equals twice the number of edges, there exists a vertex $z \in S$ that is adjacent to at most

$$\zeta np = \frac{\zeta}{1-\epsilon} np(1-\epsilon) \leq \frac{\zeta}{1-\epsilon} d(z)$$

vertices of S, since E_{deg} also occurs. We can choose ϵ small enough so that $\frac{\zeta}{1-\epsilon} < \theta$ strictly and obtain that no set containing ζn vertices is a θ-neighbourhood dominating set of G. Thus $M_n \geq \zeta n$, and from (15.3.13) we then get the lower bound in (15.3.2). This completes the proof of the theorem. □

15.3.1 Data Undersampling

In many data classification problems involving two classes, there is often a large imbalance in the size of the majority and minority classes, and this adversely affects the classwise accuracy of common predictive models like k-nearest neighbour (kNN), Random Forest, Logistic Regression, or Neural Networks [4, 7, 10]. Therefore it is important to undersample or reduce the size of the majority class and make it comparable to the size of the minority class. Many different undersampling methods have been proposed in literature like random undersampling, near miss, condensed neighbour, etc., and for more on undersampling techniques in data analysis, we refer to [11, 13].

In this chapter, we use neighbourhood domination as an undersampling methodology to obtain a balanced dataset of given size. Indeed, suppose we are given n data points from the majority class and θn data points from the minority class for some $0 < \theta < 1$. Instead of randomly selecting θn points from the majority class, we could extract a more "representative" set with the additional assumption that there is a distance metric $d(u, v)$ between two majority class data points u and v.

Henceforth any vertex represents a data point belonging to the majority class. For a vertex v, let $N_k(v)$ be the set of k-nearest neighbours of v, according to the metric d. Let H be the directed graph obtained by representing the data points as vertices and drawing a directed edge from vertex u to vertex v if $v \in N_k(u)$. The same analysis as in Theorem 2(a) holds with $\delta = \Delta = k$, and so if (15.3.1) holds, then there is a θ-neighbourhood dominating set of size at most $(\theta + 2\eta)n$.

Setting $k = M \log n$ for some large enough constant M, we see that (15.3.1) is satisfied and so there exists a θ-neighbourhood dominating set S of size$(\theta + 2\eta)n$. The advantage of using S is that any vertex v still has at least θk neighbours from the set of its k-nearest neighbours, and this would aid greatly while performing classification, say, for example, using kNN type rules.

15.4 Conclusion

In this chapter we have used a probabilistic approach toward data storage with feature selection and undersampling. We first obtained a Slepian-Wolf type result for nonstationary data and thereby demonstrated savings in data storage obtained by using information from the discarded features.

We have a couple of remarks. In (15.2.3), we have fixed thresholds ϵ_1, ϵ_2 to be arbitrary. In a practical setting, we could choose ϵ_1, ϵ_2 based on the storage constraint and retain only those data points whose estimated entropy lies within the said interval. It would be interesting to explore this for future work.

Another point to consider is the fact that we have considered X and Y to be discrete in our setup. Naturally, we are interested to know how to handle datasets with continuous entries. In this aspect, one direct solution would be to perform binning of the continuous entries. Of course, then we would have to choose the bin size optimally, and designing criteria for the same would be another future direction of research.

Next, we considered neighbourhood domination in random graphs and explained how our methodology could be used to perform constrained undersampling in imbalanced datasets.

For the future, we plan to apply our undersampling method to real-world datasets and study the resulting minority class accuracy. We would also like to develop practical encoding schemes that use information from the discarded features and therefore result in smart data storage.

From the theoretical viewpoint, it would be interesting to study the uniqueness of the undersampled set or in the alternate case estimate the number of undersampled sets satisfying the neighbourhood domination condition.

Acknowledgments I thank Professors Rahul Roy, Thomas Mountford, Federico Camia, Alberto Gandolfi, Lasha Ephremidze, and C. R. Subramanian and the referees for crucial comments that led to an improvement of the work. I also thank IMSc and IISER Bhopal for my fellowships.

References

1. Alon, N., Spencer, J.: The Probabilistic Method, 4rd edn. Wiley Interscience, New York (2008)
2. Clark, L., Johnson, D.: The independent domination number of a random graph. Discussiones Mathematicae Graph Theory **31**, 129–142 (2011)
3. Cover, T., Thomas, J.: Elements of Information Theory, 2nd edn. Wiley, New York (2006)
4. Fernández, A., del Río, S., Chawla, N.V., Herrera, F.: An insight into imbalanced big data classification: outcomes and challenges. Complex Intell. Syst. **3**, 105–120 (2017)
5. Ganesan, G.: Robust Domination in Random Graphs. In: Accepted for publication in Indian Journal of Pure and Applied Mathematics (2023). Preprint at: https://arxiv.org/abs/2301.05386
6. Glebov, R., Liebenau, A., Szabó, T.: On the concentration of the domination number of the random graph. SIAM J. Discret. Math. **29**, 1186–1206 (2012)
7. He, H., Garcia, E.A.: Learning from imbalanced data. IEEE Trans. Knowl. Data Eng. **21**, 1263–1284 (2009)
8. Hu, R., Li, J.: Exploiting Slepian-Wolf coding in wireless relay channels. In: Proceedings IEEE Signal Processing Advances for Wireless Communications (SPAWC) 2005, New York, pp. 1–6 (2005)
9. Kuhn, M., Johnson, K.: Applied Predictive Modeling, 1st edn. Springer, Berlin (2013)
10. Liu, S., Wang, Y., Zhang, J., Chen, C., Xiang, Y.: Addressing the class imbalance problem in twitter spam detection using ensemble learning. Comput. Secur. **69**, 35–49 (2017)
11. More, A.: Survey of Resampling Techniques for Improving Classification Performance in Unbalanced Datasets (2016). Preprint at: https://arxiv.org/abs/1608.06048
12. Pradhan, S.S., Kusuma, J., Ramchandran, K.: Distributed compression in a dense microsensor network. IEEE Signal Process. Mag. **19**, 51–60 (2002)
13. Sui, Y., Zhang, X., Huan, J., Hong, H.: Exploring data sampling techniques for imbalanced classification problems. In: Proceedings SPIE 11198, Fourth International Workshop on Pattern Recognition, 1119813 (2019)
14. Timo, R., Wigger, M.: Slepian-Wolf coding for broadcasting with cooperative base-stations. IEEE Trans. Commun. **63**, 1850–1866 (2015)

15. Wang, B., Xiang, K.: On $k-$tuple domination of random graphs. Appl. Math. Lett. **22**, 1513–1517 (2009)
16. Wieland, B., Godbole, A.P.: On the domination number of a random graph. Electron. J. Comb. **8**(#R37), 1–13 (2001)
17. Zheng, J., Wang, P., Li, C.: Distributed data aggregation using Slepian–Wolf coding in cluster-based wireless sensor networks. IEEE Trans. Veh. Technol. **59**, 2564–2574 (2010)

MIX
Papier aus verantwortungsvollen Quellen
Paper from responsible sources
FSC® C105338

If you have any concerns about our products,
you can contact us on
ProductSafety@springernature.com

In case Publisher is established outside the EU,
the EU authorized representative is:
Springer Nature Customer Service Center GmbH
Europaplatz 3, 69115 Heidelberg, Germany

Printed by Libri Plureos GmbH
in Hamburg, Germany